Springer Series in Information Sciences 21

Editor: Manfred R. Schroeder

Springer Series in Information Sciences

Editors: Thomas S. Huang   Teuvo Kohonen   Manfred R. Schroeder

Managing Editor: H.K.V. Lotsch

Peter Strobach

Linear Prediction Theory

A Mathematical Basis for Adaptive Systems

With 63 Figures

Springer-Verlag Berlin Heidelberg New York
London Paris Tokyo Hong Kong

Dr.-Ing. *Peter Strobach*

SIEMENS AG, Zentralabteilung Forschung und Entwicklung,
ZFE IS – Forschung für Informatik und Software,
Otto-Hahn-Ring 6, D-8000 München 83, Fed. Rep. of Germany

Series Editors:

Professor Thomas S. Huang

Department of Electrical Engineering and Coordinated Science Laboratory,
University of Illinois, Urbana, IL 61801, USA

Professor Teuvo Kohonen

Laboratory of Computer and Information Sciences, Helsinki University of Technology,
SF-02150 Espoo 15, Finland

Professor Dr. Manfred R. Schroeder

Drittes Physikalisches Institut, Universität Göttingen, Bürgerstrasse 42–44,
D-3400 Göttingen, Fed. Rep. of Germany

Managing Editor: Helmut K. V. Lotsch

Springer-Verlag, Tiergartenstrasse 17,
D-6900 Heidelberg, Fed. Rep. of Germany

ISBN-13:978-3-642-75208-7   e-ISBN-13:978-3-642-75206-3
DOI: 10.1007/978-3-642-75206-3

This work was only possible through the love, encouragement and support of my parents. It is to them that I dedicate this book.

Preface

Linear prediction theory and the related algorithms have matured to the point where they now form an integral part of many real-world adaptive systems. When it is necessary to extract information from a random process, we are frequently faced with the problem of analyzing and solving special systems of linear equations. In the general case these systems are overdetermined and may be characterized by additional properties, such as update and shift-invariance properties. Usually, one employs exact or approximate least-squares methods to solve the resulting class of linear equations.

Mainly during the last decade, researchers in various fields have contributed techniques and nomenclature for this type of least-squares problem. This body of methods now constitutes what we call the theory of linear prediction. The immense interest that it has aroused clearly emerges from recent advances in processor technology, which provide the means to implement linear prediction algorithms, and to operate them in real time. The practical effect is the occurrence of a new class of high-performance adaptive systems for control, communications and system identification applications.

This monograph presumes a background in discrete-time digital signal processing, including Z-transforms, and a basic knowledge of discrete-time random processes. One of the difficulties I have encountered while writing this book is that many engineers and computer scientists lack knowledge of fundamental mathematics and geometry. This was particularly striking when the understanding of a certain linear prediction algorithm required a sophisticated mathematical derivation, or a higher-level geometrical interpretation. Another difficulty arises from the fact that most of the material covered here is scattered through the signal processing literature without any special ordering. The many contributions to the field of linear prediction theory and algorithms generally use disparate notations and mathematical techniques, making the material less accessible for the reader unfamiliar with the subject.

It is therefore the intention of this book to bring together the most important research contributions to linear prediction theory and to present them in a logical order using a common mathematical framework. Moreover, several new and previously unpublished concepts and approaches have been added. It is hoped that the presented material may thus be assimilated more efficiently, and that the relationship between the different techniques can be better appreciated.

Munich   PETER STROBACH
February 1989

Acknowledgements

I wish to thank the series editor, Prof. *Manfred R. Schroeder*, for his interest in this work.

I would like to express my sincere gratitude to Prof. *Dieter Schütt*, chief of the information systems research department at SIEMENS AG, Munich, FRG, whose constant encouragement has been most invaluable to me throughout the preparation of this book. His tireless devotion and unfailing support of his research groups has earned my highest respect.

I thank the management of SIEMENS AG, Munich, FRG, for providing me with four weeks of paid spare time in the final phase of this work.

I am also grateful to my former M.S. student *Jörg Sokat* from the University of Duisburg, FRG, for reviewing an early version of Chap. 2.

I am most indebted to Miss *Deborah Hollis* of Springer-Verlag, Heidelberg, FRG, for constant support and much useful advice during the preparation of the final manuscript. I am most thankful to the other staff members of Springer-Verlag, Heidelberg, for their help in the production of this book.

Special thanks go to Dipl.-Ing. *M. Schielein* who has provided me with all the necessary computer equipment. How would I have done it without him ?

The beautiful drawings in this book come from Mrs. *Habecker* and her crew. Their work is greatly appreciated.

The book was typeset using the particularly useful text-processing software SIGNUM, which one of our Ph.D. students developed as a hobby (in fact, for typing his Ph.D. thesis). The software became a hit, but the thesis was never completed... Thank you, *Franz*, for your fine software !

Contents

1. Introduction

The method of least squares was discovered by the
German mathematician Carl Friedrich Gauß, in 1809,
and even to this day it has remained one of the most
powerful techniques available to statisticians

Gwilym M. Jenkins
Donald G. Watts

The topic of linear prediction that we shall study in this book treats the
problem of adapting linear systems to sets of measurements or observations. The central theme therefore is *adaptation algorithms*, which are
derived, to a great extent, from exact or approximative *least-squares
techniques* [1.1] The success of the least-squares method in this field
essentially stems from two important facts.

1. Least-squares techniques lend themselves to tractable problem formulations and efficient algorithms for their solution.

2. The Gauss statistics is a good approximation for many real-world
   processes and therefore, the least-squares method often performs
   very close to the optimal, but computationally burdensome maximum
   likelihood method.

The idea of adaptive linear prediction has found numerous applications
in the fields of adaptive filtering, parameter estimation and system identification. The linear prediction model provides a *parametric description*
of an observed process. The necessity of such a description or modelling
of random data in many technical applications becomes very obvious if
we consider the popular example of speech data. Several different records
of speech data may represent the same information, i.e., the same vowel,
but the signal curves may look considerably different to the human ob-
server, who tends to interpret the curve in a *deterministic* sense. Similar
observations can be made in the fields of sound analysis [1.2,3], adaptive
control [1.4], or in the field of failure detection in dynamic systems [1.5].

   The main goal in many of these problems is a methodology for describing larger records of data by a common set of parameters carrying
the significant process information. These parameters might be extracted
from the observed process *blockwise*, or they might be updated *recursively*, each time a new sample of the observed process is captured.

   To handle this type of problem, linear prediction algorithms for
estimating the parameters of appropriate parametric process models have

been developed. The idea of parametric process modeling in context with linear prediction methods is based on the assumption that a signal is completely determined in terms of its first-order (mean) and second-order (covariance) information when only the parameters and the excitation of the generating filter (process model) are known.

One may assume that an observed process is the output of a *non-recursive* (moving average, MA) system, or the output of a *pure recursive* (autoregressive, AR) system, or even the output of a more general *autoregressive moving average* (ARMA) system, depending on the specific application. In the case of the speech signal, the all-pole (AR) model seems to be appropriate at the first glance, since the vocal tract can be modeled physically as an all-pole resonance tube in most situations.

The speech signal is a *one-dimensional* (scalar) sequence. The considerations of this book, however, can be extended to *vector-valued* data, i.e., multichannel data, or even to the case of *multidimensional* data. For example, an image taken from a video camera can be viewed as two-dimensional data. Image sequences, e.g., a film, represent an example of three-dimensional data. The basic algorithm developments presented in this book, however, will be given for the example of a scalar AR model. In Chap. 10, and in the Appendix, it will be shown how the algorithms thus obtained can be extended to larger process models.

In the real world, the parameters of the process models introduced above can be the subject of more or less abrupt *changes*. This leads to the problem of *adaptation* or *tracking* of the model parameters when time progresses. Note at this point that our use of the terminology "time" for the independent process coordinate is due to the general understanding of a random process as a "time" series. The independent process coordinate, however, could also be a *spatial* coordinate as, e.g., in image processing. The development of appropriate algorithms for parameter estimation and tracking based on linear prediction approaches will be the main topic of this book. For the purpose of estimating the parameters of an observed process, it is sometimes useful to apply the idea of *"inverse modeling"*.

The idea of inverse modeling is basically the following. Take a linear system which has just the inverse structure of the assumed process model. Use the observed process as an excitation of this *identification filter* and adjust the system parameters such that the system output predicts the observed process in an optimum sense with respect to a certain error criterion or objective function. The parameter estimates of the underlying process model are then uniquely determined by the thus-obtained parameters of this "inverse" filter used for identification. Note that the parameters of the inverse filter are generally obtained via a procedure that primarily seeks to minimize the prediction error in some sense. In most cases, the number of error samples, which should be minimized by an appropriately chosen parameter set, is much larger than the number of coefficients or parameters of the inverse filter. This leads quite naturally to an *overdetermined* problem. It is convenient to employ least-squares methods [1.6] for calculating the desired parameter set. Least-squares methods seek a minimization of the *prediction error energy*. In the case of a least-squares solution, the system parameters are determined by a special set of linear equations termed the *Normal Equations*. The very important properties of the Normal Equations will be considered in Chap. 2 of this book. The Normal Equations are completely determined by the covariance matrix of the observed process. Furthermore, the Normal Equations have an important interpretation as the discrete form of the *Wiener-Hopf integral equation* [1.7].

Therefore, it can be shown that least-squares parameter estimates are an optimum solution of the prediction problem as long as the observed process is completely determined by its first- and second-order (mean and covariance) information, as is the case for Gaussian processes. Exact least-squares parameter estimation algorithms are sometimes still too complicated to compute. The general inversion of the Normal Equations requires a number of computations that is proportional to the cube of the model order. Such classical algorithms for inversion of the Normal Equations will be treated in Chap. 3.

A situation of particular interest arises when the parameter set of a current time step can be computed in the presence of the parameter set of the previous time step. In this case, the linear prediction problem has a *recursive* solution such that the desired actual parameter set can be computed from the previous (one time-step delayed) parameter set plus some *update information* which depends on the current (incoming) sample of the observed process. Most of these recursive least-squares algorithms exhibit the structure of a *Kalman filter* [1.8,9]. The first commonly recognized algorithm of this type was reported by Godard [1.10] in 1974 in an application of channel equalization. The earliest reference of the RLS algorithm, however, appears to be Plackett [1.11]. The recursive least-squares (RLS) algorithm, as later described by Godard, has initiated a rapid development of recursive least-squares techniques for fast parameter tracking. Godard's RLS algorithm has a computational complexity that grows with the square of the model order. The RLS algorithm can be derived using the *Sherman-Morrison identity* [1.12], which is also known as the *matrix inversion lemma* [1.13]. On the other hand, the RLS algorithm can be derived from a Kalman filter approach [1.5, 13]. This will be shown in Chap. 10. The main goal in the RLS algorithm is the *recursive updating of the inverse covariance matrix* of the observed process.

Since it can be shown (Chap. 2) that the computational complexity for the recursive updating of the covariance matrix requires only an amount of computation that grows linearly with the model order, one can assume that there must exist an algorithm that allows the recursive updating of the inverse covariance matrix, with roughly the same computational complexity, that also depends only linearly on the model order. And indeed, this is the case. An algorithm for recursive inversion of the Normal Equations with a computational complexity that depends linearly on the model order was introduced by Ljung et al. [1.14] in 1978. This algorithm became known as the *fast Kalman algorithm*. The algebraic derivation of the fast Kalman algorithm succeeds by introducing a sophisticated partitioning scheme in the Sherman–Morrison identity, leading to the *Sherman–Morrison identity for partitioned matrices*, which is treated in Chap. 5 . Until recently, this recursive law for updating an inverse partitioned covariance matrix was the only basis for the algebraic derivation of fast RLS algorithms. Several types of fast RLS algorithms have appeared in the literature. The fast a posteriori error sequential technique (FAEST) introduced by Carayannis et al. [1.15] in 1983 seems to be the fastest ever developed technique for the recursive inversion of the Normal Equations. The FAEST algorithm requires only 5p multiplications and divisions per recursion for the recursive updating of the Kalman gain vector of an order p system.

Recently, Cioffi [1.16] has shown that fast RLS techniques can also be obtained from the classical QR decomposition and Givens rotations. This approach extends the classical recursive QRLS algorithms of Gentleman [1.17] and Kung and Gentleman [1.18]. McWhirter [1.19] demonstrated how one can extract the residual from the Gentleman and Kung QRLS algorithm (Gentleman and Kung array) by a fast procedure without explicit back substitution. This will be the topic of Chap. 4. All the QRLS algorithms discussed there are based on the Givens reduction [1.20]. These algorithms do not attempt to update an inverse covariance matrix and hence, they are *unconditionally stable*.

Returning to the fast RLS algorithms based on the Sherman–Morrison identity, these algorithms seem to be very attractive due to their extremely low computational costs. Their application, however, can be cumbersome due to *numerical difficulties* that occasionally arise when fast Kalman-type algorithms are implemented with fixed-point arithmetic. The fast Kalman recursions appear to be very *ill-conditioned*. There exists a second inherent difficulty that arises with the application of RLS algorithms based on the Sherman–Morrison identity. As soon as the covariance matrix of the observed process becomes singular for possibly only a single recursion, this may cause RLS algorithms to become *unstable* and to diverge immediately. This effect arises due to the inherent assumption underlying Kalman-type algorithms, namely, that the recursively updated inverse system matrix *always exists*. This, however, is a fairly weak assumption in real-time applications with millions of iterations and quantized data. As soon as we use quantized data as input, there will always be a nonzero probability that RLS-based algorithms will become unstable in a real-time (or on-line) application. This effect presents a serious problem in the application of RLS-based algorithms. See also the discussion in Chap. 11 of this book. Another discussion on instability effects in context with the RLS algorithm was recently presented by Kubin [1.21]. He attributes the instability effects to the incompatibility of *nonpersistently exciting data* (data that contains long chains of zero samples) with the RLS update equations.

These difficulties with the application of Kalman type algorithms are probably the deeper reason for the constant interest in approximative least-squares techniques. One of the most important algorithms of this type is the *least mean squares* (LMS) algorithm introduced by Widrow and Hoff [1.22] in 1960. It can be shown that the LMS algorithm can be easily derived from the RLS algorithm of Godard, when only the Kalman gain vector is replaced by the state vector of the prediction error filter multiplied by a constant stepsize. The choice of the stepsize is critical for the *convergence rate* of the LMS algorithm. The possible values of the stepsize are fairly limited and depend on the *eigenvalue spread* of the covariance matrix of the observed process. In most cases, only a small stepsize can be used for stable operation of the LMS algorithm, and therefore the convergence rate can drop below acceptable values. On the other hand, large stepsizes generally result in very noisy parameter estimates. The reason for this unpleasant effect is that the LMS algorithm seeks to minimize only the actual prediction error sample, rather than the prediction error energy within a certain window as is the case in exact least-squares techniques. The noisy parameter estimates result from this *"local"* character of the LMS algorithm.

The RLS algorithms of the Kalman type and the LMS algorithm are based on an *algebraic solution* of the Normal Equations assuming the inverse filter is represented in a transversal (direct) structure. The term "normal", however, derives from *geometrical considerations.* Employing a geometrical interpretation, we may assume that a certain finite or infinite number of process samples constitute a signal vector. Using this geometrical model, the least-squares estimation problem can be stated much more conveniently than in the classical algebraic approaches. Assuming an all-zero prediction filter, the signal vectors which are formed by the successively delayed sequence of process samples, constitute an *oblique basis* of the so-called *subspace of past observations* of the signal. In the case that the signal covariance matrix is nonsingular, the dimension of this subspace of past observations is identical with the model order. This will be the general case and the starting point of our considerations. The problem of least-squares estimation is then equivalent with the problem of projecting the actual (incoming) process vector *orthogonally* onto the subspace of past observations. The least-squares predicted process vector is therefore the *orthogonal projection* of the input process onto the subspace of past observations and is therefore an element of the subspace of past observations. The prediction error is orthogonal with respect to the subspace of past observations and is termed the *orthogonal complement* of the predicted process vector. These very important geometrical considerations will become more apparent in Chap. 6. But it is most important to note that the solution of the Normal Equations of linear prediction leads to a set of parameters such that the linear predictor, which is determined by these parameters, performs the orthogonal projection of the observed process onto its vectors of past observations.

This very important conjunction ultimately leads to an alternative solution of the Normal Equations. Clearly, we may first successively construct an *orthogonal basis* of the subspace of past observations and, second, we may project the input process vector *successively* on the orthogonal basis vectors. This procedure was termed the *"growing order"* or *ladder* formulation [1.23] of the linear prediction problem. The required orthogonalization operations can be performed using a *Gram-Schmidt othogonalization* procedure [1.24]. One major advantage that ultimately emerges from this approach is that the model order can be varied *dynamically* depending on the prediction error or some other precision criterion. This is possible due to the *"decoupled"* nature of the orthogonal directions of approximation which are provided by the orthogonal basis vectors used in the projection process. As a consequence, the order of the Normal Equations does *not* have to be determined a priori before solving them with a ladder algorithm. We just start with the orthogonalization procedure and stop as soon as we have reached a desired accuracy of prediction or approximation. A second consequence of the orthogonalization procedure for solving the Normal Equations is that this leads to a completely new structure of the prediction error filter. Chapter 6 is devoted to this alternative structure of the prediction error filter, which has been termed the *ladder (or lattice) structure*. The ladder structure is an equivalent representation of the transversal structure, but its coefficients generally have a smaller variance than the coefficients of the transversal structure. This is one reason why ladder coefficients are *less sensitive to round-off error*. We note, however, that the round-off error sensitivity of the prediction error filter is not always the dominant part that determines the round-off error characteristic of a linear prediction algorithm. Nevertheless, least-squares ladder algorithms generally exhibit more robust characteristics when operated in numerically uncertain environments than transversal algorithms.

We may distinguish basically three classes of ladder algorithms. The first class treated in Chap. 7 are the algorithms of the *"Levinson type"*, which require the computation of the corresponding transversal predictor parameters as intermediate variables. Therefore, these algorithms rely mainly on *Levinson's recursion* [1.25] which accomplishes the order recursive conversion between the two representations (ladder and transversal) of a prediction error filter. The possibly most prominent member of this class of algorithms is the *Levinson-Durbin algorithm* [1.26] for a *block-wise* calculation of prediction error filter parameters. Levinson-type recursions, may also play an important role in the *recursive* case of linear prediction. See Chap. 7 and Sect. A.2. A second class of ladder algorithms is formed by the class of methods which seek to replace the Levinson recursion by some type of *inner-product recursions*. These methods will be discussed in Chap. 8. They also extend to the recursive case. Two different types of inner products (sometimes termed "generalized residual energies" [1.27, 28]) can be introduced, leading to different algorithms.

Similar to the transversal RLS algorithms, fast estimation algorithms can also be derived in the case of the ladder structure. They constitute the third class of ladder algorithms, treated in Chap. 9. The initial derivation of these algorithms was presented in the fundamental work of Lee [1.29] in 1980, who employed a sophisticated theory of differentials of projection operators for updating the partial projection operator at a certain stage of the algorithm as soon as a new sample of the input process is captured. The fundamentals of Lee's work can be traced back to early contributions in *perturbation theory* [1.30]. In Chap. 9, we give a detailed and instructive explanation of Lee's idea. His geometrical approach was later adopted by Cioffi to derive the transversal RLS algorithms [1.31] utilizing this geometrical framework. This approach, however, does not provide faster solutions or basically different structures of the update equations than already available with the previously discussed FAEST algorithm of Carayannis who had developed his algorithm using pure algebraic techniques. In general, one may conclude that both transversal and ladder-based RLS algorithms can be derived using either an algebraic approach based on the Sherman-Morrison identity for partitioned matrices or, equivalently, a pure geometrical procedure. Sometimes, the geometrical approach seems to be more straightforward and less "mystified" than its algebraic counterpart. In this book, we shall discuss both techniques, the algebraic approach for the RLS transversal algorithm and the geometrical derivation of the RLS ladder algorithm. This seems better suited to the historical development but also to the inherent geometrical nature of ladder forms in contrast to the more "algebraic" nature of transversal forms. Nevertheless, geometrical considerations concerning the RLS transversal algorithms may be found in the book of Honig and Messerschmitt [1.32] whereas a derivation of the fast RLS ladder algorithm by pure algebraic manipulations is given by Ljung and Söderström [1.33].

In the recursive approaches that will be discussed, one assumes that the signal model parameters are more or less a continuous function of time. The corresponding process is then termed a *nonstationary* process [1.34]. Sometimes, however, one can observe that relatively long data records of an observed process can be described with a sufficient accuracy when only a single (constant) parameter set is used. Such a process is then termed a *piecewise stationary* process. In a stationary segment, the autocorrelation coefficients will be independent of time and, as a consequence, the covariance matrix will turn into a *Toeplitz matrix*. Additionally, inside a stationary segment, there will be no need for a recursive updating of model parameters since they are constant over time. In this case, it is most appropriate to collect the data inside the stationary segment and to solve the Toeplitz-structured Normal Equations (also known als the Yule-Walker equations [1.35]) by a *nonrecursive procedure*. Such procedures, however, should be capable of exploiting the rich structural properties of Toeplitz matrices.

Toeplitz systems of linear equations have been a topic of constant interest for decades. An excellent description of the properties of Toeplitz forms is available in the work of Grenander and Szegö [1.36] published in 1958. One of the earliest and probably the most well-known algorithm for solving a Toeplitz system is the *Levinson algorithm* [1.25] introduced by Levinson in 1947. Levinson's work was later extended by Durbin [1.26] and is hence called the Levinson-Durbin algorithm. Levinson type algorithms are the subject of Chap. 7. Interestingly, the Levinson algorithm already generates the coefficients of a special ladder form with *identical coefficients* in the forward and backward predictor which is known as the *partial correlation (PARCOR) ladder form* [1.37]. This identity of the forward and backward linear predictor in the PARCOR ladder form stems from the underlying assumption of a stationary input process. The ladder structure thus obtained has an important interpretation as the digital signal processing equivalent to the *acoustic tube model* in speech processing [1.38,39] with stationary excitation. This is also the reason why the coefficients of the ladder structure are sometimes termed "*reflection*" coefficients.

The central part of the Levinson-Durbin algorithm is the Levinson recursion, which relates the reflection coefficients of the ladder form to the parameters of the transversal form prediction error filter. This famous recursion sometimes appears to be ill-conditioned and unbounded in its recursion variables. Therefore, attempts have been made to avoid the explicit computation of the transversal form predictor coefficients in order to facilitate a more robust fixed-point implementation. One of these approaches is the widely introduced LeRoux-Gueguen algorithm [1.40], published in 1977. Instead of computing the transversal prediction filter parameters directly, this algorithm uses a set of *low-variance intermediate variables with bounded dynamic range*, which have the meaning of inner products between the residual vectors of the prediction error filter and the input signal. Henceforth, we may call these important quantities *"generalized residual energies"* (GRE s). Generalized residual energies play a central role in linear prediction theory.

The Levinson-Durbin algorithm in its original version was assumed for a long time to provide the fastest possible solution of a Toeplitz system. Recently, Delsarte and Genin [1.41] have discovered, however, that even the Levinson-Durbin algorithm is redundant in complexity. It can be split into two parts, a symmetric and an antisymmetric part, only one of which needs to be processed. This new class of algorithms was called *"split Levinson"* algorithms. There are several ways of deriving split Levinson algorithms. In this book we shall discuss the original approach of Delsarte and Genin and a recently introduced formulation of Krishna [1.42, 43]. Krishna's approach will be related to the early work of Delsarte and Genin. Later, in Chap. 7, we will derive a second class of split algorithms where the Levinson type recursion, as appearing in the split Levinson algorithms of Delsarte and Krishna, is replaced by inner-product recursions [1.42, 44]. This second class of split algorithms is sometimes referred to as split Schur algorithms in honour of Schur's early work on bounded analytic functions [1.45]. Just as the LeRoux-Gueguen algorithm can be viewed as the "fixed-point counterpart" of the Levinson-Durbin algorithm, these new split Schur algorithms are the "fixed-point counterparts" of the original split Levinson algorithms of Delsarte and Krishna in that they replace the Levinson type recursion by an appropriate inner-product recursion, which, however, requires only *half the number of multiplications* for its computation when compared to the classical algorithm of LeRoux and Gueguen. These new algorithms have a better numerical accuracy, a more regular algorithm structure, and fewer recursion variables than the original split Levinson algorithms whereas the overall operations count remains roughly the same. Nevertheless, the numerical properties of split Levinson and split Schur algorithms are still a field of active research.

Having scanned some aspects of stationary linear prediction, we will now return to nonstationary processes. We have seen that the Levinson-Durbin algorithm determines a ladder form prediction error filter from the Toeplitz structured covariance matrix. This is in contrast to the fast RLS ladder algorithms, which, implicitly, determine the ladder-form prediction error filter via the time-recursively updated *inverse* covariance matrix. Both the RLS transversal algorithms and the RLS ladder algorithms exhibit a *fast Kalman type algorithm structure*. This algorithm structure is characterized by the property that it implicitly involves the updating of an inverse covariance matrix. As already mentioned in this introduction, the recursive updating of an inverse covariance matrix can

cause several inherent difficulties when the related algorithms are operated in numerically uncertain environments. In the fast Kalman type algorithm structure, round-off errors can *accumulate* in the time-recursively
updated variables. This causes the algorithms to degrade in an unpleasant
way so that they lose their fast start-up and tracking behavior and finally
perform not significantly better than simple LMS methods when operated
in finite precision environments. Another problem is that the recursions
of fast RLS algorithms prohibit the incorporation of higher-order
windows [1.46] on the data. Only simple rectangular and exponential
windows or simple modifications thereof can be used, and therefore the
advantage of a low computational complexity of such algorithms might
be compensated by the fact that the steady-state and tracking behavior
is quite limited and sometimes far from optimum. Bershad and Macchi
[1.47] recently gave an instructuve example of this limited behavior of
exponentially weighted RLS algorithms.

Therefore, it can be of great interest to develop algorithms which
compute the true least-squares ladder prediction error filter directly from
the time-recursively updated covariance matrix instead of from its inverse.
An true recursive least-squares ladder algorithm of this type was first
introduced by Strobach [1.28] in 1986. In contrast to conventional RLS-type
algorithms, the algorithm of Strobach splits the estimation problem into
a *tracking problem* and a *modeling problem*. The tracking of the time-
varying process is performed by a recursive updating of the covariance
matrix. Note that the updating of an inverse covariance matrix is *not*
employed in this approach. The true least-squares ladder reflection
coefficients are determined from the covariance information at each time
step using a pure order recursive procedure. This *puré order recursive
ladder algorithm* (PORLA) has several advantages over conventional
Kalman based RLS techiques. First, the PORLA algorithm allows the
application of higher-order recursive windows (e.g. recursive Hanning
[1.46]) to the data to improve the tracking and the steady-state estimation
behavior. Second, since the time recursion in this new method is bounded
on the calculation of the input data covariance matrix, *round-off errors
cannot propagate in time* in higher stages of the pure order recursively
constructed ladder form. This way, the superior fast start-up and tracking
behavior of the PORLA algorithm cannot be corrupted by round-off error.
The PORLA algorithm arises from an algebraic approach using generalized
residual energies. The generalized residual energies used here are somewhat different to those used in the LeRoux-Gueguen algorithm. The
generalized residual energies used in the PORLA method are defined as
inner products of residual (prediction error) *vectors*. The application of
these intermediate quantities can be traced back to the early work of
Cumani [1.27] of 1982, who first used this type of generalized residual
energies to replace the Levinson recursion in the Makhoul covariance

ladder algorithm [1.48]. In fact, Cumani's approach has motivated the search for the PORLA method. Cumani's algorithm appears as a special case of the PORLA method when only the reflection coefficients are assumed to be time-invariant and the forward and backward reflection coefficients are forced to identical values by employing *Burg's harmonic mean principle* [1.49]. Several approximate ladder algorithms can be derived from the PORLA method by a stepwise introduction of simplifying assumptions associated with stationary processes. Finally, we end up with an algorithm that has the same infinite precision behavior as the Levin-son-Durbin algorithm. This simplified PORLA algorithm can be viewed as another fixed-point implementation of the Levinson—Durbin algorithm and has, in fact, similar numerical properties when compared to the LeRoux-Gueguen algorithm but a somewhat higher computational complexity, which stems from the incorporation of different generalized residual energies in both approaches.

Now, as we have seen the possibility of deriving a fixed-point implementation of the Levinson-Durbin algorithm from the true recursive least-squares PORLA method, it should also be possible to generalize the LeRoux-Gueguen algorithm to the nonstationary case by following Strobach's idea in the opposite direction, but simply using different generalized residual energies, namely, those appearing in the LeRoux and Gueguen algorithm.

This important step of generalization has been presented by Sokat and Strobach [1.50], who demonstrated that a new class of PORLA algorithms can be established, based on the generalized residual energies of LeRoux and Gueguen, but having similar features to the original counterpart given by Strobach. The interesting point of this generalization is that the new algorithm introduced by Sokat has a lower complexity in terms of arithmetic operations than the original PORLA algorithm of Strobach. This gain in speed was achieved by a different definition of generalized residual energies which makes a better use of forward/backward predictor relationships. Another interesting discussion of faster PORLA algorithms is available in the Appendix. Here, we show how such algorithms can be derived from another intermediate quantity, termed the *"generalized covariance"*. The resulting recursions are particularly interesting for implementation on *triangular systolic arrays* [1.51]. We present the explicit systolic array structures of these new update schemes. Besides the simple scalar considerations, the algorithms are extended to the vector-valued case, both for recursive updating and block processing.

Generally, we may note that all PORLA type algorithms have a computational complexity that is proportional to the square of the model order, and thus the advantages of PORLA based algorithms must be paid for with a somewhat higher computational complexity when higher-order modeling problems are considered. On the other hand, PORLA type algorithms will always remain stable in on-line applications, even in the case of rank deficiencies of the covariance matrix or nonpersistently exciting data.

After introducing the most important concepts and algorithms of linear prediction theory for the simplest case of a scalar AR model, Chap. 10 presents ways these techniques can be extended and applied to more complicated estimation problems. Such extensions of the signal model are, e.g., the *joint process estimation case*, where a process must be modeled with respect to the subspace of past observations of a *related* process. The joint process approach has a direct application in *FIR system identification* and *noise cancelling*. Other extensions deal with the case of a *vector autoregressive* (VAR) process model which can be used in modeling *multichannel* data. The vector-valued case of linear prediction is also the basis for many *ARMA system identification algorithms*. The ARMA system identification problem can be *embedded* in the larger VAR estimation problem. In the stationary case, those concepts may apply also to the problem of *parametric ARMA spectral estimation*.

Chapter 11 discusses several aspects of implementation, such as considerations about numerical accuracy and stability. A large number of references are provided in Chap. 11 to enable the interested reader to find links to the special application-oriented literature.

Some last-minute results have been summarized in the Appendix. There, we develop a new class of covariance ladder algorithms of the PORLA type. These new algorithms are based on a clever time-update property of the error covariance. The new algorithms are particularly useful for a VLSI implementation. The explicit VLSI structures are presented. Moreover, the new algorithms have been extended to the block processing case and to the multichannel case. A discussion of practical problems arising in the computation of large reflection matrices in the multichannel case and their solution using the Penrose pseudoinverse concept has also been added.

2. The Linear Prediction Model

One of the most powerful techniques for describing time-varying processes arises from the assumption that an observed process $\{x(t)\}$, represented by an ensemble of L independent measurements $x_0(t)$, $x_1(t)$, ..., $x_{L-1}(t)$ at time step t

$$\mathbf{x}(t) = [x_0(t),\ x_1(t),\ ...\ ,\ x_{L-1}(t)]^T, \qquad (2.1)$$

can be predicted by a *weighted linear combination* of previous measurements available in the data vectors $\mathbf{x}(t-1)$, $\mathbf{x}(t-2)$, ..., $\mathbf{x}(t-p)$, where

$$\mathbf{x}(t-m) = [x_0(t-m),\ x_1(t-m),\ ...\ ,\ x_{L-1}(t-m)]^T\ ; \quad 1 \le m \le p \qquad (2.2)$$

and p denotes the *model order* . Introducing the associated *weighting coefficients* $a_1(t)$, $a_2(t)$, ..., $a_p(t)$, one can write the linear combination as

$$\hat{\mathbf{x}}^f(t) = \sum_{m=1}^{P} a_m(t)\mathbf{x}(t-m) \qquad (2.3)$$

where $\hat{\mathbf{x}}^f(t)$ is the resulting *forward predicted process vector*. The parameter vector

$$\mathbf{a}(t) = [\ a_1(t),\ a_2(t),\ ...\ ,\ a_p(t)\ ]^T \qquad (2.4)$$

should be determined in a sense such that the forward predicted process vector $\hat{\mathbf{x}}^f(t)$ becomes as "close" as possible to the true process vector $\mathbf{x}(t)$. Introducing the matrix of *past observations* $\mathbf{X}(t)$

$$\mathbf{X}(t) = [\ \mathbf{x}(t-1),\ \mathbf{x}(t-2),\ ...\ ,\ \mathbf{x}(t-p)\ ] \qquad (2.5)$$

we may express the linear combination or *predictor equation* (2.3) as

$$\hat{\mathbf{x}}^f(t) = \mathbf{X}(t)\,\mathbf{a}(t)\ . \qquad (2.6)$$

The predictor determined by the expressions (2.3) or (2.6) can be inter-

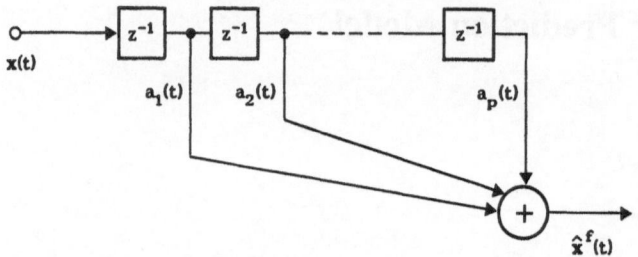

Fig. 2.1. All-zero forward prediction filter of order p

preted as an all-zero or moving average (MA) filter structure as shown in Fig. 2.1.

The evaluation of the *"closeness"* of the forward predicted process vector $\hat{\mathbf{x}}^f(t)$ to the true observation $\mathbf{x}(t)$ makes it necessary to define a *prediction error vector* $\mathbf{e}(t)$ as

$$\mathbf{e}(t) = \mathbf{x}(t) - \hat{\mathbf{x}}^f(t) \quad . \tag{2.7}$$

A substitution of the prediction filter equation (2.6) into the definition of the prediction error vector (2.7) yields another filter equation defining what is called the *prediction error filter* of order p:

$$\mathbf{e}(t) = \mathbf{x}(t) - \mathbf{X}(t)\, \mathbf{a}(t) \quad . \tag{2.8}$$

The structure of the prediction error filter is illustrated in Fig. 2.2.

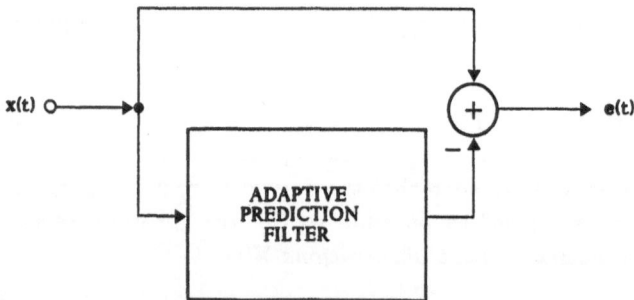

Fig. 2.2. All-zero prediction error filter of order p

2.1 The Normal Equations of Linear Prediction

Expression (2.8) reveals that the prediction error is a *linear* function of the parameter set $a(t)$; hence the terminology *"linear prediction"*. The parameter or coefficient set $a(t)$ must be adjusted such that the prediction error vector $e(t)$ is minimized in some sense.

For the case $L > p$, i.e., the case where the number of measurements exceeds the model order, relation (2.8) turns into an overdetermined system of linear equations for the desired parameter vector $a(t)$. In principle, there is a wide variety of possible methods for solving (2.8).

A special case of interest, however, arises when the parameters $a(t)$ are determined in a sense such that the prediction error *energy* $E(t)$

$$E(t) = e^T(t)\, e(t) = \sum_{n=0}^{L-1} e_n^2(t) \tag{2.9}$$

is minimized by solving the least-squares problem

$$E(t) \overset{!}{=} min \longrightarrow \hat{a}(t) \tag{2.10}$$

where $\overset{!}{=}$ indicates that the minimum of $E(t)$ is associated with the parameter set $\hat{a}(t)$ which is called the *least-squares estimate* of $a(t)$. Throughout this book we will restrict our considerations to the broad class of algorithms which solve the problem (2.10).

A quick inspection of expression (2.9) reveals that the prediction error energy $E(t)$ has an important *geometrical interpretation* in that it is equivalent to the *Euclidean length* of the error vector $e(t)$. Obviously, the solution of the least-squares problem (2.10) yields a set of parameters $\hat{a}(t)$ which, in turn, force the prediction error filter to produce an error vector of least Euclidean length or minimal "energy".

In order to find a solution to the least-squares problem (2.10), we may take advantage of the fact that the prediction error energy $E(t)$ can be expressed as a *quadratic scalar function* of the prediction filter parameter set $a(t)$. This is accomplished by substituting (2.8) into (2.9), yielding

$$E(t,a(t)) = x^T(t)\, x(t) - a^T(t)\, X^T(t)\, x(t) - x^T(t)\, X(t)\, a(t)$$
$$+ a^T(t)\, X^T(t)\, X(t)\, a(t) \overset{!}{=} min . \tag{2.11}$$

The functional $E(t,a(t))$ determines a *performance surface* in the $(p+1)$-dimensional space. This performance surface has the *convex* regular shape of a p-dimensional parabola with *one distinct minimum*. The situation is illuminated by the order $p = 2$ example of Fig. 2.3.

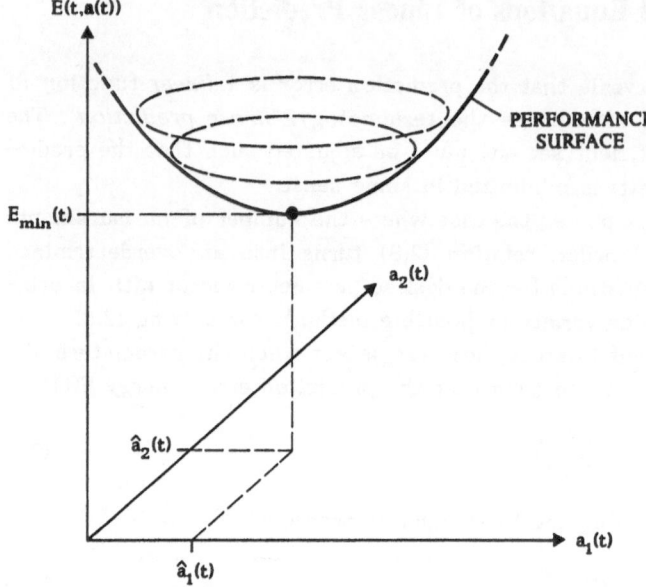

Fig. 2.3. Convex performance surface as determined by the functional $E(t,a(t))$. Example shown for the order $p = 2$ problem

From these considerations, it is obvious that the minimum of $E(t,a(t))$ can be determined by setting the partial derivatives of $E(t,a(t))$ with respect to $a^T(t)$ to zero,

$$\partial E(t,a(t)) / \partial a^T(t) = -2X^T(t)x(t) + 2X^T(t)X(t)a(t) \tag{2.12}$$

$$= \begin{bmatrix} 0 \ldots 0 \end{bmatrix}^T,$$

which gives the result

$$X^T(t)X(t)\hat{a}(t) = X^T(t)x(t). \tag{2.13}$$

The system (2.13) constitutes the *Normal Equations of linear prediction* for determining the parameter estimate $\hat{a}(t)$ in the least-squares sense from the observed process vector $x(t)$ and the matrix of past observations $X(t)$.

The history of the Normal Equations can be traced back to Levinson [2.1], who formulated the Wiener filtering problem in discrete time. The solution to the Wiener filtering problem, or the explicit formula for the optimum estimate in the continuous time case, requires the solution of an integral equation known as the *Wiener-Hopf integral equation* (Wiener and Hopf 1931) [2.2]. In 1947, Levinson showed that in the case of discrete time signals the Wiener–Hopf integral equation takes on a matrix form and this matrix form is identical to the *Normal Equations* (2.13).

2.2 Geometrical Interpretation of the Normal Equations

The term "normal" reveals that the Normal Equations have a geometrical interpretation. In fact, it will be shown in the following that the solution of the Normal Equations yields a parameter set $\hat{\mathbf{a}}(t)$ which – in a geometrical sense – provides the *orthogonal projection* of the process vector $\mathbf{x}(t)$ onto what is called the *subspace of past observations*, which is spanned by the matrix of past observations $\mathbf{X}(t)$. In this geometrical model, the process vectors of past observations, which form the matrix $\mathbf{X}(t)$, are interpreted as *oblique basis vectors* of the subspace of past observations. Assuming that the system matrix $\mathbf{X}^T(t)\mathbf{X}(t)$ of the Normal Equations is *nonsingular*, i.e., the column vectors of $\mathbf{X}(t)$ are all *nonzero* and *linearly independent*, the dimension of this subspace is identical with the model order p. In this case, the Normal Equations have the solution

$$\hat{\mathbf{a}}(t) = \left(\mathbf{X}^T(t)\mathbf{X}(t)\right)^{-1}\mathbf{X}^T(t)\mathbf{x}(t). \tag{2.14}$$

In the case of a *rank deficient* system matrix, i.e., when the dimension of the subspace of past observations drops below the predefined model order, the inverse matrix in (2.14) is not defined. In this case, the simple yet powerful concept of the inverse no longer holds. See Chap. 3 for a detailed discussion of this case of a singular system matrix of the Normal Equations. Throughout this book, we will assume that $\left(\mathbf{X}^T(t)\mathbf{X}(t)\right)^{-1}$ exists, and we shall add appropriate comments as soon as a violation of this condition causes particular problems.

Using the least-squares parameter estimate, as defined by (2.14), and substituting it into the prediction filter (2.6) yields the following expression for the forward predicted process vector $\hat{\mathbf{x}}^f(t)$ in the least-squares sense:

$$\hat{\mathbf{x}}^f(t) = \mathbf{X}(t)\left(\mathbf{X}^T(t)\mathbf{X}(t)\right)^{-1}\mathbf{X}^T(t)\mathbf{x}(t) . \tag{2.15}$$

According to the prediction filter equation (2.6), the forward predicted process $\hat{\mathbf{x}}^f(t)$ is always a linear combination of weighted basis vectors of the subspace of past observations. Consequently, $\hat{\mathbf{x}}^f(t)$ is always an element of the subspace of past observations. From these considerations, it becomes obvious that $\hat{\mathbf{x}}^f(t)$ can be obtained as the *projection* of the actual observation $\mathbf{x}(t)$ onto the subspace of past observations spanned by $\mathbf{X}(t)$. Introducing the *projection operator* $\mathbf{P}(t)$

$$\mathbf{P}(t) = \mathbf{X}(t)\left(\mathbf{X}^T(t)\mathbf{X}(t)\right)^{-1}\mathbf{X}^T(t) \quad ; \qquad \mathbf{P}(t) \in \mathbb{R}^{L \times L}, \tag{2.16}$$

we may interpret the operation (2.15) as a projection operator applied on the actual observation $\mathbf{x}(t)$, hence (2.15) reduces to

$$\hat{\mathbf{x}}^f(t) = \mathbf{P}(t)\mathbf{x}(t) \tag{2.17}$$

and the corresponding prediction error vector $\mathbf{e}_{min}(t)$ is determined by

$$\mathbf{e}_{min}(t) = \Big(\mathbf{I} - \mathbf{P}(t)\Big)\mathbf{x}(t) = \mathbf{P}^\perp(t)\mathbf{x}(t) \ , \tag{2.18}$$

where $\mathbf{I}$ is the identity matrix of appropriate dimension and $\mathbf{P}^\perp(t)$ is the *orthogonal complement* of $\mathbf{P}(t)$.

We may now state two important properties of the projection operator $\mathbf{P}(t)$.

(1) *SYMMETRY*
From definition (2.16), it can readily be seen that

$$\mathbf{P}^T(t) = \mathbf{P}(t). \tag{2.19}$$

(2) *IDEMPOTENCE*
More important than the symmetric property of the projection operator $\mathbf{P}(t)$ is its property of *idempotence*. In fact, it can be demonstrated utilizing definition (2.16) that the following property holds:

$$\mathbf{P}(t)\mathbf{x}(t) = \mathbf{P}(t)\ \mathbf{P}(t)\ ...\ \mathbf{P}(t)\mathbf{x}(t). \tag{2.20}$$

Expression (2.20) [idempotence property of the projection operator $\mathbf{P}(t)$] can be interpreted as follows. Once a vector has been obtained as the projection onto the subspace $\mathbf{X}(t)$ using $\mathbf{P}(t)$, it can never be modified by any further application of $\mathbf{P}(t)$.

In the next step, we give a proof that $\mathbf{P}(t)$ provides the *orthogonal* projection of the observed process $\mathbf{x}(t)$ onto the subspace of past observations. For this purpose, we have to demonstrate that $\mathbf{e}_{min}(t)$ is orthogonal with respect to the oblique basis $\mathbf{X}(t)$ of the subspace of past observations, and hence, we must show that

$$\mathbf{X}^T(t)\mathbf{e}_{min}(t) = \Big[0...0\Big]^T \tag{2.21}$$

holds.

Proof. Substituting (2.18) into (2.21) and taking definition (2.16) into consideration gives

$$\left[\mathbf{X}^T(t) - \mathbf{X}^T(t)\, \mathbf{P}(t) \right] \mathbf{x}(t)$$

$$= \left[\mathbf{X}^T(t) - \mathbf{X}^T(t)\, \mathbf{X}(t) \left(\mathbf{X}^T(t)\, \mathbf{X}(t) \right)^{-1} \mathbf{X}^T(t) \right] \mathbf{x}(t)$$

$$= \left[0 \ldots 0 \right]^T . \tag{2.22}$$

End of proof.

From (2.21), it follows directly that

$$\hat{\mathbf{x}}^{f\,T}(t)\, \mathbf{e}_{min}(t) = 0 \ , \tag{2.23}$$

$$\mathbf{x}^T(t)\, \mathbf{e}_{min}(t) = \left(\hat{\mathbf{x}}^{f\,T}(t) + \mathbf{e}^T_{min}(t) \right) \mathbf{e}_{min}(t)$$

$$= \mathbf{e}^T_{min}(t)\, \mathbf{e}_{min}(t) = E_{min}(t) \ . \tag{2.24}$$

Figure 2.4 illustrates the resulting geometrical relationships. As a consequence of these considerations, we may note that the demand for minimal Euclidean length of the error process $\mathbf{e}(t)$ and the demand for orthogonality of the error process $\mathbf{e}(t)$ with respect to the subspace of past observations $\mathbf{X}(t)$ are *equivalent formulations* of the same least-squares error criterion.

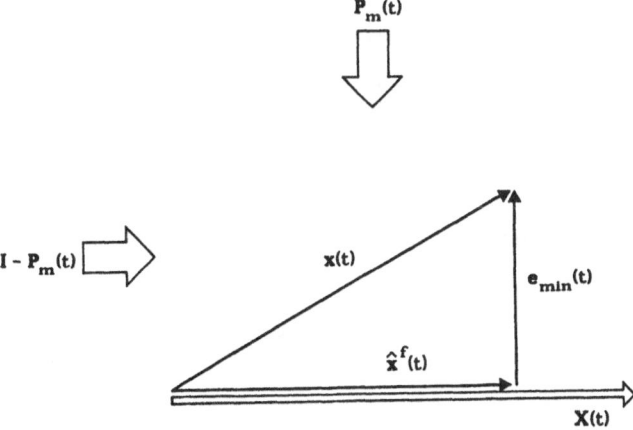

Fig. 2.4. Decomposition of the process vector $\mathbf{x}(t)$ into the forward predicted process $\hat{\mathbf{x}}^f(t)$ and its orthogonal complement represented by the prediction error vector $\mathbf{e}_{min}(t)$ of least Euclidean length

2.3 Statistical Interpretation of the Normal Equations

Besides the geometrical interpretation, there exists a meaningful *statistical* interpretation of the Normal Equations. For this purpose, we shall define the unnormalized covariance matrix of time-shifted observations $\{x(t-i), 0 \le i \le p\}$ of the process $\{x(t)\}$ as

$$\Phi(t) = \left[\Phi_{i,j}(t) \right] \tag{2.25}$$

$$\Phi_{i,j}(t) = x^T(t-i)\, x(t-j) \;; \qquad 0 \le i,j \le p \;, \tag{2.26}$$

equivalently,

$$\Phi(t) = \left[x(t) \quad X(t) \right]^T \left[x(t) \quad X(t) \right] \;. \tag{2.27}$$

Utilizing (2.27), the following partitioning scheme of the covariance matrix can be introduced:

$$\Phi(t) = \begin{bmatrix} x^T(t)\, x(t) & x^T(t)\, X(t) \\ X^T(t)\, x(t) & X^T(t)\, X(t) \end{bmatrix} \;. \tag{2.28}$$

Viewing (2.28), it is easy to check that the Normal Equations of linear prediction (2.13) can alternatively be expressed as

$$\Phi(t) \begin{bmatrix} -1 \\ \hat{a}(t) \end{bmatrix} = \begin{bmatrix} x^T(t)\, X(t)\, \hat{a}(t) - x^T(t)\, x(t) \\ 0 \\ \vdots \\ 0 \end{bmatrix} \;. \tag{2.29}$$

Substitution of (2.14) into (2.11) reveals that

$$E_{min}(t) = E(t,\hat{a}(t)) = x^T(t)\, x(t) - x^T(t)\, X(t)\, \hat{a}(t) \tag{2.30}$$

and hence, (2.29) reduces to

$$\Phi(t) \begin{bmatrix} -1 \\ \hat{a}(t) \end{bmatrix} = \begin{bmatrix} -E_{min}(t) \\ 0 \\ \vdots \\ 0 \end{bmatrix} \;. \tag{2.31}$$

Through a leftside multiplication of (2.31) with the expression $\left[-1 \quad \hat{a}^T(t) \right]$, we conclude that the prediction error energy can be computed from the covariance matrix of the observed process and the least-squares parameter estimates as follows:

$$E_{min}(t) = \begin{bmatrix} -1 & \hat{a}^T(t) \end{bmatrix} \Phi(t) \begin{bmatrix} -1 \\ \hat{a}(t) \end{bmatrix} . \tag{2.32}$$

Equation (2.32) describes the *linear transformation* of a multivariate Gaussian process described by its covariance matrix $\Phi(t)$ into a Gaussian output process of variance $E_{min}(t)$ [2.3]. Clearly, the linear system performing this transformation is the prediction error filter (2.8) having the least-squares parameter estimates $\hat{a}(t)$ as coefficients. This expression (2.32) also reveals that least-squares parameter estimation considers only the first- and second-order (mean and covariance) information in the optimization process, and hence the method of least squares is only optimal for Gaussian processes, since a Gaussian process is completely determined by its mean and covariance information. For non-Gaussian processes, the least-squares estimate is worse than the optimal, but computationally burdensome maximum likelihood method [2.4, 5].

A perfect prediction in the sense of a minimal prediction error energy with a given pth order prediction error filter is possible when the observed process is the output of an *autoregressive process model* excited by a *white Gaussian noise process*. The autoregressive (AR) process model is defined as the *inverse* structure of the pth order prediction error filter.

Assuming time-invariant coefficients of the prediction error filter (2.8) for the moment, we shall state the z-transform of the prediction error filter as

$$H_p(z) = 1 - \sum_{m=1}^{p} z^{-m} a_m . \tag{2.33}$$

The z-transform of the AR process model $H_M(z)$ is then defined as the *inverse* of the z-transform of the prediction error filter, hence

$$H_M(z) = 1 \left/ \left(1 - \sum_{m=1}^{p} z^{-m} a_m \right) \right. \tag{2.34}$$

determines the corresponding AR process model. Taking the inverse z-transform of (2.34), we obtain the corresponding filter structure of the AR process model

$$x(t) = y(t) + \sum_{m=1}^{p} a_m(t) x(t-m) , \tag{2.35}$$

where we have reintroduced time-varying coefficients $\{a_m(t), 1 \le m \le p\}$ and $\{y(t)\}$ is a white Gaussian noise process. The inverse correspondence between the z-transform of the AR process model and the z-transform of the prediction error filter is why the prediction error filter is sometimes called an "*inverse filter*". It just "compensates" the correlation that a white Gaussian noise process has gained by being fed through the

AR process model. The coefficients $a_1(t)$, $a_2(t)$, . . . , $a_p(t)$ are frequently termed AR parameters, because they are the coefficients of the AR model which is assumed to be the source of the observed *Gaussian AR process* $\{x(t)\}$

The Gaussian AR process of order p can be perfectly identified by the all-zero prediction error filter of order p with coefficients adjusted in the least-squares sense. The prediction error is then again a white

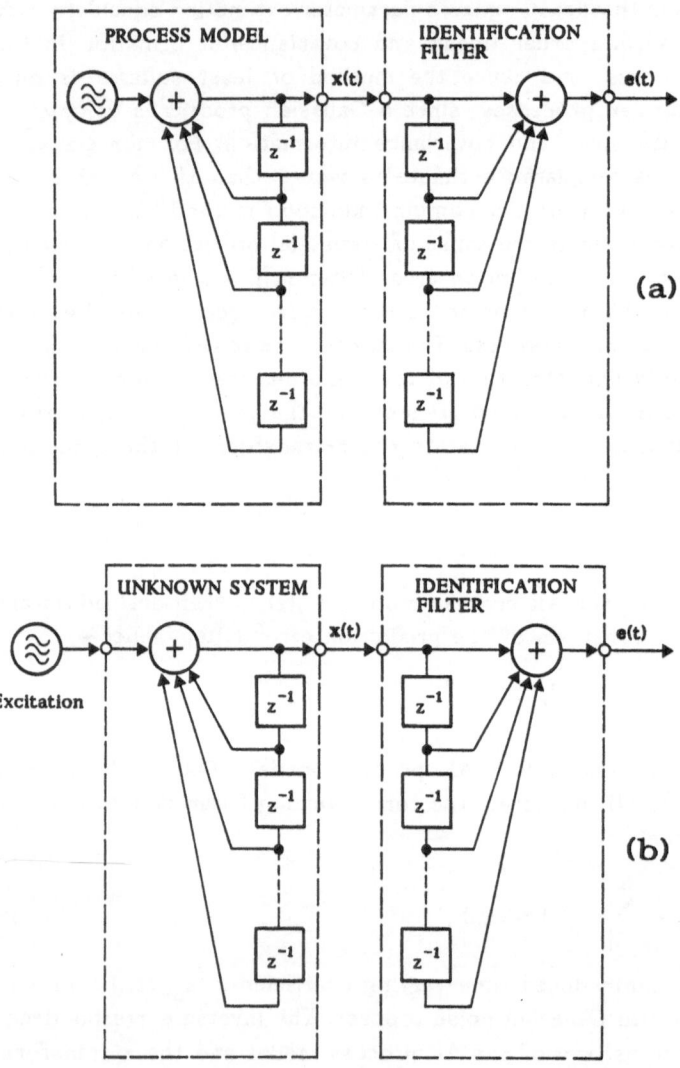

Fig. 2.5 a, b. Principle of autoregressive (AR) process and system identification. (a) AR modeling of an unknown signal. (b) AR modeling of an unknown linear system excited by a white Gaussian noise process

process with Gaussian amplitude statistics. This gave rise to the terminology "*whitening filter*" as an alternative expression for the all-zero
prediction error filter.

Figure 2.5a illustrates the relationship between the AR process model
and the MA prediction error filter. Figure 2.5b shows that the concept
underlying the AR process model can be applied to the problem of
identifying an unknown system with time-varying parameters when this
system is excited by a white Gaussian noise source.

2.4 The Problem of Signal Observation

Until now, we have assumed that - at each time step - a number L of
measurements of the observed process $x(t)$ are available, where L > p,
leading to the least-squares solution for the desired parameters. Unfortunately, in a real-time application, we will be faced with the problem
that - at each time step - we will have access to only *one* sample
measurement of the observed process. In this case, the process vector
$x(t)$ (2.1) , defined by an *ensemble* of measurements at the same time step,
reduces to a scalar quantity and the least-squares approach can no longer
be applied. A way to circumvent this difficulty is to form a process
vector from the measurements of *subsequent* time steps. The time interval of the measurements contributing to the process vector is usually
called a *window.* We conclude that one distinguishes between process
vectors formed from *ensembles* of measurements and *time series* of
measurements.

(1) Ensemble definition of a process vector [see also (2.1)]:

$$\mathbf{x}(t) = [x_0(t), x_1(t), ..., x_{L-1}(t)]^T .$$ (2.36)

(2) Time definition of a process vector:

$$\mathbf{x}(t) = [x(t), x(t-1), ..., x(t-L+1)]^T .$$ (2.37)

Figure 2.6 illustrates the two different definitions of the process
vector $x(t)$. In real-time applications, one is frequently faced with the
problem that the ensemble definition of the process vector (2.36) cannot
be applied since there is only a single realization (observable component)
of the process vector available at one time step. In those cases, the time
definition of the process vector (2.37) is used as an *approximation* of the
true process vector defined by (2.36). This approximation is justified
under the assumption that the statistics of the observed process does
not change significantly over a time interval equivalent to the length L
of the process vector or "data window".

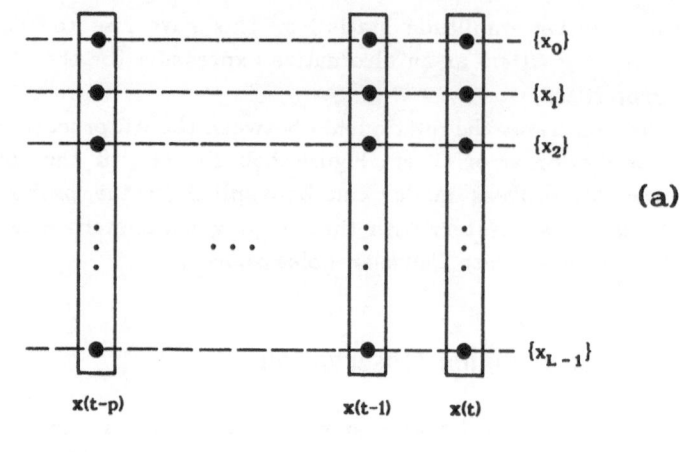

(a)

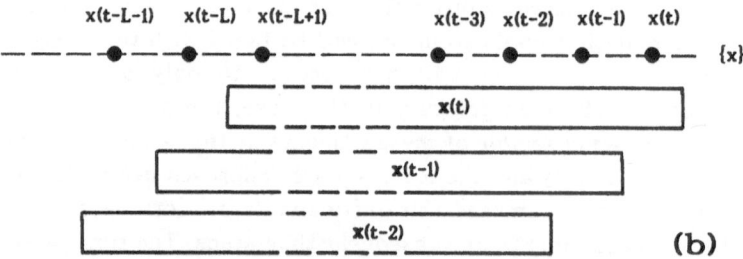

(b)

Fig. 2.6a,b. Different definitions of the process vector $\mathbf{x}(t)$. (a) Ensemble definition. (b) Time definition

The substitution of the time definition for the ensemble definition of process vectors can cause a serious degradation of the tracking performance of the estimation algorithm when the dynamics of the variations of the parameters of the observed process have a small time constant compared to the window length L. Large values of the window length L make it more difficult to follow rapid changes of system or signal parameters. On the other hand, large windows improve the *steady-state behavior* of an estimation algorithm since the uncertainty of a single measurement inside the process vector has a smaller influence on the overall estimation result. In practice, a tradeoff must be found between a desired tracking performance and an acceptable steady-state behavior.

Until now, we have only considered the case of a simple *rectangular window*, that is, we have assumed that all measurements inside the window are *weighted* with a factor of one, whereas measurements outside the window are weighted with zero. A considerable improvement of both the steady-state and the tracking behavior can be achieved if we apply *higher-order weighting functions* (higher-order windows) to the data. Here the general idea is that measurements at the boundaries of the process vector are weighted with successively smaller factors than measurements in the center of the process vector. This way, the influence of an incoming data value can be limited to an acceptable level while keeping the process vector length as small as possible for fast tracking. A detailed discussion of windowing techniques is given in Sect. 2.8.

2.5 Recursion Laws of the Normal Equations

The tracking of nonstationary processes requires a calculation of the process parameters at each time step. This gives rise to the question whether it is necessary to compute the solution completely anew for each new (incoming) sample of the observed process, or whether we can exploit knowledge about the Normal Equations of the *previous* time step in order to obtain faster algorithms for solving the Normal Equations, i.e., algorithms that require *fewer operations* for calculating a desired parameter set. For this purpose, we wish to discover the *recursive properties* of the Normal Equations determined by the recursion laws of the covariance matrix according to

$$\Phi(t-1) \xrightarrow{\ ?\ } \Phi(t) \ . \tag{2.38}$$

These recursion laws will later be useful in deriving fast algorithms for a time-recursive calculation or *updating* of the AR parameters as posed in the following problem relation:

$$a(t-1) \xrightarrow{\ ?\ } a(t) \ . \tag{2.39}$$

First, we are interested in (2.38). We see from definition (2.26) that the covariance matrix of the observed process is always *symmetric:*

$$\Phi_{i,j}(t) = \Phi_{j,i}(t) \ ; \qquad 0 \le i,j \le p \ . \tag{2.40}$$

Furthermore, we note that the process vectors of subsequent time steps are *shifted versions* of each other. Consequently, the covariance matrix of the observed process satisfies the following *shift invariance property:*

$$\Phi_{i,j}(t-1) = \mathbf{x}^{T}(t-1-i)\mathbf{x}(t-1-j) = \Phi_{i+1,j+1}(t) \ ; \qquad 0 \le i,j \le p-1 \ . \tag{2.41}$$

Figure 2.7 illustrates this very useful shift invariance property of the covariance matrix. If we consider the symmetric property in conjunction with the shift invariance property, it becomes obvious that the covariance matrix of an observed process can be updated, at each recursion, by computing just the first row or column vector of the covariance matrix.

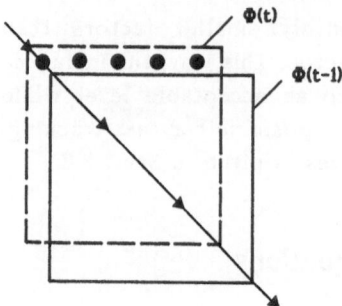

Fig. 2.7. Time-recursive updating of the covariance matrix of an observed process. Only the dotted elements must be computed completely anew at each recursion

Interestingly, this recursion law of the covariance matrix holds for both the ensemble and time definitions of the process vector. Normal Equations at subsequent time steps can therefore always be interpreted as *shifted versions* of each other.

Another recursion law for updating the covariance matrix can be directly derived from definition (2.26) as follows:

$$\Phi_{i,j}(t) = \Phi_{i,j}(t-1) + \left[x(t) \ \mathbf{z}^T(t) \right]^T \left[x(t) \ \mathbf{z}^T(t) \right]$$
$$- \left[x(t-L) \ \mathbf{z}^T(t-L) \right]^T \left[x(t-L) \ \mathbf{z}^T(t-L) \right] \ , \tag{2.42}$$

where $\mathbf{z}(t)$ is the state vector of the prediction error filter defined as

$$\mathbf{z}(t) = \left[x(t-1), \ x(t-2), \ ..., \ x(t-p) \right]^T \ . \tag{2.43}$$

From these considerations, we can state that the updating of the covariance matrix requires a computational complexity of only $O(p)$ operations per recursion, that is, the computational complexity for a recursive calculation of the covariance matrix grows *linearly* with the model order. This gives rise to the important question whether the *inverse* covariance matrix could also be updated via a procedure that requires only $O(p)$ operations per recursion. This question leads to the fast recursive least-squares (RLS) algorithms, which are the topic of Chap. 5 in this book.

2.6 Stationarity – A Special Case of Linear Prediction

An important case of linear prediction arises when the observed process is *stationary*. We shall define the *autocorrelation vector* of the observed process $\{x(t)\}$ as the first row vector of the covariance matrix:

$$\Phi_{0,j}(t) = \mathbf{x}^T(t)\, x(t-j) \ ; \qquad 0 \le j \le p \ . \tag{2.44}$$

The process $\{x(t)\}$ is called stationary if the corresponding autocorrelation vector defined by (2.44) is *time invariant*, i.e., satisfies

$$\Phi_{0,j}(t-k) = \Phi_{0,j}(t) \quad ; \qquad -\infty \leq k \leq +\infty \quad . \tag{2.45}$$

Given a stationary process, the covariance matrix turns into a *Toeplitz* matrix, as can be easily seen from (2.45). The Toeplitz covariance matrix has the structure

$$\Phi = \begin{bmatrix} \Phi_{0,0} & \Phi_{0,1} & \cdots\cdots & \Phi_{0,p-1} & \Phi_{0,p} \\ \Phi_{0,1} & \Phi_{0,0} & \Phi_{0,1} & & \Phi_{0,p-1} \\ \vdots & & \ddots & & \vdots \\ \Phi_{0,p-1} & & \ddots & \Phi_{0,0} & \Phi_{0,1} \\ \Phi_{0,p} & \Phi_{0,p-1} & \cdots\cdots & \Phi_{0,1} & \Phi_{0,0} \end{bmatrix} . \tag{2.46}$$

In the literature, one can frequently find the *normalized* version of the Toeplitz covariance matrix. Each element of (2.46) might be normalized to the process energy $\Phi_{0,0}$ to obtain the autocorrelation coefficients c_i,

$$c_i = \Phi_{0,i} \Big/ \Phi_{0,0} \quad ; \qquad 0 \leq i \leq p \quad . \tag{2.47}$$

Utilizing (2.47), the normalized Toeplitz Normal Equations can be stated as

$$\begin{bmatrix} 1 & c_1 & \cdots\cdots\cdots & c_{p-1} \\ c_1 & 1 & c_1 & \vdots \\ \vdots & & \ddots & \\ & & & c_1 \\ c_{p-1} & \cdots\cdots\cdots & c_1 & 1 \end{bmatrix} \begin{bmatrix} a_1 \\ a_2 \\ \vdots \\ \\ a_p \end{bmatrix} = \begin{bmatrix} c_1 \\ c_2 \\ \vdots \\ \\ c_p \end{bmatrix} . \tag{2.48}$$

The system of linear equations (2.48) is sometimes called the *Yule-Walker equations* [2.6].

2.7 Covariance Method and Autocorrelation Method

In practice, an observed process $\{x(t)\}$ is never completely stationary. If one wants to work with the Yule-Walker equations, the process must be *"forced"* to stationarity. A convenient way to achieve this is to extend the observation vectors by zero samples in the following intervals:

$$x(\tau) = 0 \qquad for \qquad \begin{cases} t+1 \leq \tau \leq \infty \; , \\ -\infty \leq \tau \leq t-L \; . \end{cases} \qquad (2.49)$$

The process samples outside a certain interval are set to zero. This case is also known as the *autocorrelation method of* linear prediction or the *windowed case* in contrast to the general case, which is known as the *covariance method* of linear prediction or the *unwindowed case* [2.7]. The definition of the covariance matrix (2.26) reveals that the "windowing" operation (2.49) automatically turns the covariance matrix into a Toeplitz matrix since each of the shifted data vectors $\{\mathbf{x}(t-m), 0 \leq m \leq p\}$ contains the same section of nonzero data, which just appears in a *shifted position* in consecutively shifted data vectors within an observation interval $\{-\infty \leq \tau \leq \infty\}$.

A windowing operation like (2.49) can also be interpreted as an element-wise multiplication of the process $\{x(t)\}$ with a window function $\mathbf{w}$. In the case of (2.49), the window function is $\{w(k) = 1; 0 \leq k \leq L-1\}$, and zero outside (rectangular window). According to classical system theory, the windowing operation corresponds to a *convolution* of the process spectrum with the spectrum of the window function. This can cause a serious degradation of the estimated parameters when only a simple rectangular window is used. The distortions can be reduced if we replace the rectangular window by more appropriate window functions which are characterized by a more *smoothed* descent at the window boundaries. Such windows can improve the estimation accuracy dramatically. A detailed discussion of appropriate window functions can be found in the classical signal processing literature [2.8].

2.8 Recursive Windowing Algorithms

In the previous section, we discussed the case of *nonrecursive* windowing of the process $\{x(t)\}$ in the autocorrelation method of linear prediction. Windowing algorithms, however, can also be important for the *recursive* updating of a covariance matrix in order to reduce the influence of incoming data points when the process vector length L is small for fast tracking. This leads to the problem of *recursive* windowing. Recursive windowing algorithms operate on the *lag sequence* rather than on the sequence of process samples. The lag sequence $\{z_{0,j}(t)\}$ is defined as

$$z_{0,j}(t) = x(t)x(t-j) \; ; \qquad\qquad 0 \leq j \leq p \; . \qquad (2.50)$$

The windowed unnormalized autocorrelation vector is then completely specified by

$$\Phi_{0,j}(t) = \sum_{k=0}^{L-1} w(k) \, z_{0,j}(t-k) \; ; \quad 0 \leq j \leq p \; , \qquad (2.51)$$

where

$$\mathbf{w} = [w(0), \ w(1), \ w(2), \dots , \ w(L-1)]^{T} \qquad (2.52)$$

is an appropriate window function.

Note that (2.52) is a *nonrecursive* formulation of the windowing problem. The most important windowing functions, however, can be brought into a *recursive* formulation [2.9]. Table 2.1 contains a summary of the most important recursive windowing algorithms. The windowing algorithms of Table 2.1 can be classified into two categories: first algorithms with a *finite* window length L, also called *sliding windows*, and second, *infinite* or *growing window* algorithms. Sliding window algorithms generally require the storage of (L+p+1) samples of the lag sequence, a fact that can cause serious problems in some applications where the window length L tends to larger values.

The second class of windowing algorithms, namely, the infinite (or growing) windows, is characterized by the fact that, generally, all past data samples are incorporated in the autocorrelation estimate. Therefore, growing window algorithms cannot adapt as rapidly as sliding window algorithms. Growing window algorithms, however, do not require a large data buffer as do in their finite duration counterparts.

A closer look at the sliding window algorithms of Table 2.1a reveals that these algorithms can be interpreted as *fixed recursive filters* having the lag sequence $\{z_{0,j}(t)\}$ as input and the respective values of the first row vector of the covariance matrix as output. Interestingly, the poles of these filters are located exactly on the unit circle and hence these windowing algorithms are only *wide-sense stable*. It can be shown that these algorithms can only be operated stably in an arithmetic environment which causes no round-off error in the loop addition in order to avoid an accumulation of round-off noise in the recursive unity gain loops of the algorithms. Clearly, this demand is satisfied only for fixed-point arithmetic, and it is not commonly known that each of the algorithms of Table 2.1a are *numerically unstable* when floating-point arithmetic is used, since floating-point arithmetic causes round-off errors in both addition and multiplication.

The growing window algorithms of Table 2.1b have their poles *inside* the unit circle, and therefore these algorithms are *unconditionally stable*, independent of the type of arithmetic involved in the computations.

A compromise between the typical behavior of sliding and growing window algorithms is obtained when the modified Barnwell window given in Table 2.1b is used. This windowing algorithm was introduced by Strobach in [2.10] and is the lag-recursive counterpart of an early proposal by Barnwell [2.11] who first used a window of this second-order type in the autocorrelation method of linear prediction. This modified Barnwell

Table 2.1. Recursive windowing algorithms for updating of the autocorrelation vector at each time step where $z_{O,j}(t) = x(t)x(t-j)$, λ is the exponential weighting factor and β is the modified Barnwell weighting factor. (a) Recursive finite impulse response (FIR, sliding) windows. (b) Recursive infinite impulse response (IIR, growing) windows

WINDOW (FIR)	WINDOW FUNCTION	RECURSIVE WINDOWING ALGORITHM
Rectangular	$w(k) = 1$ for $0 \le k \le L\text{-}1$, $w(k) = 0$ otherwise	$\Phi_{O,j}(t) = \Phi_{O,j}(t\text{-}1) + z_{O,j}(t) - z_{O,j}(t\text{-}L)$
Triangular	$w(k) = 1 - k/L$ for $0 \le k \le L\text{-}1$, $w(k) = 0$ otherwise	$\Phi_{O,j}(t) = \Phi_{O,j}(t\text{-}1) + z_{O,j}(t) - (1/L)\,X_{O,j}(t\text{-}1)$ $X_{O,j}(t) = X_{O,j}(t\text{-}1) + z_{O,j}(t) - z_{O,j}(t\text{-}L)$
Bartlett	$w(k) = 2k/(L\text{-}1)$ for $0 \le k \le (L\text{-}1)/2$, $w(k) = 2(L\text{-}1\text{-}k)/(L\text{-}1)$ for $(L\text{-}1)/2 \le k \le L\text{-}1$, $w(k) = 0$ otherwise	$\Phi_{O,j}(t) = \Phi_{O,j}(t\text{-}1)$ $\qquad + 2/(L\text{-}1)\big[X_{O,j}(t) - X_{O,j}(t\text{-}(L\text{-}1)/2)\big]$ $X_{O,j}(t) = X_{O,j}(t\text{-}1) + z_{O,j}(t) - z_{O,j}(t\text{-}(L\text{-}1)/2)$
Hanning	$w(k) = 0.5 - 0.5\cos\big(2\pi k/(L\text{-}1)\big)$ for $0 \le k \le L\text{-}1$, $w(k) = 0$ otherwise	$\Phi_{O,j}(t) = 0.5\,X_{O,j}(t) - 0.5\,Y_{O,j}(t)$ $X_{O,j}(t) = X_{O,j}(t\text{-}1) + z_{O,j}(t) - z_{O,j}(t\text{-}L)$ $Y_{O,j}(t) = -Y_{O,j}(t\text{-}2) + z_{O,j}(t) - z_{O,j}(t\text{-}L\text{-}1)$ $+\cos\big(2\pi/(L\text{-}1)\big)\big(2Y_{O,j}(t\text{-}1) - z_{O,j}(t\text{-}1) - z_{O,j}(t\text{-}L)\big)$

(a)

WINDOW (IIR)	WINDOW FUNCTION	RECURSIVE WINDOWING ALGORITHM
Exponential	$w(k) = \lambda^k$ for $\lambda \le 1$, $k \ge 0$ $w(k) = 0$ otherwise	$\Phi_{O,j}(t) = \lambda\,\Phi_{O,j}(t\text{-}1) + z_{O,j}(t)$
Modified Barnwell	$w(k) = (1+k)\,\beta^k$ for $\beta \le 1$, $k \ge 0$ $w(k) = 0$ otherwise	$\Phi_{O,j}(t) = \beta\,\Phi_{O,j}(t\text{-}1) + X_{O,j}(t)$ $X_{O,j}(t) = \beta\,X_{O,j}(t\text{-}1) + z_{O,j}(t)$

(b)

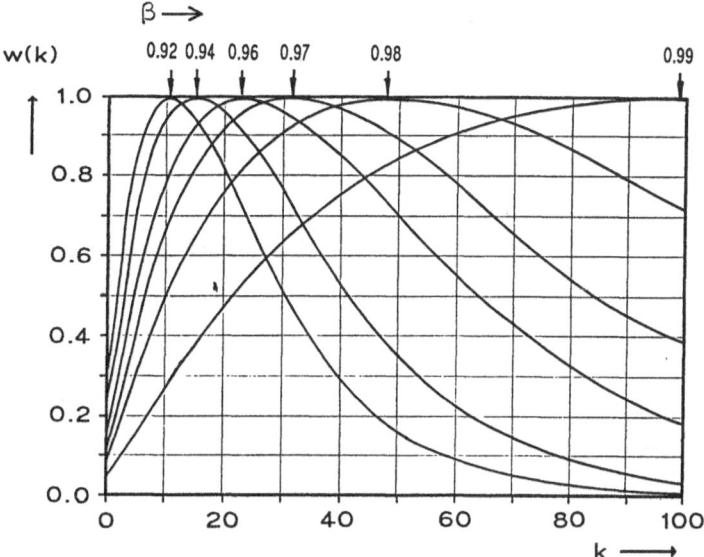

Fig. 2.8. The recursive modified Barnwell window for several values of the weighting factor β. [Window functions normalized to $max\{w(k)\}$]

window performs quite similarly to the recursive Hanning window, but needs no large data buffer. Figure 2.8 shows the modified Barnwell window for several values of the weighting factor β.

To conclude this section, we give some guidelines for a meaningful choice of the window parameter β in the case of the modified Barnwell window. As already discussed, this window function causes an algorithm to behave very similarly to a finite duration window estimation algorithm using, e.g., the Hanning window. It is therefore of interest to determine a characteristic quantity from which an "effective window length" can be deduced. Such parameters can help in the meaningful determination of β, since they give a quick estimate of what will be the approximate number of past process samples contributing to the actual parameter estimate.

Clearly, one such parameter is the location of the maximum of $w(k)$ as a function of β. To determine the location of this maximum, it is sufficient to evaluate the function

$$\frac{\partial}{\partial k}(1+k)\beta^k = 0 \ , \tag{2.53}$$

which gives the location of the maximum k_{max} according to

$$k_{max} = -\frac{1+\ln(\beta)}{\ln(\beta)} \tag{2.54}$$

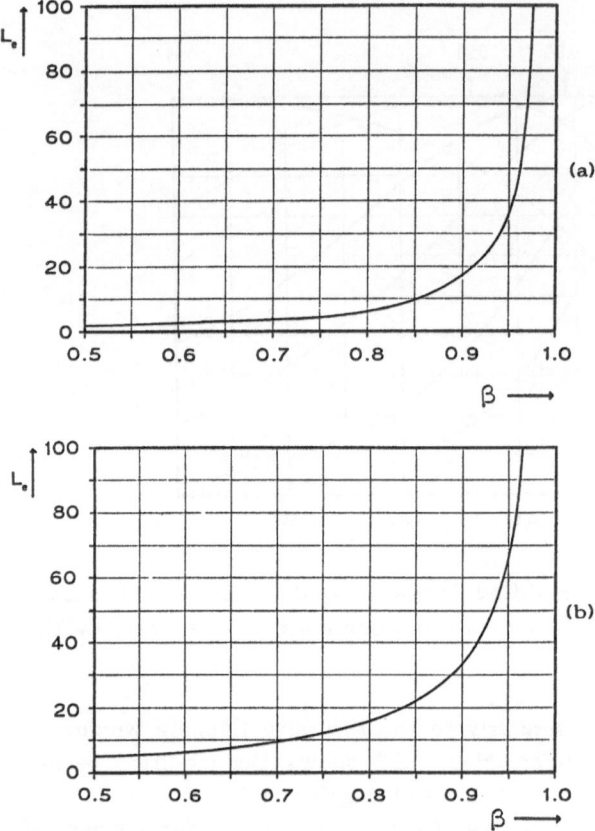

Fig. 2.9a,b. Determination of effective window length L_e as a function of β. (a) $L_e = 2k_{max}$ according to (2.55) (definition based on location of maximum). (b) $L_e = 2n$ according to (2.59) (area-based definition)

or, conversely,

$$\beta = \exp\left(-\frac{1}{1 + k_{max}}\right) . \tag{2.55}$$

Thus, a first possible definition of the effective window length might be, for example, two times the value of k_{max}. This measure of the effective window length is displayed in Fig. 2.9a as a function of β.

In a second approach, we might take advantage of the fact that the area covered by the window function for the modified Barnwell window satisfies a closed expression as follows:

$$\sum_{k=0}^{n}(1+k)\beta^k = \frac{1 + \beta^{(n+1)}\left(\beta(n+1) - n - 2\right)}{(1 - \beta)^2} . \tag{2.56}$$

Then it can be shown that the window function covers a total area

$$\sum_{k=0}^{\infty} (1+k) \beta^k = \frac{1}{(1-\beta)^2} .$$ (2.57)

So the effective window length 2n in this case can be defined as twice the specific value of n for which the window function covers half of its total area, hence

$$\frac{1}{2(1-\beta)^2} = \frac{1 + \beta^{(n+1)}\big(\beta(n+1) - n - 2\big)}{(1-\beta)^2}$$ (2.58)

or, equivalently,

$$1 + 2\beta^{(n+1)}\big(\beta(n+1) - n - 2\big) = 0 ,$$ (2.59)

which is the desired relationship between n and β, where $2n = L_e$ is the effective window length. Figure 2.9b shows this definition of the effective window length as a function of the window parameter β.

2.9 Backward Linear Prediction

Until now, we have only treated the case of *forward linear prediction*, where an *actual observation* $x(t)$ is predicted from its *subspace of past observations* spanned by $x(t-1), x(t-2), \ldots, x(t-p)$. It is important to note, however, that a counterpart problem exists in predicting the *past observation* $x(t-p)$ from its *subspace of previous observations* $x(t), x(t-1), \ldots, x(t-p+1)$. This case is called *backward linear prediction*. The problems of forward and backward linear prediction are in many ways *dual*.

Throughout this book, it will become apparent that the derivation of linear prediction algorithms generally requires the simultaneous solution of both the forward and backward linear prediction subproblems, at least in the nonstationary case. In the strictly stationary (Toeplitz) case, it will turn out that forward and backward linear prediction will become *equivalent*. We shall introduce the basic relationships of backward linear prediction and we shall relate them to the already developed relationships of forward linear prediction. A more complete summary and comparison of forward/backward linear prediction relationships may be found in Appendix A.1.

Consider the *past observation* $x(t-p)$,

$$\mathbf{x}(t-p) = \Big[x(t-p), x(t-p-1), \ldots, x(t-p-L+1))\Big]^T ,$$ (2.60)

and the *subspace of previous observations* $\mathbf{X}(t+1)$,

$$\mathbf{X}(t+1) = \Big[\mathbf{x}(t),\ \mathbf{x}(t-1), \ldots, \mathbf{x}(t-p+1) \Big] \ . \qquad (2.61)$$

Then, the *predicted process vector* $\hat{\mathbf{x}}^{b}(t)$ generated by backward linear prediction is determined by

$$\hat{\mathbf{x}}^{b}(t) = \mathbf{X}(t+1)\,\mathbf{b}(t) \ , \qquad (2.62)$$

where $\mathbf{b}(t)$ is the *backward linear prediction parameter vector*. The *backward linear prediction error vector* $\mathbf{r}(t)$ is then defined as

$$\mathbf{r}(t) = \mathbf{x}(t-p) - \hat{\mathbf{x}}^{b}(t) \ . \qquad (2.63)$$

Clearly, (2.62) is the counterpart relationship to (2.6), and (2.63) is the counterpart to (2.7). See also Fig. 2.10 for an illustration of forward/backward linear prediction relationships.

Following this line of thought, we may define the *backward linear prediction residual energy* $R(t)$ as the inner product

$$R(t) = \mathbf{r}^{T}(t)\mathbf{r}(t) \qquad (2.64)$$

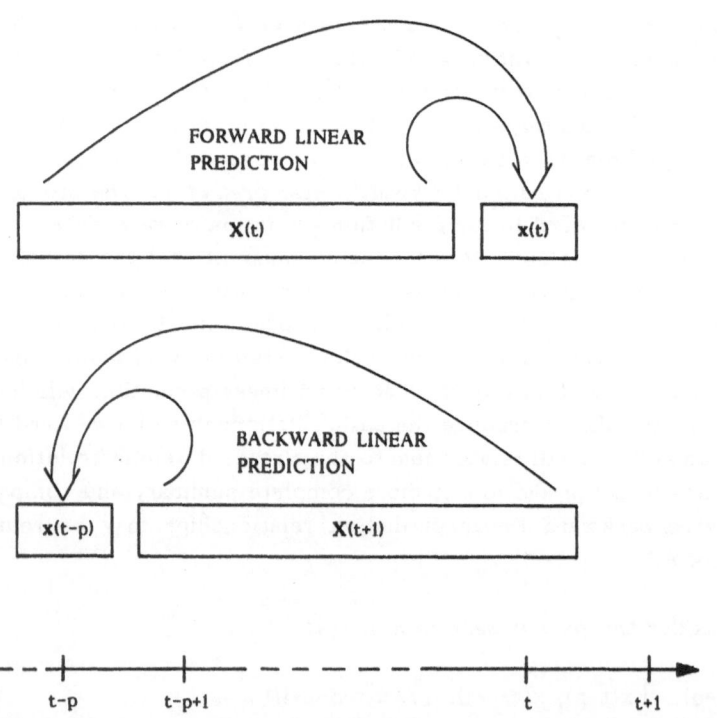

Fig. 2.10. Illustration of the dual problems of forward and backward linear prediction

and, obviously, we may determine the least-squares estimate of $\mathbf{b}(t)$, denoted $\hat{\mathbf{b}}(t)$ for the moment, by a minimization of $R(t)$ according to

$$R(t) \overset{!}{=} min \longrightarrow \hat{\mathbf{b}}(t) \ . \tag{2.65}$$

Clearly,

$$R(t,\mathbf{b}(t)) = \mathbf{x}^T(t-p)\mathbf{x}(t-p) - \mathbf{b}^T(t)\mathbf{X}^T(t+1)\mathbf{x}(t-p)$$

$$- \mathbf{x}^T(t-p)\mathbf{X}(t+1)\mathbf{b}(t) + \mathbf{b}^T(t)\mathbf{X}^T(t+1)\mathbf{X}(t+1)\mathbf{b}(t) \ , \tag{2.66}$$

which constitutes again a *convex performance surface* whose minimum might be determined by setting the gradient of $R(t,\mathbf{b}(t))$ to zero, yielding

$$\partial R(t,\mathbf{b}(t)) \big/ \partial \mathbf{b}^T(t) = - 2\mathbf{X}^T(t+1)\mathbf{x}(t-p)$$

$$+ 2\mathbf{X}^T(t+1)\mathbf{X}(t+1)\mathbf{b}(t) = \begin{bmatrix} 0 \dots 0 \end{bmatrix}^T, \tag{2.67}$$

which finally constitutes the *Normal Equations of backward linear prediction*:

$$\mathbf{X}^T(t+1)\mathbf{X}(t+1)\hat{\mathbf{b}}(t) = \mathbf{X}^T(t+1)\mathbf{x}(t-p) \ . \tag{2.68}$$

The relationship (2.68) is the counterpart to the forward linear prediction Normal Equations as stated in (2.13). A more complete comparison and summary of important forward/backward linear prediction relationships can be found in Appendix A.1.

2.10 Chapter Summary

In Chap. 2, we have developed the Normal Equations of linear prediction that determine the least-squares estimate of the autoregressive process model parameters. The assumption of a process model with autoregressive (AR) structure corresponds to a predictor that has a pure nonrecursive (MA) structure. This concept leads to Normal Equations that can be formed from the covariance matrix of the observed process. The different cases of time and ensemble definitions of the observed process have been discussed. Besides the algebraic definition, we paid much attention on the geometrical interpretation of the Normal Equations. Special cases such as *stationarity* and *recursivity* of the Normal Equations have been considered. An understanding of these fundamental properties of the Normal Equations is essential for a better assimilation of the following chapters, where we will derive the most important algorithms for solving the Normal Equations.

Interestingly, it turns out that in the derivation of linear prediction algorithms, we generally have to solve the two subproblems of forward and backward linear prediction. These two subproblems are in many ways *dual*. For these reasons, the optimal exploitation of forward/backward linear prediction relationships plays a central role in linear prediction algorithms.

3. Classical Algorithms for Symmetric Linear Systems

This chapter presents an overview of the most important algorithms for solving symmetric linear systems as appearing in the *nonrecursive covariance case* of linear prediction and several other useful techniques for matrix computations which will be required throughout this book. The techniques presented here have also a wide range of applications in the field of numerical analysis [3.1-4].

3.1 The Cholesky Decomposition

Consider the covariance Normal Equations of the form (2.13)

$$\mathbf{X}^T(t)\mathbf{X}(t)\mathbf{a}(t) = \mathbf{X}^T(t)\mathbf{x}(t) \ , \tag{3.1}$$

where $\hat{\mathbf{a}}(t)$ is replaced by $\mathbf{a}(t)$ for convenience. Using the shorthand notation

$$\mathbf{A}\mathbf{a} = \mathbf{h} \ , \tag{3.2}$$

where

$$\mathbf{A} = \mathbf{X}^T(t)\mathbf{X}(t) \qquad \text{and} \tag{3.3a}$$

$$\mathbf{h} = \mathbf{X}^T(t)\mathbf{x}(t) \ , \tag{3.3b}$$

from (3.3a) it follows that the system matrix is symmetric

$$\mathbf{A}^T = \mathbf{A} \tag{3.4}$$

and can therefore be expressed as the product of an upper and a lower *triangular* matrix as

$$\mathbf{A} = \mathbf{L}\mathbf{L}^T \ , \tag{3.5}$$

where $\mathbf{L}$ is of the form

$$\mathbf{L} = \begin{bmatrix} \begin{array}{ccc} 0 \ldots \ldots 0 \\ \ddots \ \vdots \\ 0 \end{array} \end{bmatrix} \ . \tag{3.6}$$

Assuming first that we succeed in finding this decomposition (3.5), the linear system (3.2) attains the form

$$\mathbf{L}\mathbf{L}^T\mathbf{a} = \mathbf{h} \ . \tag{3.7}$$

Now, the solution vector **a** can be obtained by a forward and back substitution as follows:

$$L v = h \longrightarrow v \; , \tag{3.8a}$$

$$L^T a = v \longrightarrow a \; . \tag{3.8b}$$

As the next step, we have to provide the triangular matrix **L**. A recursion relation for computation of the decomposition matrix **L** can be derived directly from the formula for matrix-by-matrix multiplication. Writing the matrix product LL^T in a more explicit form, we have

$$A_{i,j} = \sum_{k=1}^{p} L_{i,k} L^T_{k,j} = \sum_{k=1}^{p} L_{i,k} L_{j,k} \; , \tag{3.9}$$

where p denotes the dimension of the system matrix **A**. Since **L** was assumed to be of lower triangular form, the summation in (3.9) need not be computed for all elements of **L**. Omitting zero multiplications, (3.9) can be rewritten with an upper summation index that is a function of the row and column index pair (i,j):

$$A_{i,j} = \sum_{k=1}^{min(i,j)} L_{i,k} L_{j,k} \; . \tag{3.10}$$

From (3.10) it follows directly that

$$A_{1,1} = L^2_{1,1} \tag{3.11}$$

or, equivalently,

$$L_{1,1} = \left(A_{1,1} \right)^{1/2} \; ; \qquad\qquad A_{1,1} > 0 \; . \tag{3.12}$$

An evaluation of the first column of **A**

$$A_{i,1} = L_{i,1} L_{1,1} \tag{3.13}$$

leads to a computation of the first column of **L** as

$$L_{i,1} = A_{i,1} \Big/ L_{1,1} \; . \tag{3.14}$$

Rewriting (3.10) in a slightly different form,

$$A_{i,j} = L_{i,j} L_{j,j} + \sum_{k=1}^{j-1} L_{i,k} L_{j,k} \; , \tag{3.15}$$

equation (3.15) can be rearranged, yielding the desired recursion for the elements of the jth column vector of **L**:

$$L_{i,j} = \left(A_{i,j} - \sum_{k=1}^{j-1} L_{i,k} L_{j,k}\right) / L_{j,j} \quad \text{for} \quad i > j \tag{3.16}$$

where

$$A_{j,j} = \sum_{k=1}^{j} L_{j,k}^2 = \sum_{k=1}^{j-1} L_{j,k}^2 + L_{j,j}^2 \quad , \tag{3.17a}$$

$$L_{j,j} = \left(A_{j,j} - \sum_{k=1}^{j-1} L_{j,k}^2\right)^{1/2} \quad . \tag{3.17b}$$

Next it is of importance to state the condition which must be satisfied to guarantee the stability of the decomposition determined by (3.13-17). In other words, we are seeking for the condition on **A** that guarantees that the roots in the computation of (3.17b) are all real valued. It can be shown that this demand can be satisfied by a *positive definite* system matrix **A**.

A real matrix **A** is called **positive definite** iff the following two conditions are satisfied:

(1) $\mathbf{A}^T = \mathbf{A}$, $\tag{3.18a}$

(2) $\mathbf{x}^T \mathbf{A} \mathbf{x} > 0$ *for all* $\mathbf{x} > \begin{bmatrix} 0 \ldots 0 \end{bmatrix}^T$. $\tag{3.18b}$

In Table 3.1 we summarize the algorithm derived above for factorizing a symmetric matrix **A** into a product of a lower triangular matrix **L** and its transpose $\mathbf{L}^T$.

Table 3.1. Cholesky decomposition of a real symmetric positive definite matrix **A** of rank p

---

Initialize: **A**   (Matrix to be factorized)

          p   (Dimension of **A**)

FOR j = 1, 2, . . . , p

$$L_{j,j} = \left(A_{j,j} - \sum_{k=1}^{j-1} L_{j,k}^2\right)^{1/2}$$

  FOR i = j+1, j+2, . . . , p

$$L_{i,j} = \left(A_{i,j} - \sum_{k=1}^{j-1} L_{i,k} L_{j,k}\right) / L_{j,j}$$

The application of the Cholesky decomposition to signal processing problems was discussed by Dickinson [3.5, 6] and Gibson [3.7].

3.2 The QR Decomposition

The Cholesky decomposition, as described, operates on the Normal Equations given in an *explicit* form (3.2). From an experiment, one normally obtains the data represented by the matrix of past observations $\mathbf{X}(t)$ and the actual process vector $\mathbf{x}(t)$. From this information, the system matrix $\mathbf{A}(t)$ and the right-hand-side vector $\mathbf{h}(t)$ of the Normal Equations must be computed by evaluating (3.1) before the Cholesky decomposition can be applied.

This raises the question whether there exist methods of factorizing $\mathbf{X}(t)$ *directly*, leading to a solution of the Normal Equations that does not require the explicit computation of $\mathbf{A}$ and $\mathbf{h}$. In fact, it can be shown that $\mathbf{X}$ can be decomposed into an *orthogonal* matrix $\mathbf{Q}$ and a *triangular* matrix $\mathbf{R}$. Such a decomposition is commonly referred to as *QR decomposition* [3.1]. Algorithms based on the QR decomposition generally have a better numerical accuracy than algorithms which require the explicit calculation of $\mathbf{A}$ and $\mathbf{h}$.

We consider the Normal Equations in a shorthand notation:

$$\mathbf{X}^T \mathbf{X} \, \mathbf{a} = \mathbf{X}^T \mathbf{x} \; ; \qquad \mathbf{X} \in \mathbb{R}^{L \times P} , \quad \mathbf{a} \in \mathbb{R}^P , \quad \mathbf{x} \in \mathbb{R}^L . \tag{3.19}$$

The QR decomposition is based on the assumption that the following factorization of the matrix of past observations exists:

$$\mathbf{X} = \mathbf{Q} \, \mathbf{R}' \; ; \qquad \mathbf{Q} \in \mathbb{R}^{L \times L}, \quad \mathbf{R}' \in \mathbb{R}^{L \times P} \; ; \tag{3.20}$$

where $\mathbf{R}'$ satisfies the following partitioning scheme:

$$\mathbf{R}' = \begin{bmatrix} \mathbf{R} \\ \mathbf{0} \end{bmatrix} \; ; \qquad \mathbf{R} \in \mathbb{R}^{P \times P} , \quad \mathbf{0} \in \mathbb{R}^{(L-P) \times P} \; ; \tag{3.21}$$

$\mathbf{R}$ is an upper-right triangular matrix and $\mathbf{0}$ is the null matrix

$$\mathbf{R} = \begin{bmatrix} 0 & \diagdown \\ \vdots & \diagdown \\ 0 \cdots & 0 \end{bmatrix} , \tag{3.22a}$$

$$\mathbf{0} = \begin{bmatrix} 0. & \cdots & .0 \\ \vdots & & \vdots \\ 0. & \cdots & .0 \end{bmatrix} . \tag{3.22b}$$

Assuming that $\mathbf{Q}$ is an *orthogonal* matrix satisfying the orthogonality relation

$$\mathbf{Q}^T\mathbf{Q} = \mathbf{I} \tag{3.23}$$

and utilizing this property together with (3.20 and 21), we have

$$\mathbf{Q}^T\mathbf{X} = \mathbf{Q}^T\mathbf{Q}\begin{bmatrix} \mathbf{R} \\ \mathbf{0} \end{bmatrix} = \begin{bmatrix} \mathbf{R} \\ \mathbf{0} \end{bmatrix} . \tag{3.24}$$

Accordingly, the Normal Equations can be rewritten in terms of the QR decomposition

$$\mathbf{X}^T\mathbf{X}\mathbf{a} = \begin{bmatrix} \mathbf{R} \\ \mathbf{0} \end{bmatrix}^T \mathbf{Q}^T\mathbf{Q}\begin{bmatrix} \mathbf{R} \\ \mathbf{0} \end{bmatrix}\mathbf{a} = \begin{bmatrix} \mathbf{R} \\ \mathbf{0} \end{bmatrix}^T \mathbf{Q}^T\mathbf{x} = \mathbf{X}^T\mathbf{x} . \tag{3.25}$$

Introducing the partitioning scheme

$$\mathbf{Q}^T\mathbf{x} = \begin{bmatrix} \mathbf{d}_1 \\ \mathbf{d}_2 \end{bmatrix} ; \qquad \mathbf{d}_1 \in \mathbb{R}^P , \qquad \mathbf{d}_2 \in \mathbb{R}^{L-P} , \tag{3.26}$$

the right-hand-side vector of the Normal Equations can be rewritten in terms of the auxiliary vectors $\mathbf{d}_1$ and $\mathbf{d}_2$ and the upper triangular matrix $\mathbf{R}$ as

$$\mathbf{X}^T\mathbf{x} = \begin{bmatrix} \mathbf{R} \\ \mathbf{0} \end{bmatrix}^T \begin{bmatrix} \mathbf{d}_1 \\ \mathbf{d}_2 \end{bmatrix} = \mathbf{R}^T\mathbf{d}_1 + \mathbf{0}^T\mathbf{d}_2 = \mathbf{R}^T\mathbf{d}_1 . \tag{3.27}$$

Utilizing again the orthogonality relation (3.23) together with (3.27), equation (3.25) simplifies to

$$\mathbf{X}^T\mathbf{X}\mathbf{a} = \mathbf{R}^T\mathbf{R}\mathbf{a} = \mathbf{R}^T\mathbf{d}_1$$

$$= \mathbf{X}^T\mathbf{x} \longrightarrow \mathbf{d}_1 \qquad \textit{(forward substitution)} . \tag{3.28}$$

Since $\mathbf{R}$ is an upper triangular matrix, the solution of the Normal Equations can be obtained from a back substitution of the system

$$\mathbf{R}\mathbf{a} = \mathbf{d}_1 \longrightarrow \mathbf{a} \qquad \textit{(back substitution)} . \tag{3.29}$$

Note at this point that the explicit computation of the orthogonal matrix

Q is *not* required when the QR decomposition is used to solve the Normal Equations.

Next we consider the problem of computing the matrices **Q** and **R** from the observation matrix **X**. For this purpose, we discuss two prominent methods: first the *Givens reduction* and second the *Householder reduction*.

3.2.1 The Givens Reduction

In the Givens reduction [3.8], the desired matrices **Q** and **R** are generated by successive coordinate (Givens) rotations. Consider a vector $[x_1, x_2]^T$ with an Euclidean length r and angle ψ in the two-dimensional space, see Fig. 3.1. A related vector $[x_1', x_2']^T$ with the same Euclidean length r but an angle $\varphi + \psi$ can be obtained via the operation of an *anticlockwise rotation* with a rotation angle φ, as illustrated in Fig. 3.1. Let

$$\begin{bmatrix} x_1 \\ x_2 \end{bmatrix} = r \begin{bmatrix} \sin \psi \\ \cos \psi \end{bmatrix} \qquad \text{and} \qquad (3.30a)$$

$$\begin{bmatrix} x_1' \\ x_2' \end{bmatrix} = r \begin{bmatrix} \sin(\psi + \varphi) \\ \cos(\psi + \varphi) \end{bmatrix} \qquad (3.30b)$$

be the original and rotated vectors, respectively. We wish to derive the rotation operator as a function of the rotation angle φ.

Taking into account the trigonometric theorem

$$\begin{bmatrix} \sin(\psi + \varphi) \\ \cos(\psi + \varphi) \end{bmatrix} = \begin{bmatrix} \cos\psi \sin\varphi + \sin\psi\cos\varphi \\ \cos\psi \cos\varphi - \sin\psi \sin\varphi \end{bmatrix} , \qquad (3.31)$$

we may express $[x_1', x_2']^T$ as a function of the rotation angle φ as follows:

$$\begin{bmatrix} x_1' \\ x_2' \end{bmatrix} = r \begin{bmatrix} \cos\varphi & \sin\varphi \\ -\sin\varphi & \cos\varphi \end{bmatrix} \begin{bmatrix} \sin \psi \\ \cos \psi \end{bmatrix} . \qquad (3.32)$$

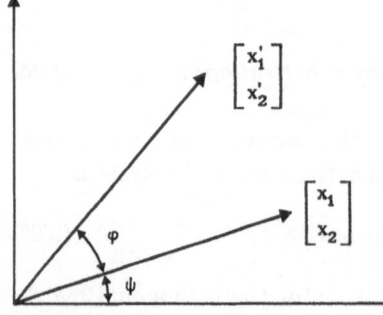

Fig. 3.1. Two vectors $[x_1, x_2]^T$ and $[x_1', x_2']^T$ related by rotation. φ is the rotation angle

From (3.32) and in consideration of (3.30a,b), we obtain the rotations

$$\begin{bmatrix} x_1' \\ x_2' \end{bmatrix} = \begin{bmatrix} \cos\varphi & \sin\varphi \\ -\sin\varphi & \cos\varphi \end{bmatrix} \begin{bmatrix} x_1 \\ x_2 \end{bmatrix} , \tag{3.33a}$$

$$\begin{bmatrix} x_1 \\ x_2 \end{bmatrix} = \begin{bmatrix} \cos\varphi & -\sin\varphi \\ \sin\varphi & \cos\varphi \end{bmatrix} \begin{bmatrix} x_1' \\ x_2' \end{bmatrix} . \tag{3.33b}$$

The rotation matrix $\mathbf{D}$

$$\mathbf{D} = \begin{bmatrix} \cos\varphi & \sin\varphi \\ -\sin\varphi & \cos\varphi \end{bmatrix} = \begin{bmatrix} c & s \\ -s & c \end{bmatrix} \tag{3.34}$$

is *orthogonal* and hence satisfies the condition

$$\mathbf{D}^T \mathbf{D} = \mathbf{I} \tag{3.35}$$

where $\mathbf{I}$ is the identity matrix of dimension 2×2.

Next we consider the special case of a rotation where the rotation matrix $\mathbf{D}$ is such that the second component of the rotated vector is zero. Without loss of generality we assume $x_2' = 0$ is achieved through a rotation of $[x_1, x_2]^T$. Hence, from the expression

$$\begin{bmatrix} x_1' \\ 0 \end{bmatrix} = \begin{bmatrix} \cos\varphi & \sin\varphi \\ -\sin\varphi & \cos\varphi \end{bmatrix} \begin{bmatrix} x_1 \\ x_2 \end{bmatrix} \tag{3.36}$$

we can determine the rotation angle that satisfies (3.36) by means of

$$-x_1 \sin\varphi + x_2 \cos\varphi = 0 , \tag{3.37}$$

$$\varphi = \tan^{-1}\left(x_2 / x_1\right) , \tag{3.38}$$

as well as the elements of the rotation matrix $\mathbf{D}$

$$c = \cos\varphi = \cos\left[\tan^{-1}\left(x_2 / x_1\right)\right] = x_1 \left(x_1^2 + x_2^2\right)^{-1/2} , \tag{3.39a}$$

$$s = \sin\varphi = \sin\left[\tan^{-1}\left(x_2 / x_1\right)\right] = x_2 \left(x_1^2 + x_2^2\right)^{-1/2} . \tag{3.39b}$$

With these preliminary considerations, we can now derive the Givens reduction. Suppose that

$$\mathbf{X} = \mathbf{Q}_{2,1} \mathbf{R}_{2,1}' \qquad \text{with} \tag{3.40}$$

$$\mathbf{Q}_{2,1} = \mathbf{I} , \tag{3.41}$$

$$\mathbf{R}_{2,1}' = \mathbf{X} \tag{3.42}$$

is an initial equation from which the QR decomposition (3.20) can be computed. Then it is the goal in the Givens reduction to apply *successive rotations* to the system (3.40) in order to generate zero elements in the auxiliary matrix $\mathbf{R}'_{i,j}$ until the final matrix $\mathbf{R}'$

$$\mathbf{R}' = \begin{bmatrix} \diagdown \\ \\ 0 \end{bmatrix} = \begin{bmatrix} \mathbf{R} \\ \\ 0 \end{bmatrix} \tag{3.43}$$

is reached, where the index pair (i,j) denotes the element in the auxiliary matrix $\mathbf{R}'_{i,j}$ that is set to zero by a rotation. The rotation for setting the element (i,j) in the auxiliary matrix $\mathbf{R}'_{i,j}$ to zero is accomplished by substituting the orthogonal rotation matrix $\mathbf{G}_{i,j} \in \mathbb{R}^{L \times L}$ into (3.40), yielding

$$\mathbf{X} = \mathbf{Q}_{i,j}\,\mathbf{G}_{i,j}^{-1}\,\mathbf{G}_{i,j}\,\mathbf{R}'_{i,j}\;; \qquad\qquad i > j \;. \tag{3.44}$$

From (3.44), we obtain two recurrence relations for updating the $\mathbf{Q}$ and $\mathbf{R}'$ matrices as follows:

$$\mathbf{Q}_{i+1,j} = \mathbf{Q}_{i,j}\,\mathbf{G}_{i,j}^{-1} \;, \tag{3.45a}$$

$$\mathbf{R}'_{i+1,j} = \mathbf{G}_{i,j}\,\mathbf{R}'_{i,j} \;. \tag{3.45b}$$

The equations (3.45a,b) assume a row-by-row generation of zero elements in each column of the matrix $\mathbf{R}'_{i,j}$.

With our previous considerations about vector rotations, we can now easily establish the required Givens rotation matrix $\mathbf{G}_{i,j}$. According to (3.34 and 36), $\mathbf{G}_{i,j}$ will be of the form

where all unspecified elements in $\mathbf{G}_{i,j}$ are zero.

We note that the application of the Givens rotation matrix $\mathbf{G}_{i,j}$ in the sense of (3.34) produces a rotation as in (3.36). The values of c and s can be determined according to (3.39a,b). Specifically, we have

$$
\begin{bmatrix}
R'_{i+1,j}(j,j) \\
R'_{i+1,j}(i,j) = 0
\end{bmatrix}
=
\begin{bmatrix}
c & s \\
-s & c
\end{bmatrix}
\begin{bmatrix}
R'_{i,j}(j,j) \\
R'_{i,j}(i,j)
\end{bmatrix}
. \tag{3.47}
$$

From (3.47), it follows that

$$
c = R'_{i,j}(j,j) \big/ q \ , \tag{3.48a}
$$

$$
s = R'_{i,j}(i,j) \big/ q \ , \tag{3.48b}
$$

where

$$
q = \left(R'^{2}_{i,j}(i,j) + R'^{2}_{i,j}(j,j) \right)^{1/2} . \tag{3.49}
$$

Considering (3.46), and having in mind that $c^2 + s^2 = 1$, it is easy to check that $\mathbf{G}_{i,j}$ satisfies the orthogonality relation

$$
\mathbf{G}^{T}_{i,j} \, \mathbf{G}_{i,j} = \mathbf{I} \ , \tag{3.50}
$$

or equivalently,

$$
\mathbf{G}^{-1}_{i,j} = \mathbf{G}^{T}_{i,j} \ , \tag{3.51}
$$

which simplifies the evaluation of (3.45a). At this point it is important to verify whether the orthogonality of $\mathbf{Q}$ is invariant with respect to the application of a Givens rotation matrix. More specifically, we have to check whether the operation

$$
\mathbf{Q}_{i+1,j} = \mathbf{Q}_{i,j} \, \mathbf{G}^{T}_{i,j} \tag{3.52}
$$

violates the orthogonality of $\mathbf{Q}$. Recall that the orthogonality of $\mathbf{Q}$ through all steps of the decomposition was a prerequisite in the concept of a QR decomposition. Assuming that $\mathbf{Q}_{i,j}$ was an orthogonal matrix, then we can state that

$$
\mathbf{Q}_{i+1,j} \, \mathbf{Q}^{T}_{i+1,j} = \mathbf{Q}_{i,j} \, \mathbf{G}^{T}_{i,j} \mathbf{G}_{i,j} \, \mathbf{Q}^{T}_{i,j} = \mathbf{Q}_{i,j} \, \mathbf{Q}^{T}_{i,j} = \mathbf{I} \ , \tag{3.53}
$$

which proves that the orthogonality of $\mathbf{Q}$ holds through all Givens rotations when the reduction process starts with initial conditions as stated in (3.41 and 42).

A summary of the Givens reduction is provided in Table 3.2. From the Givens algorithm of this table, we obtain the desired matrices $\mathbf{Q}$ and $\mathbf{R}$ as follows:

$$\begin{bmatrix} \mathbf{R} \\ \mathbf{0} \end{bmatrix} = \begin{bmatrix} \mathbf{R}'_{L,p} \end{bmatrix} , \tag{3.54a}$$

$$\mathbf{Q} = \mathbf{Q}_{L,p} . \tag{3.54b}$$

Now, from (3.46) it is clear that $\mathbf{G}_{i,j}$ is a sparse matrix and the matrix multiplications in Table 3.2 reduce to four scalar multiplications. This way, a more efficient computation of the Givens reduction is obtained. Such an algorithm is summarized in Table 3.3.

Table 3.2. QR decomposition of an observation matrix $\mathbf{X}$ using the Givens reduction. Note that the computation of $\mathbf{Q}$ is not needed for solving a least-squares problem

---

Initialize :

$\mathbf{R}'_{2,1} = \mathbf{X}$; $\mathbf{Q}_{2,1} = \mathbf{I}$

FOR   j = 1, 2, . . . , p
$\begin{bmatrix} \text{FOR   i = j+1, j+2, . . . , L} \\ \begin{bmatrix} \textit{calculate } \mathbf{G}_{i,j} \textit{ according to: } (3.46, 48a,b \text{ and } 49) \\ \mathbf{R}'_{i+1,j} = \mathbf{G}_{i,j} \mathbf{R}'_{i,j} \\ \mathbf{Q}_{i+1,j} = \mathbf{Q}_{i,j} \mathbf{G}^T_{i,j} \end{bmatrix} \end{bmatrix}$

The triangularization of an observation matrix $\mathbf{X}$ of dimension $L \times p$ using the Givens reduction of Table 3.3 necessitates the computation of $Lp - 0.5p^2 - 0.5p$ square roots.

This large number of square roots complicates the computation of the Givens reduction on conventional multiplier–accumulator based signal processing hardware. The solution to this problem can be a special–purpose hardware based, for example, on the CORDIC principle (see Sect. 4.4 for details). On the other hand it is possible to derive *square-root-free* versions of the Givens reduction. An early discussion of Givens transformations without square roots was given by Gentleman [3.9] and Hammarling [3.10]. More recent considerations were made by Nash and Hansen [3.11] and Götze and Schwiegelshohn [3.12].

Table 3.3. Efficient implementation of the Givens reduction. The quantities X(j,k) and Q(k,j) need to be stored twice to avoid overwriting. The desired matrix **R'** arises in **X** through rotations. As the computation of the orthogonal matrix **Q** is not a part of the recursion, its computation remains *optional*. This algorithm involves division. Whenever the divisor is small, set $c = 1$ and $s = 0$ (identity rotor)

FOR k = 1, 2, . . . , L *initialize:*

$\quad$ Q(k,k) = 1 ;$\qquad$ Q(k,j) = 0 $\quad \forall \quad$ j ≠ k

FOR j = 1, 2, . . . , p

$\quad$ FOR i = j+1, j+2, . . . , L

$\qquad q = \left(X^2(j,j) + X^2(i,j) \right)^{1/2}$

$\qquad c = X(j,j) \big/ q$

$\qquad s = X(i,j) \big/ q$

$\qquad$ FOR k = j, j+1, . . . , p : *rotate* **X**

$\qquad\quad$ X(j,k) = $\quad$ c X(j,k) + s X(i,k)

$\qquad\quad$ X(i,k) = $\;$ -s X(j,k) + c X(i,k)

$\qquad$ FOR k = 1, 2, . . . , L :$\quad$ *rotate* **Q**

$\qquad\quad$ Q(k,j) = $\quad$ c Q(k,j) + s Q(k,i) $\qquad\Big\}$ *optional*

$\qquad\quad$ Q(k,i) = $\;$ -s Q(k,j) + c Q(k,i)

3.2.2 The Householder Reduction

Besides the Givens reduction, the so-called *Householder reduction* [3.13] has received much attention. The method has several similarities with the Givens reduction; however, we use the Householder matrix **H** instead of the Givens rotation matrix **G**. The Householder matrix **H** is defined as

$$H = I - 2ww^T ; \qquad H \in \mathbb{R}^{L \times L} , \tag{3.55}$$

where

$$w^T w = 1 ; \qquad w \in \mathbb{R}^L . \tag{3.56}$$

The Householder matrix **H** is orthogonal, as is the Givens rotation matrix, and therefore satisfies the orthogonality relation

$$H^T H = I . \tag{3.57}$$

Additionally, the Householder matrix satisfies the *involutary property*

$$H H = I . \tag{3.58}$$

As shown in Sect. 3.2.1, the property of orthogonality is an indispensable requirement for the use of a transform matrix to calculate a QR decomposition. Recall relation (3.53). We interpreted the recurrent application of Givens matrices as a sequence of rotations. In a similar way, it can be instructive to interpret the Householder matrix as a *projection operator.*

To pick up this idea, we recall the definition of an orthogonal projection operator as given in (2.16). Since $\mathbf{w}$ has a Euclidean norm of unity, it is clear that we can write

$$\mathbf{w}\,\mathbf{w}^T = \mathbf{w}\left(\mathbf{w}^T\mathbf{w}\right)^{-1}\mathbf{w}^T \ . \tag{3.59}$$

The operator (3.59) can be interpreted as a projection operator that projects arbitrary vectors as elements of $\mathbb{R}^L$ *orthogonally* onto $\mathbf{w} \in \mathbb{R}^L$. See again the definition (2.16) of an orthogonal projection operator and the corresponding illustration in Fig. 2.4. In the following step, it becomes apparent that the Householder matrix $\mathbf{H}$ represents a special case of a projection operator. From its definition (3.55), we may deduce that $\mathbf{H}$ is a projection operator that *reflects* an arbitrary vector $\mathbf{x} \in \mathbb{R}^L$ on the orthogonal subspace of $\mathbf{w}$. This statement is illustrated in Fig. 3.2.

Now we would like to apply the Householder matrix to the problem of computing a QR decomposition. In contrast to the Givens reduction, which generates only one zero element in each rotation, the Householder matrix offers the possibility of generating column vectors in $\mathbf{R}'$ that have *any desired* number of zero elements. This way, the Householder reduction can generate the desired matrices $\mathbf{Q}$ and $\mathbf{R}$ by only a number p of transformation steps.

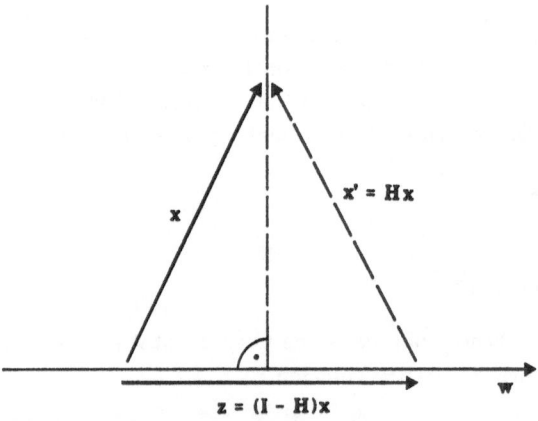

Fig. 3.2. Geometrical interpretation of the Householder matrix $\mathbf{H}$ as a projection operator that generates a reflected image $\mathbf{x}'$ with respect to the orthogonal subspace of $\mathbf{w}$

We consider the following initial situation:

$$Q_1 = I , \tag{3.60a}$$

$$R_1' = X \tag{3.60b}$$

and the QR decomposition

$$X = Q R' . \tag{3.61}$$

Combining (3.60a,b and 61), we obtain the initial equation

$$X = Q_1 R_1' = Q_1 X . \tag{3.62}$$

Taking into account the involutary property (3.58) of the Householder matrix, we may express (3.62) in the form

$$X = Q_j H\, HR_j' ; \qquad 1 \le j \le p , \tag{3.63}$$

which ultimately yields the recurrence relations

$$Q_{j+1} = Q_j H \tag{3.64a}$$

$$\left. \begin{array}{l} \end{array} \right\} \quad 1 \le j \le p ,$$

$$R'_{j+1} = H R'_j \tag{3.64b}$$

where

$$H = I - 2ww^T , \tag{3.65}$$

$$w = \left[0 \ldots 0,\ w_j^T \right]^T$$
$$= \left[0 \ldots 0,\ w_j(1),\ w_j(2),\ \ldots ,\ w_j(L\!-\!j\!+\!1) \right]^T . \tag{3.66}$$

Utilizing (3.66), the Householder matrix H can be partitioned as

$$H = \begin{bmatrix} I & 0 \\ 0 & H_j \end{bmatrix} , \tag{3.67}$$

where I is of dimension $(j\!-\!1) \times (j\!-\!1)$ and H is of dimension $(L\!-\!j\!+\!1) \times (L\!-\!j\!+\!1)$, hence

$$H_j = I - 2w_j w_j^T . \tag{3.68}$$

Now, we are interested in determining the elements of $\mathbf{w}_j$ such that the elements of the jth column vector of the matrix $\mathbf{R}'_j$, namely, the elements from $R'_j(j+1,j)$ to $R'_j(L,j)$, are transformed to zero by application of $\mathbf{H}_j$. Introducing the following column vector notation

$$\mathbf{r}_j = \Big[R'_j(j,j),\ R'_j(j+1,j),\ \ldots,\ R'_j(L,j) \Big]^T \tag{3.69}$$

and taking account of (3.64b), we obtain the desired transformation vector $\mathbf{w}_j$ that satisfies the condition

$$\Big(\mathbf{I} - 2\mathbf{w}_j\mathbf{w}_j^T\Big)\mathbf{r}_j = \mathbf{r}_{j+1} = \Big[R'_{j+1}(j,j),\ 0\ldots 0 \Big]^T . \tag{3.70}$$

From condition (3.70), it can be readily seen that

$$\mathbf{r}_{j+1}^T\,\mathbf{r}_{j+1} = R'^{2}_{j+1}(j,j) = \mathbf{r}_j^T\mathbf{H}_j\mathbf{H}_j\mathbf{r}_j = \mathbf{r}_j^T\mathbf{r}_j , \tag{3.71}$$

which gives

$$R'_{j+1}(j,j) = \pm\Big(\mathbf{r}_j^T\mathbf{r}_j \Big)^{1/2} . \tag{3.72}$$

Only the *absolute value* of the element $R'_{j+1}(j,j)$ is uniquely determined by (3.72). Later in this context we will see that the freedom in choosing the sign of $R'_{j+1}(j,j)$ can be used to obtain a numerically more accurate implementation of the Householder algorithm.

Premultiplying (3.70) by $\mathbf{w}_j^T$ allows a second important relationship to be derived, namely

$$\mathbf{w}_j^T\mathbf{r}_j - 2\mathbf{w}_j^T\mathbf{w}_j\mathbf{w}_j^T\mathbf{r}_j = \mathbf{w}_j^T\mathbf{r}_{j+1} . \tag{3.73}$$

Note $\mathbf{w}_j^T\mathbf{w}_j = 1$, and therefore

$$-\mathbf{w}_j^T\mathbf{r}_j = \mathbf{w}_j^T\mathbf{r}_{j+1} = w_j(1)\,R'_{j+1}(j,j) . \tag{3.74}$$

Introducing the auxiliary quantity h_j,

$$h_j = 2\mathbf{w}_j^T\mathbf{r}_j = -2\,w_j(1)\,R'_{j+1}(j,j) , \tag{3.75}$$

we can split (3.70) into

$$R'_j(j,j) - h_j\,w_j(1) = R'_{j+1}(j,j) \tag{3.76a}$$

and the set of equations

$$R'_j(k,j) - h_j\,w_j(k-j+1) = 0 \quad for \quad k = j+1,\ j+2,\ \ldots,\ L . \tag{3.76b}$$

Utilizing (3.75), one may now rewrite (3.76 a) to obtain

$$R_j'(j,j) + 2 w_j^2(1) R_{j+1}'(j,j) = R_{j+1}'(j,j) \quad , \tag{3.77}$$

which gives

$$w_j(1) = \left[0.5 \left(1 - R_j'(j,j) \big/ R_{j+1}'(j,j) \right) \right]^{1/2} \quad . \tag{3.78}$$

An evaluation of (3.78) reveals that this expression can be ill-conditioned through a subtraction of two almost identical numbers as soon as

$$R_j'(j,j) \big/ R_{j+1}'(j,j) \approx 1 \quad . \tag{3.79}$$

This difficulty can be circumvented by exploiting the fact that $R_{j+1}'(j,j)$ is only predetermined in its absolute value due to (3.72). Thus, a choice of the sign of $R_{j+1}'(j,j)$ according to the rule

$$sign\left(R_{j+1}'(j,j) \right) = - sign\left(R_j'(j,j) \right) \tag{3.80}$$

guarantees that

$$- R_j'(j,j) \big/ R_{j+1}'(j,j) > 0 \quad , \tag{3.81}$$

and thus avoids the subtraction of two almost identical values.

Next we require the calculation steps for determining the quantities $\{ w_j(k), j+1 \le k \le L \}$. Combining (3.72) with (3.78) and taking into account (3.75), we find h_j is uniquely determined by

$$h_j = - 2 w_j(1) R_{j+1}'(j,j) \quad , \tag{3.82}$$

which, in consideration of (3.76b), yields the desired rule for calculating the w's as follows:

$$w_j(k-j+1) = R_j'(k,j) \big/ h_j \qquad for \qquad k = j+1, j+2, \ldots, L \quad . \tag{3.83}$$

Table 3.4 provides a summary of the Householder algorithm.

A good review of Householder schemes in signal processing was given by Steinhardt [3.14] and Rader and Steinhardt [3.15].

3.2.3 Calculation of Prediction Error Energy

Once we have solved the Normal Equations, it is sometimes of interest to check the quality of the least-squares approximation by evaluating the prediction error energy E_{min}. At first glance, it seems that the prediction error energy can be determined by simply evaluating the expression

Table 3.4. Householder reduction of a real matrix $\mathbf{X} \in \mathbb{R}^{L \times p}$. The rotation of $\mathbf{Q}$ remains *optional*. The quantity $R'_{j+1}(j,j)$ needs to be stored twice to avoid "overwriting". This algorithm involves division. Whenever the divisor is small, set $1/x = x$

Initialize:

$\mathbf{R}'_1 = \mathbf{X}$; $\mathbf{Q}_1 = \mathbf{I}$;

FOR $j = 1, 2, \ldots , p$

$\quad \mathbf{r}_j = \left[R'_j(j,j), \ldots , R'_j(L,j) \right]^T$

$\quad R'_{j+1}(j,j) = \left(\mathbf{r}_j^T \mathbf{r}_j \right)^{1/2}$

$\quad$ IF $(R'_j(j,j) \geq 0)$ THEN

$\qquad R'_{j+1}(j,j) = - R'_{j+1}(j,j)$ $\qquad\qquad$ *change sign*

$\quad$ ENDIF

$\quad w_j(1) = \left[0.5 \left(1 - R'_j(j,j) / R'_{j+1}(j,j) \right) \right]^{1/2}$

$\quad h_j = - 2 R'_{j+1}(j,j) \, w_j(1)$

$\quad$ FOR $k = j+1, j+2, \ldots , L$

$\qquad w_j(k-j+1) = R'_j(k,j) / h_j$

$\quad \mathbf{H}_j = \mathbf{I}_{L-j+1} - 2 \mathbf{w}_j \mathbf{w}_j^T$

$\quad \mathbf{R}'_{j+1} = \begin{bmatrix} \mathbf{I} & \mathbf{0} \\ \mathbf{0} & \mathbf{H}_j \end{bmatrix} \mathbf{R}'_j$ $\qquad\qquad$ *rotate* $\mathbf{R}'_j$

$\quad \mathbf{Q}_{j+1} = \mathbf{Q}_j \begin{bmatrix} \mathbf{I} & \mathbf{0} \\ \mathbf{0} & \mathbf{H}_j \end{bmatrix}$ $\qquad\qquad$ *rotate* $\mathbf{Q}_j$ (*optional*)

$\begin{bmatrix} \mathbf{R} \\ \mathbf{0} \end{bmatrix} = \mathbf{R}'_{p+1}$; $\qquad\qquad \mathbf{Q} = \mathbf{Q}_{p+1}$

$$E_{min} = e_{min}^T e_{min} ,$$ (3.84)

where the prediction error vector e_{min} can be obtained from the predictor equation (2.8) when the parameter vector $\hat{a}$ is the least-squares solution to a given problem. In the case of an accurate prediction, however, it is clear that

$$x \approx X\hat{a} ,$$ (3.85)

which leads to a subtraction of two large but almost identical values in the computation of the prediction error vector e_{min} . The computation of the prediction error energy using the straightforward formula (3.84) can therefore be the subject of large numerical errors, which appear as a *bias* on the true least-squares prediction error energy.

The solution to this problem is a *direct computation* of the prediction error energy from auxiliary quantities that we obtain during a QR decomposition. Starting with the known relation for the prediction error energy

$$E(a) = e^T e = x^T x - x^T X a - a^T X^T x + a^T X^T X a ,$$ (3.86)

we may substitute the Normal Equations (2.13) into (3.86) to obtain

$$E_{min} = x^T x - x^T X\hat{a} ,$$ (3.87)

where we have reintroduced the notation $\hat{a}$ for the moment in order to distinguish the general weighting parameter vector a appearing in the functional $E(a)$ from the least-squares estimate of a denoted by $\hat{a}$. Using the QR decomposition, we note that [recall (3.26)]

$$x^T x = x^T Q Q^T x = d_1^T d_1 + d_2^T d_2$$ (3.88)

and, similarly,

$$x^T X\hat{a} = x^T Q \begin{bmatrix} R \\ 0 \end{bmatrix} \hat{a} = \begin{bmatrix} d_1 \\ d_2 \end{bmatrix}^T \begin{bmatrix} R\hat{a} \\ 0 \end{bmatrix} = d_1^T d_1 ,$$ (3.89)

which [from (3.87)] ultimately yields the useful result

$$E_{min} = d_2^T d_2 .$$ (3.90)

The relation (3.90) enables a numerically much more accurate determination of the prediction error energy than is possible with an explicit computation of the prediction error energy as the inner product of residual vectors (3.84).

3.3 Some More Principles for Matrix Computations

This section reviews some useful classical principles for matrix computations which allow us to analyze and solve special problems, for example, the computation of $K = X^{-1} Y$, where X and Y are square matrices of possibly very large dimension. This type of problem arises in the determination of reflection matrices in the vector-valued case of linear prediction. See also Appendix A.3. In particular, we shall introduce the *singular value decomposition* (SVD) [3.1, 16] as a special case of an orthogonal decomposition of an observation matrix. We show how the Normal Equations can be analyzed and solved from the elements of an SVD. The important concept of the *Penrose pseudoinverse* [3.17] is also discussed.

3.3.1 The Singular Value Decomposition

An arbitrary observation matrix $X \in \mathbb{R}^{L \times P}$ of dimension $L \times p$ has a singular value decomposition (SVD) [3.16]

$$X = Q S V^T \; ; \qquad Q \in \mathbb{R}^{L \times L}, \quad S \in \mathbb{R}^{L \times P}, \quad V \in \mathbb{R}^{P \times P}, \qquad (3.91)$$

where Q and V are *orthogonal* matrices and S is a *diagonal* matrix with k *nonnegative* elements of the form

$$S = \begin{bmatrix} & & & 0 \cdots 0 \\ & S^{(k)} & & \vdots \quad \vdots \\ & & & 0 \cdots 0 \\ 0 \cdots\cdots 0 & 0 \cdots 0 \\ \vdots & \vdots \quad \vdots \\ \vdots & \vdots \quad \vdots \\ 0 \cdots\cdots 0 & 0 \cdots 0 \end{bmatrix} , \qquad (3.92a)$$

$$S^{(k)} = \begin{bmatrix} S_1^{(k)} & & & \\ & S_2^{(k)} & & \\ & & \ddots & \\ & & & S_k^{(k)} \end{bmatrix} . \qquad (3.92b)$$

The elements $S_1^{(k)}, S_2^{(k)}, \ldots, S_k^{(k)}$ are the *singular values* of X. Note that this set of numbers uniquely determines the observation matrix X. The value of k is called the *rank* of X:

$$rank \; (X) = k \; . \qquad (3.93)$$

The matrix X is called *singular* when k < *min*(L,p), else the matrix X is called *nonsingular*.

Let $X = QSV^T$ be the singular value decomposition of X, then the *eigenvalue decomposition* of the system matrix of the Normal Equations $A = X^TX$ is given by

$$A = X^TX = V\left(S^TS\right)V^T = V\Omega V^T .$$ (3.94)

The matrix product $\Omega = S^TS$ forms another diagonal matrix, and its elements are called the *eigenvalues* of A.

The singular value decomposition of a given observation matrix X can be conveniently determined by the following procedure, employing either Givens or Householder reductions. For this purpose, the problem of computing $X = QSV^T$ is split into two subproblems:

$$R = SV^T ,$$ (3.95a)

$$X = QR = QSV^T .$$ (3.95b)

The system (3.95b) is treated first. We wish to generate the orthogonal matrix Q and an upper triangular form in R, hence

$$\begin{bmatrix} X \end{bmatrix} = \begin{bmatrix} Q \end{bmatrix}\begin{bmatrix} \begin{matrix} 0 \cdot \\ \vdots \cdot \cdot \cdot 0 \\ \vdots \\ 0 \cdot \cdot \cdot 0 \end{matrix} \end{bmatrix} .$$ (3.96)

$$R$$

Employing, for example, Givens rotations, this QR decomposition can be accomplished by setting

$$X = Q_{i,j}G_{i,j}^TG_{i,j}R_{i,j} ; \qquad\qquad i > j .$$ (3.97)

Recall (3.44). We shall obtain the recurrence relations

$$Q_{i+1,j} = Q_{i,j}G_{i,j}^T ,$$ (3.98a)

$$R_{i+1,j} = G_{i,j}R_{i,j} ,$$ (3.98b)

where $G_{i,j}$ is the Givens rotation matrix of recursion (i,j), as defined by (3.46). We may use the algorithm of Table 3.3, or the algorithm of Table 3.4 when the Householder reduction was chosen for computing the QR decomposition (3.96).

In a second step, one may *diagonalize* $\mathbf{R}^T$ such that

$$\mathbf{R}^T = \left(\mathbf{S}\,\mathbf{V}^T \right)^T = \mathbf{V}\,\mathbf{S}^T \tag{3.99}$$

or, equivalently,

$$\left[\begin{array}{c} \begin{smallmatrix} 0 \cdots \cdots \cdots 0 \\ \ddots \vdots \\ 0 \cdots 0 \end{smallmatrix} \end{array} \right] = \left[\begin{array}{c} \mathbf{V} \end{array} \right] \left[\begin{array}{c} \begin{smallmatrix} 0 \cdots \cdots \cdots 0 \\ 0 \ddots \vdots \\ 0 \cdot 0 \quad 0 \cdots 0 \end{smallmatrix} \end{array} \right] . \tag{3.100}$$

$$\quad\quad \mathbf{R}^T \quad\quad\quad\quad\quad\quad\quad \mathbf{S}^T$$

Again, the diagonalization of $\mathbf{R}^T$ may be accomplished via the approach

$$\mathbf{R}^T = \mathbf{V}_{i,j}\mathbf{H}^*_{i,j}\mathbf{H}^*_{i,j}\mathbf{S}^T_{i,j} \quad, \tag{3.101}$$

and consequently, one obtains the recursions

$$\mathbf{V}_{i+1,j} = \mathbf{V}_{i,j}\,\mathbf{H}^*_{i,j} \quad, \tag{3.102a}$$

$$\mathbf{S}^T_{i+1,j} = \mathbf{H}^*_{i,j}\,\mathbf{S}^T_{i,j} \tag{3.102b}$$

for the computation of the second orthogonal matrix $\mathbf{V}$ and the finally diagonal matrix $\mathbf{S}$ where $\mathbf{H}^*$ denotes a sequence of special rotations that accomplish the diagonalization of $\mathbf{R}$ by a *two-step procedure*. First, a sequence of Householder transformations is applied to generate a *bidiagonal* matrix. The bidiagonal form is then split into two independent parts that can be further diagonalized by a sequence of special rotations. See Chap. 18 of [3.1] for a detailed discussion about the explicit computation of a singular value decomposition. Note that Lawson and Hanson [3.1] provided a complete FORTRAN coded routine for the explicit computation of the singular value decomposition.

3.3.2 Solving the Normal Equations by Singular Value Decomposition

Suppose now that we have given the SVD of an observation matrix $\mathbf{X}$ as

$$\mathbf{X} = \mathbf{Q}\,\mathbf{S}\,\mathbf{V}^T . \tag{3.103}$$

Then, we shall describe how one may compute the solution of the Normal Equations (2.13) by means of singular value decomposition [3.18]. Recall that the Normal Equations derive from the least-squares problem

$$\mathbf{e} = \mathbf{x} - \mathbf{X}\mathbf{a} \quad, \tag{3.104a}$$

$$E = \mathbf{e}^T\mathbf{e} \stackrel{!}{=} min \longrightarrow \hat{\mathbf{a}} \quad. \tag{3.104b}$$

We shall recall the partitioning scheme (3.26)

$$Q^T x = d = \begin{bmatrix} d_1 \\ d_2 \end{bmatrix} \begin{matrix} \} k \\ \} L-k \end{matrix} \qquad (3.105)$$

Additionally, we introduce a second partitioning scheme of the form

$$V^T a = y = \begin{bmatrix} y_1 \\ y_2 \end{bmatrix} \begin{matrix} \} k \\ \} L-k \end{matrix} \qquad (3.106)$$

Note that the vectors d_1 and d_2 are uniquely determined by the orthogonal matrix Q and the observation vector x, whereas y_1 and y_2 are still unknown. A substitution of the SVD into (3.104a) gives

$$e = x - Q S V^T a \quad . \qquad (3.107)$$

Note

$$E = \left| x - Q S V^T a \right|^2 = \left| Q^T x - S V^T a \right|^2 \quad . \qquad (3.108)$$

Using the partitioning schemes (3.105 and 106), the prediction error energy E may be expressed as

$$E = \left| \begin{bmatrix} d_1 \\ d_2 \end{bmatrix} - S \begin{bmatrix} y_1 \\ y_2 \end{bmatrix} \right|^2 \quad . \qquad (3.109)$$

Employing the partitioning scheme of S (3.92a,b), we may rewrite (3.109) as

$$E = \left| \begin{bmatrix} d_1 \\ d_2 \end{bmatrix} - S^{(k)} y_1 \right|^2 \quad . \qquad (3.110)$$

We note that E is *independent* of the choice of y_2 , hence we may set $y_2 = \begin{bmatrix} 0 \dots 0 \end{bmatrix}^T$. This choice of y_2 has an important interpretation in that nonexistent dimensions in the subspace of past observations have *zero weight* when the signal covariance matrix is singular. This important consequence will be discussed in more detail later.

An evaluation of (3.110) reveals that E is composed of two additive terms as

$$E = \left| S^{(k)} y_1 - d_1 \right|^2 + \left| d_2 \right|^2 \quad . \qquad (3.111)$$

Clearly, E attains its minimum value when

$$\mathbf{S}^{(k)}\mathbf{y}_1 - \mathbf{d}_1 = \begin{bmatrix} 0 \ldots 0 \end{bmatrix}^T \longrightarrow \hat{\mathbf{y}}_1 , \tag{3.112}$$

which is a convenient way of determining the optimal vector $\hat{\mathbf{y}}_1$.

As a consequence of (3.112), we may check that

$$E_{min} = \left| \mathbf{d}_2 \right|^2 = \mathbf{d}_2^T \mathbf{d}_2 , \tag{3.113}$$

which has brought us back to (3.90). The intermediate vector $\hat{\mathbf{y}}$ is now completely determined via

$$\hat{\mathbf{y}} = \begin{bmatrix} \hat{\mathbf{y}}_1 \\ 0 \\ \vdots \\ 0 \end{bmatrix} . \tag{3.114}$$

We may substitute this solution (3.114) into the partitioning scheme (3.106) to obtain a uniquely determined linear system of equations for computing the optimal parameter vector $\hat{\mathbf{a}}$

$$\mathbf{V}^T\mathbf{a} - \hat{\mathbf{y}} = \begin{bmatrix} 0 \ldots 0 \end{bmatrix}^T \longrightarrow \hat{\mathbf{a}} , \tag{3.115}$$

which is the desired solution of the Normal Equations. The prediction error energy E_{min} can be computed via (3.113) and the associated prediction error vector $\mathbf{e}_{min}$ may be obtained as follows:

$$\mathbf{e}_{min} = \mathbf{x} - \mathbf{X}\hat{\mathbf{a}} = \mathbf{x} - \mathbf{Q}\mathbf{S}\mathbf{V}^T\hat{\mathbf{a}} = \mathbf{Q}\begin{bmatrix} \mathbf{d}_1 \\ \mathbf{d}_2 \end{bmatrix} - \mathbf{Q}\mathbf{S}\begin{bmatrix} \mathbf{y}_1 \\ 0 \\ \vdots \\ 0 \end{bmatrix} . \tag{3.116}$$

As we have assumed $\mathbf{S}^{(k)}\mathbf{y}_1 = \mathbf{d}_1$ [recall (3.112)], we shall find

$$\mathbf{e}_{min} = \mathbf{Q}\begin{bmatrix} 0 \\ \vdots \\ 0 \\ \mathbf{d}_2 \end{bmatrix} . \tag{3.117}$$

Finally, we note that the choice $\mathbf{y}_2 = \begin{bmatrix} 0 \ldots 0 \end{bmatrix}^T$ changes the partitioning scheme (3.106) into

$$\mathbf{V}^T\hat{\mathbf{a}} = \begin{bmatrix} \hat{\mathbf{y}}_1 \\ 0 \\ \vdots \\ 0 \end{bmatrix} . \tag{3.118}$$

3.3.3 The Penrose Pseudoinverse

We shall define the *Penrose pseudoinverse* [3.17] of the diagonal matrix S, denoted by S^+, as

$$
S^+ = \begin{bmatrix} \left[S^{(k)} \right]^{-1} & \begin{matrix} 0 \cdots 0 \\ \vdots \quad \vdots \\ 0 \cdots 0 \end{matrix} \\ \begin{matrix} 0 \cdots \cdots 0 \\ \vdots \qquad \vdots \\ 0 \cdots \cdots 0 \end{matrix} & \begin{matrix} 0 \cdots 0 \\ \vdots \quad \vdots \\ 0 \cdots 0 \end{matrix} \end{bmatrix} \quad . \tag{3.119}
$$

From (3.119), one sees that the pseudoinverse S^+ is simply the *transposed* matrix S with *inverted* diagonal elements of the nonsingular submatrix $S^{(k)}$.

When a matrix X has a singular value decomposition $X = QSV^T$, then the pseudoinverse X^+ is defined as

$$
X^+ = V S^+ Q^T \quad . \tag{3.120}
$$

We shall give a meaningful interpretation of definition (3.120). For this purpose, consider the system of linear equations (3.112). This system can be rewritten as

$$
\hat{y}_1 = \left[S^{(k)} \right]^{-1} d_1 \quad , \tag{3.121}
$$

and in terms of the pseudoinverse we may write

$$
\hat{y} = \begin{bmatrix} \hat{y}_1 \\ 0 \\ \vdots \\ 0 \end{bmatrix} = S^+ \begin{bmatrix} d_1 \\ d_2 \end{bmatrix} \quad , \tag{3.122}
$$

or, using the partitioning scheme (3.105),

$$
\hat{y} = S^+ Q^T x \quad . \tag{3.123}
$$

Premultiplication of (3.123) by the second orthogonal matrix V reveals

$$
\hat{a} = V \begin{bmatrix} \hat{y}_1 \\ 0 \\ \vdots \\ 0 \end{bmatrix} = V S^+ Q^T x = X^+ x \quad . \tag{3.124}
$$

This result leads to an interpretation of the pseudoinverse as the *general* solution of the least-squares problem (3.104a,b). The best coefficient vector $\hat{\mathbf{a}}$ is determined by $\mathbf{X}^+$ and the observation $\mathbf{x}$ such that nonexisting dimensions in the subspace of past observations attain a *zero weight*. Clearly, we might interpret the fact that the specific choice of $\mathbf{y}_2$ had *no effect* on the overall prediction error energy [recall (3.110 and 111)] as that any finite linear combination of zero basis vectors will again give the zero vector, and hence these *"nonexciting"* dimensions in the sub-space of past observations will not contribute or improve the overall prediction error energy.

In other words, $\hat{\mathbf{a}}$ is determined via the pseudoinverse such that the optimal prediction will be determined only from *nonzero* and *linearly independent* basis vectors of the subspace of past observations. This fact has an important practical consequence. Imagine a fixed predictor has been trained with a data set with an incomplete basis. Furthermore, suppose that the predictor thus obtained determined by $\hat{\mathbf{a}}$ is operated on a data set where just these "untrained" dimensions are suddenly present. In this case the predictor designed via the pseudoinverse approach (3.124) will provide a misadjusted, but still *fairly reasonable* output because of the choice of $\mathbf{y}_2 = \begin{bmatrix} 0 \dots 0 \end{bmatrix}^{\mathsf{T}}$.

A further note about practical computing seems to be in order. Clearly, in practice all diagonal elements of $\mathbf{S}$ will show a small nonzero value due to roundoff errors. For this reason, a singular value *analysis* may be added before the computation of $\mathbf{S}^+$. In this analysis, we may check whether a considered element S_j is considerably smaller than the smallest expected singular value for the given problem. In this case we may set $S_j^+ = S_j$. On the other hand, if the value of S_j is reasonably large, we compute $S_j^+ = S_j^{-1}$.

3.3.4 The Problem of Computing $X^{-1}Y$

The computation of $\mathbf{X}^{-1}\mathbf{Y}$ is sometimes required in linear prediction and related least-squares problems. A prominent example is the computation of *reflection matrices* of multichannel (vector-valued) ladder forms, which requires the explicit quantification of expressions of the type

$$\mathbf{K} = \mathbf{X}^{-1}\mathbf{Y} . \tag{3.125}$$

See Chap. 10 and Appendix A.3 for a more detailed discussion. In many such cases, particularly when the dimension of the involved matrices is large, the matrix $\mathbf{X}$ is likely to be singular, and hence $\mathbf{K}$ is not uniquely determined by the expression

$$\mathbf{X}\mathbf{K} = \mathbf{Y} . \tag{3.126}$$

From the infinite multitude of possible solutions for $\mathbf{K}$ that satisfy

(3.126), we may determine $\mathbf{K}$ in terms of the pseudoinverse $\mathbf{X}^+$ such that

$$\mathbf{K}_{opt} = \mathbf{X}^+\mathbf{Y} \ . \tag{3.127}$$

With this choice, $\mathbf{K}_{opt}$ attains reasonable values. Particularly when the so-trained multichannel predictors are operated in an environment where previously zero dimensions suddenly turn up in the data, the predictor will react well-behaved in terms of the input/output behavior. See also Appendix A.3, where these aspects concerning the application of the pseudoinverse in the computation of reflection matrices are discussed.

3.4 Chapter Summary

In this chapter, we have presented the most important algorithms for solving the Normal Equations in the nonrecursive covariance case of linear prediction. We have distinguished two important cases. First, the case where the Normal Equations were explicitly given, but the measurement matrix $\mathbf{X}$ and the process vector $\mathbf{x}$ are no longer accessible. In this case, we have, at this early stage of the considerations, introduced the Cholesky decomposition as an appropriate method. The Cholesky decomposition provides a factorization of the system matrix of the Normal Equations, hence facilitating an easy solution of the factorized system by a simple forward/backward substitution procedure. Another interesting case appears when the measurement matrix $\mathbf{X}$ and the vector $\mathbf{x}$ are available. In this case, it is better not to compute the Normal Equations explicitly, but to apply a QR decomposition to $\mathbf{X}$, which leads to a convenient way of solving the Normal Equations without computing their system matrix and the right-hand-side vector explicitly. Two methods of providing a QR decomposition have been discussed: The Givens and Householder reductions. Both methods have similar numerical properties, but the Householder reduction has a computationally more efficient processing scheme, i.e., it requires fewer calculations. Generally, the Givens and Householder schemes have a better numerical accuracy than the Cholesky decomposition. A drawback of QR-decomposition schemes, however, is that their computational complexity grows with the length L of the measurement matrix $\mathbf{X}$, whereas the computational cost of the Cholesky decomposition is independent of L.

We have introduced the singular value decomposition (SVD) as an important tool for analyzing and solving the Normal Equations. Specifically, the SVD allows the determination of the number of linearly independent basis vectors of the subspace of past observations, which is equivalent to the number of nonzero singular values of the observation matrix $\mathbf{X}$. When this number is smaller than the model order, the system matrix of the Normal Equations is *rank deficient*. In this case, we have a multitude of possible parameter vectors and we may employ the useful concept of the Penrose pseudoinverse to obtain a meaningful solution for the desired parameter vector.

Generally, all methods presented in this chapter are *batch processing schemes* for application to a given set of data. Later in this book, we will see how one can extend the discussed techniques to solve the *recursive case*, where the solution must be provided at each time step, by taking into account the solution and status of the recursive algorithm at the previous time step.

4. Recursive Least-Squares Using the QR Decomposition

In Chap. 3, we treated the *nonrecursive* solution of the Normal Equations as a batch processing least-squares problem. In many applications, like recursive identification and adaptive filtering, we are interested in a *recursive* solution of the Normal Equations such that given the solution at time t-1 , we may compute the updated version of this solution at time t upon the arrival of new data. Algorithms of this type are known as *recursive least-squares* (RLS) *algorithms.* As mentioned in Chap. 2, we can exploit the recursive properties of the Normal Equations, which we have discussed, to derive recursive $O(p^2)$ or even *fast* recursive $O(p)$ algorithms for calculating the updated solution of the Normal Equations at consecutive time steps of observation.

There are basically four approaches to obtaining recursive least-squares algorithms. In this chapter, we extend the classical QR decomposition [4.1] of Chap. 3 to the recursive case by exploiting the update properties of the QR decomposition of an augmented observation matrix **X**. One of these methods is the recursive $O(p^2)$ (triangular array) Givens reduction. Several versions of this algorithm are derived and their implementation on triangular systolic arrays is discussed. The origin of this class of RLS algorithms can be traced back to the early work of Gentleman [4.2] who considered the Givens reduction of an augmented observation matrix (however in a much different context than the RLS problem). Gentleman pointed out that the Givens reduction [4.3] is a very efficient way of updating the QR decomposition of an augmented observation matrix. Kung and Gentleman [4.4] presented a triangular systolic array for the recursive Givens reduction. McWhirter [4.5] subsequently showed how one can obtain the residual from this array by a simple processing scheme.

4.1 Formulation of the Growing-Window Recursive Least-Squares Problem

So far, we have formulated the least-squares problem based on an observed data set of L samples. In the derivation of recursive least-squares techniques, it is sometimes useful to consider *growing windows* on the data. In this case, the process vector $\mathbf{x}(t)$ contains all data samples from an initial time step $t = 0$ up to the current time step. As time progresses, the length of the vector $\mathbf{x}(t)$ of serialized process samples grows as well:

$$\mathbf{x}(t) = [x(t), x(t-1), x(t-2), \ldots, x(1), x(0)]^T . \qquad (4.1)$$

All past data beyond the initial time step $t = 0$ are assumed to be zero and hence we may write

$$x(t) = 0 \qquad for \qquad t < 0 , \qquad (4.2)$$

which is called the *"pre-windowing"* condition on the data. With the pre-windowing condition and serialized data, the observation matrix $\mathbf{X}(t) = \begin{bmatrix} \mathbf{x}(t-1), \mathbf{x}(t-2), \ldots, \mathbf{x}(t-p) \end{bmatrix}$ has the form

$$\mathbf{X}(t) = \begin{bmatrix} x(t-1) & x(t-2) & \ldots\ldots\ldots & x(t-p) \\ & & & \vdots \\ x(t-2) & x(t-3) & & \\ & & & \vdots \\ x(t-3) & & & \\ \vdots & & & \vdots \\ \vdots & & & \\ x(p-1) & x(p-2) & \ldots\ldots\ldots & x(0) \\ x(p-2) & & & 0 \\ \vdots & & & \vdots \\ \vdots & & & \\ x(1) & x(0) & & \vdots \\ x(0) & 0 & \ldots\ldots\ldots\ldots & 0 \end{bmatrix} . \qquad (4.3)$$

In this case of a growing window, the length of the data vector $\mathbf{x}(t)$ and the observation matrix $\mathbf{X}(t)$ increases by one as soon as a new process sample is captured.

In order to keep the energy of the considered signal bounded, it is necessary to impose an *exponential decay* on the data. Table 2.1b provides the exponentially weighted updating procedure of the covariance matrix of the observed process. This recursive procedure of Chap. 2 is equivalent to weighting the observation matrix $\mathbf{X}(t)$ with the square-root of the exponential weighting matrix $\Lambda(t)$:

$$
\Lambda(t) = \begin{bmatrix} 1 & 0 & \cdots\cdots\cdots & 0 \\ 0 & \lambda & & \vdots \\ \vdots & & \lambda^2 & \vdots \\ \vdots & & & 0 \\ 0 & \cdots\cdots & 0 & \lambda^{t-1} \end{bmatrix} \quad ; \qquad \lambda \leq 1 \ . \tag{4.4}
$$

4.2 Recursive Least–Squares Based on the Givens Reduction

With the exponential decay (4.4), we may rewrite (3.24) to express the QR decomposition of the exponentially weighted observation at time step t–1 as

$$
\mathbf{Q}^T(t-1)\,\Lambda^{1/2}(t-1)\,\mathbf{X}(t-1) = \Lambda^{1/2}(t-1) \begin{bmatrix} \mathbf{R}(t-1) \\ \mathbf{0} \end{bmatrix} = \begin{bmatrix} \mathbf{R}_\Lambda(t-1) \\ \mathbf{0} \end{bmatrix}, \tag{4.5}
$$

where $\mathbf{R}_\Lambda(t-1)$ is the exponentially weighted counterpart of the triangular matrix $\mathbf{R}(t-1)$. Similarly to (4.5), the partitioning scheme (3.26) can be rewritten in an exponentially weighted form

$$
\mathbf{Q}^T(t-1)\,\Lambda^{1/2}(t-1)\,\mathbf{x}(t-1) = \Lambda^{1/2}(t-1) \begin{bmatrix} \mathbf{d}_1(t-1) \\ \mathbf{d}_2(t-1) \end{bmatrix} = \begin{bmatrix} \mathbf{d}_{\Lambda 1}(t-1) \\ \mathbf{d}_{\Lambda 2}(t-1) \end{bmatrix} . \tag{4.6}
$$

The vectors $\mathbf{d}_{\Lambda 1}(t-1)$ and $\mathbf{d}_{\Lambda 2}(t-1)$ are the exponentially weighted counterparts of the transformed data vectors $\mathbf{d}_1(t-1)$ and $\mathbf{d}_2(t-1)$.

Introducing the rank-increased matrix $\overline{\mathbf{Q}}(t-1)$

$$
\overline{\mathbf{Q}}(t-1) = \begin{bmatrix} 1 & 0\ldots\ldots 0 \\ 0 & \\ \vdots & \mathbf{Q}(t-1) \\ 0 & \end{bmatrix} \tag{4.7}
$$

and the weighting matrix at time step t satisfying the partitioning scheme

$$
\Lambda(t) = \begin{bmatrix} 1 & 0\ldots\ldots 0 \\ 0 & \\ \vdots & \lambda\Lambda(t-1) \\ 0 & \end{bmatrix} \tag{4.8}
$$

and taking into account that

$$\mathbf{X}(t) = \begin{bmatrix} \mathbf{z}^T(t) \\ \mathbf{X}(t-1) \end{bmatrix} \tag{4.9}$$

where

$$\mathbf{z}(t) = \begin{bmatrix} x(t-1),\ x(t-2),\ \dots\ ,\ x(t-p) \end{bmatrix}^T , \tag{4.10}$$

we can express the updated version of (4.5) as

$$\overline{\mathbf{Q}}^T(t-1)\,\mathbf{\Lambda}^{1/2}(t)\,\mathbf{X}(t) = \mathbf{\Lambda}^{1/2}(t) \begin{bmatrix} \mathbf{z}^T(t) \\ \mathbf{R}(t-1) \\ \mathbf{0} \end{bmatrix} = \begin{bmatrix} \mathbf{z}^T(t) \\ \lambda^{1/2}\mathbf{R}_\Lambda(t-1) \\ \mathbf{0} \end{bmatrix} \tag{4.11}$$

or, more explicitly,

$$\begin{bmatrix} 1 & 0 \dots 0 \\ 0 & \\ \vdots & \mathbf{Q}^T(t-1) \\ 0 & \end{bmatrix} \begin{bmatrix} 1 & 0 \dots\dots\dots 0 \\ 0 & \\ \vdots & \lambda^{1/2}\mathbf{\Lambda}^{1/2}(t-1) \\ 0 & \end{bmatrix} \begin{bmatrix} \mathbf{z}^T(t) \\ \mathbf{X}(t-1) \end{bmatrix}$$

$$= \begin{bmatrix} \mathbf{z}^T(t) \\ \lambda^{1/2}\,\mathbf{Q}^T(t-1)\,\mathbf{\Lambda}^{1/2}(t-1)\,\mathbf{X}(t-1) \end{bmatrix} = \begin{bmatrix} \mathbf{z}^T(t) \\ \lambda^{1/2}\mathbf{R}_\Lambda(t-1) \\ \mathbf{0} \end{bmatrix} . \tag{4.12}$$

Using a sequence of Givens rotations $\mathbf{T}(t)$

$$\mathbf{T}(t) = \prod_{j=1}^{P} \mathbf{G}_{j+1,j}(t) , \tag{4.13}$$

we may triangularize the right-hand side of (4.11) as

$$\mathbf{T}(t)\,\overline{\mathbf{Q}}^T(t-1)\,\mathbf{\Lambda}^{1/2}(t)\,\mathbf{X}(t) = \mathbf{T}(t)\,\mathbf{\Lambda}^{1/2}(t) \begin{bmatrix} \mathbf{z}^T(t) \\ \mathbf{R}(t-1) \\ \mathbf{0} \end{bmatrix}$$

$$= \mathbf{T}(t) \begin{bmatrix} \mathbf{z}^T(t) \\ \lambda^{1/2}\mathbf{R}_\Lambda(t-1) \\ \mathbf{0} \end{bmatrix} , \tag{4.14}$$

where

$$\mathbf{Q}^T(t) = \mathbf{T}(t)\,\overline{\mathbf{Q}}^T(t-1) = \mathbf{T}(t) \begin{bmatrix} 1 & 0 \dots 0 \\ 0 & \\ \vdots & \mathbf{Q}^T(t-1) \\ 0 & \end{bmatrix} . \tag{4.15}$$

Taking into account that

$$\mathbf{Q}^T(t)\,\mathbf{\Lambda}^{1/2}(t)\,\mathbf{X}(t) = \mathbf{\Lambda}^{1/2}(t)\begin{bmatrix} \mathbf{R}(t) \\ \mathbf{0} \end{bmatrix} = \begin{bmatrix} \mathbf{R}_\Lambda(t) \\ \mathbf{0} \end{bmatrix}, \tag{4.16}$$

it follows from a comparison of (4.16) with (4.14 and 15) that

$$\begin{bmatrix} \mathbf{R}_\Lambda(t) \\ \mathbf{0} \end{bmatrix} = \mathbf{T}(t)\begin{bmatrix} \mathbf{z}^T(t) \\ \lambda^{1/2}\mathbf{R}_\Lambda(t-1) \\ \mathbf{0} \end{bmatrix}. \tag{4.17}$$

Making use of the update property of the process vector $\mathbf{x}(t)$

$$\mathbf{x}(t) = \begin{bmatrix} x(t) \\ \mathbf{x}(t-1) \end{bmatrix} \tag{4.18}$$

and taking into account (4.7, 8 and 15), we may compute the update of the left–hand side of (4.6) as

$$\mathbf{Q}^T(t)\,\mathbf{\Lambda}^{1/2}(t)\,\mathbf{x}(t)$$

$$= \mathbf{T}(t)\begin{bmatrix} 1 & 0 \dots 0 \\ 0 & \\ \vdots & \mathbf{Q}^T(t-1) \\ 0 & \end{bmatrix}\begin{bmatrix} 1 & 0 \dots\dots\dots 0 \\ 0 & \\ \vdots & \lambda^{1/2}\mathbf{\Lambda}^{1/2}(t-1) \\ 0 & \end{bmatrix}\begin{bmatrix} x(t) \\ \mathbf{x}(t-1) \end{bmatrix}$$

$$= \mathbf{T}(t)\begin{bmatrix} x(t) \\ \lambda^{1/2}\,\mathbf{Q}^T(t-1)\,\mathbf{\Lambda}^{1/2}(t-1)\,\mathbf{x}(t-1) \end{bmatrix}. \tag{4.19}$$

But, according to (4.6), the partitioning scheme for the transformed data vector also holds for the current time step,

$$\mathbf{Q}^T(t)\,\mathbf{\Lambda}^{1/2}(t)\,\mathbf{x}(t) = \mathbf{\Lambda}^{1/2}(t)\begin{bmatrix} \mathbf{d}_1(t) \\ \mathbf{d}_2(t) \end{bmatrix} = \begin{bmatrix} \mathbf{d}_{\Lambda 1}(t) \\ \mathbf{d}_{\Lambda 2}(t) \end{bmatrix}. \tag{4.20}$$

A comparison of (4.20, 19 and 6) reveals that

$$\begin{bmatrix} \mathbf{d}_{\Lambda 1}(t) \\ \mathbf{d}_{\Lambda 2}(t) \end{bmatrix} = \mathbf{T}(t)\begin{bmatrix} x(t) \\ \lambda^{1/2}\mathbf{d}_{\Lambda 1}(t-1) \\ \lambda^{1/2}\mathbf{d}_{\Lambda 2}(t-1) \end{bmatrix}, \tag{4.21}$$

which, together with (4.17), constitutes a complete recursion for the updating of the triangular matrix $\mathbf{R}_\Lambda(t)$ and the partitioned transformed data vector consisting of the components $\mathbf{d}_{\Lambda 1}(t)$ and $\mathbf{d}_{\Lambda 2}(t)$.

The required sequence of Givens rotations, as stated in (4.13), has the explicit form

$$
\mathbf{T}(t) = \begin{bmatrix} c_1 & s_1 & & & \\ -s_1 & c_1 & & & \\ & & 1 & & \\ & & & \ddots & \\ & & & & 1 \end{bmatrix} \begin{bmatrix} 1 & & & & \\ & c_2 & s_2 & & \\ & -s_2 & c_2 & & \\ & & & 1 & \\ & & & & \ddots \\ & & & & & 1 \end{bmatrix} \cdots
$$

$$
\cdots \begin{bmatrix} 1 & & & & \\ & \ddots & & & \\ & & 1 & & \\ & & c_p & s_p & \\ & & -s_p & c_p & \\ & & & & 1 \\ & & & & & \ddots \\ & & & & & & 1 \end{bmatrix} , \tag{4.22}
$$

where the values for c_* and s_* can be calculated according to (3.48a,b and 49). Table 4.1 summarizes this QR decomposition of the recursively updated observation matrix $\mathbf{X}(t)$, assuming an *exponential decay*. Note that the array provided for $\mathbf{X}(t)$ need only be of length p+1, if we assume that the QR decomposition of the previous time step is already available. Note again that the orthogonal part $\mathbf{Q}(t)$ is only *implicitly* required during the derivation. It is not explicitly required if we are interested in using the QR decomposition for solving a least-squares problem.

In contrast to Table 3.3 (nonrecursive Givens reduction) we recognize that the second loop over the loop variable i has been replaced by setting i = j+1 in the recursions of Table 4.1, since only the elements in a diagonal below the main diagonal of $\mathbf{R}_\Lambda(t)$ must be annihilated. See also Fig. 4.9 for an illustration of the annihilation process of this recursive Givens algorithm. Furthermore, we find that the algorithm of Table 4.1 has a computational complexity of $O(p^2)$ operations per recursion.

As time progresses, the dimension of the matrix $\mathbf{Q}$ grows to infinity. This does not lead to a problem since we have seen that the explicit computation of $\mathbf{Q}$ is *not required* in this context. Nevertheless, the matrix $\mathbf{Q}$ is a *"DFT-like"* matrix and when applied to the data vector has some effects that are quite similar to what we know from the *Fourier transform*. For example, it is interesting to see that some kind of a *"Parseval's" relation* [4.6] holds in that

Table 4.1. Recursive Givens reduction assuming an exponential decay of the observation matrix $\mathbf{X}(t)$. The variable $X_{j,k}(t)$ needs to be stored twice to avoid overwriting. This algorithm involves division. Whenever the divisor is small, set $c = 1$ and $s = 0$ (identity rotor)

FOR each time step $t = 0, 1, 2, 3, \ldots$ do the following:

available at time t: $\quad \mathbf{z}(t)$, $\mathbf{R}_\Lambda(t-1)$, λ

set: $\quad \mathbf{X}(t) = \begin{bmatrix} \mathbf{z}^T(t) \\ \lambda^{1/2}\mathbf{R}_\Lambda(t-1) \end{bmatrix}$

FOR $j = 1, 2, \ldots, p$

$q = \left(X_{j,j}^2(t) + X_{j+1,j}^2(t) \right)^{1/2}$

$c = X_{j,j}(t) / q$

$s = X_{j+1,j}(t) / q$

FOR $k = j, j+1, j+2, \ldots, p$: $\quad$ rotate $\mathbf{X}(t)$

$X_{j,k}(t) \;\;= \;\; c\, X_{j,k}(t) + s\, X_{j+1,k}(t)$

$X_{j+1,k}(t) = -s\, X_{j,k}(t) + c\, X_{j+1,k}(t)$

$\begin{bmatrix} \mathbf{R}_\Lambda(t) \\ 0 \ldots 0 \end{bmatrix} = \mathbf{X}(t)$

$$\mathbf{x}^T \mathbf{Q}\, \mathbf{Q}^T \mathbf{x} \;=\; \mathbf{x}^T \mathbf{x} \;=\; \begin{bmatrix} \mathbf{d}_1^T & \mathbf{d}_2^T \end{bmatrix} \begin{bmatrix} \mathbf{d}_1 \\ \mathbf{d}_2 \end{bmatrix} . \tag{4.23}$$

The effect of applying $\mathbf{Q}$ to a data vector $\mathbf{x}$ is therefore sometimes loosely interpreted as the *"least-squares frequency transform"* of the data vector $\mathbf{x}$.

The actual solution $\mathbf{a}(t)$ can be obtained by first computing $\mathbf{d}_{\Lambda 1}(t)$ and $\mathbf{d}_{\Lambda 2}(t)$ recursively using (4.21) followed by a back substitution of the triangular system

$$\mathbf{R}_\Lambda(t)\, \mathbf{a}(t) = \mathbf{d}_{\Lambda 1}(t) . \tag{4.24}$$

Table 4.2. Recursive least-squares algorithm using the Givens reduction and an exponential windowing of the input data. The variables $X_{j,k}(t)$ and $d_j(t)$ need to be stored twice to avoid overwriting. The rank of $R_\Lambda(t)$ is equal to $min(t,p)$. This algorithm involves division. Whenever the divisor is small, set $c = 1$ and $s = 0$ (identity rotor)

Initialize: $R_\Lambda(-1) = 0$; $d_{\Lambda 1}(-1) = \begin{bmatrix} 0 \ldots 0 \end{bmatrix}^T$

FOR each time step t = 0, 1, 2, 3, . . . do the following:

    available at time t: $x(t)$, $z(t)$, $R_\Lambda(t-1)$, $d_{\Lambda 1}(t-1)$, λ

    set: $X(t) = \begin{bmatrix} z^T(t) \\ \lambda^{1/2} R_\Lambda(t-1) \end{bmatrix}$; $d(t) = \begin{bmatrix} x(t) \\ \lambda^{1/2} d_{\Lambda 1}(t-1) \end{bmatrix}$

      FOR j = 1, 2, . . . , p

        $q = \left(X_{j,j}^2(t) + X_{j+1,j}^2(t) \right)^{1/2}$

        $c = X_{j,j}(t) / q$

        $s = X_{j+1,j}(t) / q$

        FOR k = j, j+1, j+2, . . . , p: *rotate* $X(t)$

          $X_{j,k}(t) \ \ = c\, X_{j,k}(t) + s\, X_{j+1,k}(t)$

          $X_{j+1,k}(t) = -s\, X_{j,k}(t) + c\, X_{j+1,k}(t)$

        $d_j(t) \ \ = c\, d_j(t) + s\, d_{j+1}(t)$ *rotate* $d(t)$

        $d_{j+1}(t) = -s\, d_j(t) + c\, d_{j+1}(t)$

    $\begin{bmatrix} R_\Lambda(t) \\ 0 \ldots 0 \end{bmatrix} = X(t)$; $\begin{bmatrix} d_{\Lambda 1}(t) \\ 0 \end{bmatrix} = d(t)$

    $R_\Lambda(t)\, a(t) = d_{\Lambda 1}(t) \longrightarrow a(t)$ *through back substitution.*

Table 4.2 summarizes the complete recursive least-squares algorithm based on Givens reductions. From the definition of the observation matrix $X(t)$ in (4.3), we can see that the pre-windowing condition (4.2) causes the lower part of $X(t)$ to be triangular. In the start-up phase of the algorithm this fact must be considered for exact initialization since the lower triangular part of $\mathbf{X}(t)$ has just the *opposite* triangular shape to $\mathbf{R}(t)$. Before starting the algorithm, the pre-windowing condition (4.2) reveals that we can set

$$\mathbf{R}_\Lambda(-1) = \mathbf{0} , \qquad (4.25a)$$

$$\mathbf{d}_{\Lambda 1}(-1) = \begin{bmatrix} 0 \ldots 0 \end{bmatrix}^{\mathrm{T}} , \qquad (4.25b)$$

$$\mathbf{d}_{\Lambda 2}(-1) = \begin{bmatrix} 0 \ldots 0 \end{bmatrix}^{\mathrm{T}} , \qquad (4.25c)$$

where $\mathbf{0}$ is the null matrix of appropriate dimension. Recalling the structure of the lower part of $\mathbf{X}(t)$

$$\mathbf{X}(t) = \begin{bmatrix} x(2) & x(1) & x(0) & 0 \ldots \ldots \\ x(1) & x(0) & 0 & 0 \ldots \ldots \\ x(0) & 0 & 0 & 0 \ldots \ldots \\ 0 & 0 & 0 & 0 \ldots \ldots \end{bmatrix} = \begin{bmatrix} \mathbf{z}^{\mathrm{T}}(3) \\ \mathbf{z}^{\mathrm{T}}(2) \\ \mathbf{z}^{\mathrm{T}}(1) \\ \mathbf{z}^{\mathrm{T}}(0) \end{bmatrix} , \qquad (4.26)$$

we see that the time step $t = 0$ requires no rotations, but only the collection of the input data $\mathbf{z}(0)$ as follows:

$$\begin{bmatrix} \mathbf{R}_\Lambda(0) \\ \mathbf{0} \end{bmatrix} = \begin{bmatrix} \mathbf{z}^{\mathrm{T}}(0) \\ \mathbf{0} \end{bmatrix} , \qquad (4.27a)$$

$$\begin{bmatrix} \mathbf{d}_{\Lambda 1}(0) \\ \mathbf{d}_{\Lambda 2}(0) \end{bmatrix} = \begin{bmatrix} x(0) \\ 0 \\ \vdots \\ 0 \end{bmatrix} . \qquad (4.27b)$$

At time step $t = 1$, the first rotation is required to annihilate the second nonzero element of $\mathbf{d}_\Lambda$:

$$\begin{bmatrix} \mathbf{R}_\Lambda(1) \\ \mathbf{0} \end{bmatrix} = \begin{bmatrix} \mathbf{z}^{\mathrm{T}}(1) \\ \mathbf{0} \end{bmatrix} , \qquad (4.28a)$$

$$\begin{bmatrix} \mathbf{d}_{\Lambda 1}(1) \\ \mathbf{d}_{\Lambda 2}(1) \end{bmatrix} = \mathbf{G}_{2,1}(1) \begin{bmatrix} x(1) \\ \lambda^{1/2} x(0) \\ 0 \\ \vdots \\ 0 \end{bmatrix} . \qquad (4.28b)$$

At time step $t = 2$, the first rotation is necessary to annihilate the element in the second row, first column, of the updated matrix $\mathbf{R}_\Lambda$:

$$
\begin{bmatrix} \mathbf{R}_\Lambda(2) \\ \mathbf{0} \end{bmatrix} = \mathbf{G}_{2,1}(2) \begin{bmatrix} \mathbf{z}^T(2) \\ \lambda^{1/2}\mathbf{R}_\Lambda(1) \\ \mathbf{0} \end{bmatrix} , \tag{4.29a}
$$

$$
\begin{bmatrix} \mathbf{d}_{\Lambda 1}(2) \\ \mathbf{d}_{\Lambda 2}(2) \end{bmatrix} = \mathbf{G}_{2,1}(2)\, \mathbf{G}_{3,2}(2) \begin{bmatrix} x(2) \\ \lambda^{1/2}\mathbf{d}_{\Lambda 1}(1) \\ \lambda^{1/2}\mathbf{d}_{\Lambda 2}(1) \end{bmatrix} . \tag{4.29b}
$$

In the next time step, $t = 3$, two Givens rotations will be needed for rotation of $\mathbf{R}_\Lambda$, and three rotations are applied to $\mathbf{d}_\Lambda$, hence

$$
\begin{bmatrix} \mathbf{R}_\Lambda(3) \\ \mathbf{0} \end{bmatrix} = \mathbf{G}_{2,1}(3)\, \mathbf{G}_{3,2}(3) \begin{bmatrix} \mathbf{z}^T(3) \\ \lambda^{1/2}\mathbf{R}_\Lambda(2) \\ \mathbf{0} \end{bmatrix} , \tag{4.30a}
$$

$$
\begin{bmatrix} \mathbf{d}_{\Lambda 1}(3) \\ \mathbf{d}_{\Lambda 2}(3) \end{bmatrix} = \mathbf{G}_{2,1}(3)\, \mathbf{G}_{3,2}(3)\, \mathbf{G}_{4,3}(3) \begin{bmatrix} x(3) \\ \lambda^{1/2}\mathbf{d}_{\Lambda 1}(2) \\ \lambda^{1/2}\mathbf{d}_{\Lambda 2}(2) \end{bmatrix} , \tag{4.30b}
$$

and finally, one sees that

$$
\begin{bmatrix} \mathbf{R}_\Lambda(t) \\ \mathbf{0} \end{bmatrix} = \prod_{j=1}^{min(t-1,p)} \mathbf{G}_{j+1,j}(t) \begin{bmatrix} \mathbf{z}^T(t) \\ \lambda^{1/2}\mathbf{R}_\Lambda(t-1) \\ \mathbf{0} \end{bmatrix} , \tag{4.31a}
$$

$$
\begin{bmatrix} \mathbf{d}_{\Lambda 1}(t) \\ \mathbf{d}_{\Lambda 2}(t) \end{bmatrix} = \prod_{j=1}^{min(t,p)} \mathbf{G}_{j+1,j}(t) \begin{bmatrix} x(t) \\ \lambda^{1/2}\mathbf{d}_{\Lambda 1}(t-1) \\ \lambda^{1/2}\mathbf{d}_{\Lambda 2}(t-1) \end{bmatrix} , \tag{4.31b}
$$

which concludes our consideration of the exact initialization procedure. It can be shown that in the case of pre-windowed data (4.3) this exact initialization is automatically obtained when one only sets $c = 1$ and $s = 0$ (identity rotor) in the case that the quantity q falls below a certain threshold. Table 4.2 provides a summary of this recursive least-squares algorithm using the Givens reduction.

4.3 Systolic Array Implementation

The algorithm of Table 4.2 can be implemented on a two-dimensional (triangular) systolic array [4,4,7,8]. We will next develop the principal structure of this array by evaluating the algorithm of Table 4.2.

Writing the outer loop for $j = 1, 2, 3, \ldots$ explicitly, we obtain the following sequence of rotations applied on the elements of column k of the observation matrix $\mathbf{X}(t)$:

$$X'_{1,k}(t) = c_1 X_{1,k}(t) + s_1 X_{2,k}(t) \tag{4.32a}$$

$$X'_{2,k}(t) = -s_1 X_{1,k}(t) + c_1 X_{2,k}(t) \tag{4.32b}$$

$$X''_{2,k}(t) = c_2 X'_{2,k}(t) + s_2 X_{3,k}(t) \tag{4.32c}$$

$$X'_{3,k}(t) = -s_2 X'_{2,k}(t) + c_2 X_{3,k}(t) \tag{4.32d}$$

$$X''_{3,k}(t) = c_3 X'_{3,k}(t) + s_3 X_{4,k}(t) \tag{4.32e}$$

$$X'_{4,k}(t) = -s_3 X'_{3,k}(t) + c_3 X_{4,k}(t) \tag{4.32f}$$

$$X''_{4,k}(t) = c_4 X'_{4,k}(t) + s_4 X_{5,k}(t) \tag{4.32g}$$

$$X'_{5,k}(t) = -s_4 X'_{4,k}(t) + c_4 X_{5,k}(t) \tag{4.32h}$$

$$\vdots \qquad \vdots \qquad \vdots$$

The elements denoted by X' are intermediate variables in the rotation process. The elements X'' are the finally obtained elements of column k in the rotated (output) observation matrix.

These rotations, as performed for a fixed column k, may be visualized by a linear array as shown in Fig. 4.1. In this figure, the "rotational element" is defined by

$$\begin{bmatrix} u_o \\ v_o \end{bmatrix} = \begin{bmatrix} c & s \\ -s & c \end{bmatrix} \begin{bmatrix} u_i \\ v_i \end{bmatrix} . \tag{4.33}$$

Now, stating the shift condition (time update)

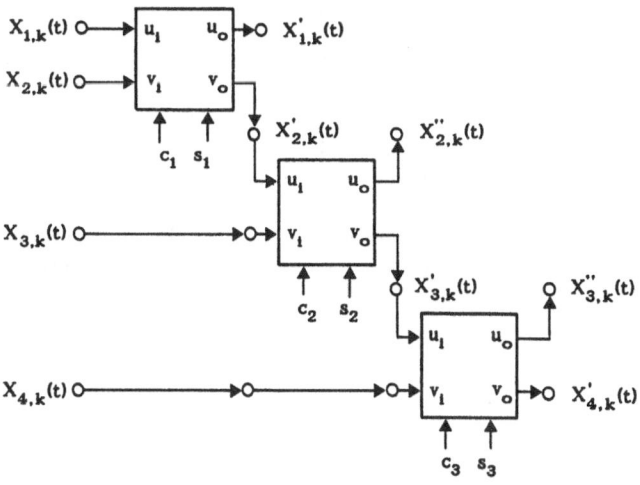

Fig. 4.1. Interrelationships of rotational elements in subsequent rows in a fixed column k

$$X_{j,k}(t) = \lambda^{1/2} X''_{j-1,k}(t-1) ; \qquad j = 3, 4, 5, \ldots , \tag{4.34a}$$

$$X_{2,k}(t) = \lambda^{1/2} X'_{1,k}(t-1) , \tag{4.34b}$$

$$X_{1,k}(t) = z_k(t) = x(t-k) , \tag{4.34c}$$

we see that there exists a "feedback" in that the rotated elements are fed back into the input of the rotational elements with a delay of one clock cycle. Additionally, the gain in this feedback loop is identical with the exponential weighting factor $\lambda^{1/2}$. This also reveals that we have to add a storage element denoted by z^{-1} to each rotational unit. In practice, each of these storage elements contains an element of the upper triangular form $R_\Lambda(t)$ in $X(t)$. This gives rise to the structure shown in Fig. 4.2.

After these deliberations it is now easy to compose the final array for rotation of the upper right triangular form in $X(t)$. Figure 4.3 shows an example for p = 3. The array shown in this figure provides the following operation:

$$\begin{bmatrix} z_1(t) & z_2(t) & z_3(t) \\ \lambda^{1/2}X'_{1,1}(t-1) & \lambda^{1/2}X'_{1,2}(t-1) & \lambda^{1/2}X'_{1,3}(t-1) \\ 0 & \lambda^{1/2}X''_{2,2}(t-1) & \lambda^{1/2}X''_{2,3}(t-1) \\ 0 & 0 & \lambda^{1/2}X''_{3,3}(t-1) \end{bmatrix}$$

$$\Rightarrow \begin{bmatrix} X'_{1,1}(t) & X'_{1,2}(t) & X'_{1,3}(t) \\ 0 & X''_{2,2}(t) & X''_{2,3}(t) \\ 0 & 0 & X''_{3,3}(t) \\ 0 & 0 & 0 \end{bmatrix} . \tag{4.35}$$

For each column, the rotation parameters c and s must be determined such that the lower diagonal elements are transformed to zero. This is accomplished by "angle units" determined by the following set of equations:

$$q = \left(u^2 + v^2\right)^{1/2} , \tag{4.36a}$$

$$c = u/q , \tag{4.36b}$$

$$s = v/q . \tag{4.36c}$$

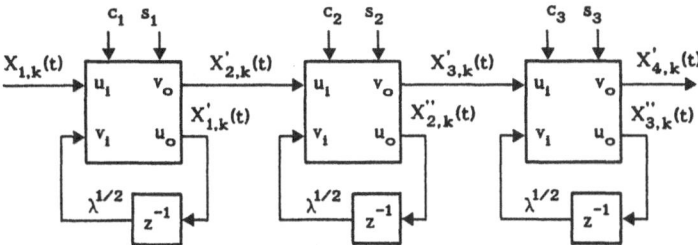

Fig. 4.2. Interrelationship of rotational elements of a fixed column in the recursive case

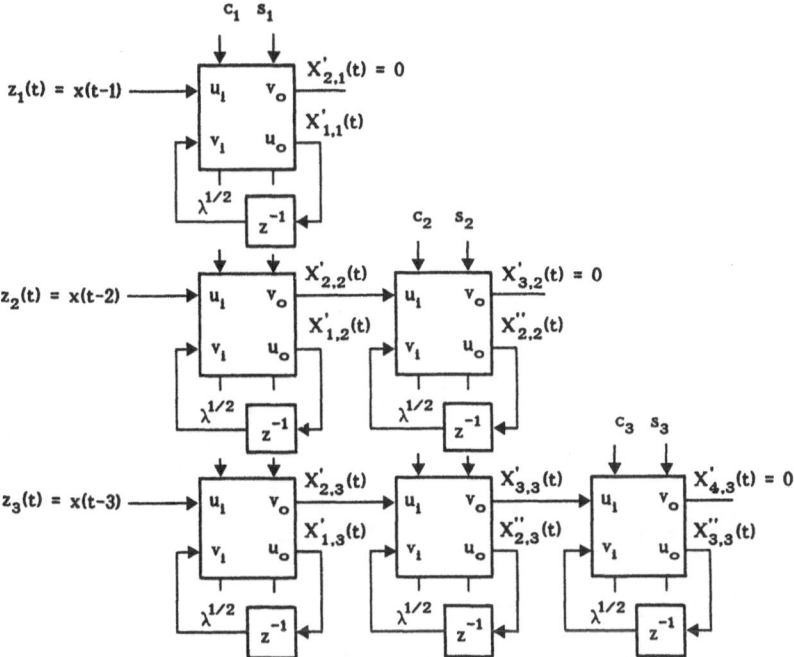

Fig. 4.3. Triangular array of dimension $p = 3$ for annihilation of the lower diagonal elements in the updated triangular form in $\mathbf{X}(t)$ [first matrix in (4.35)]

Figure 4.4 illustrates the angle unit while Fig. 4.5 shows how the angle units interact with the triangular array section. Additionally, we may use a linear array section for rotation of the vector $\mathbf{d}(t)$. This linear array section is located below the triangular array in Fig. 4.5. After the rotations of the updated observation matrix $\mathbf{X}(t)$ and the vector $\mathbf{d}(t)$ have been completed, one requires a second (interleaved) array for the back-substitution process which is required to compute the desired parameter vector $\mathbf{a}(t)$.

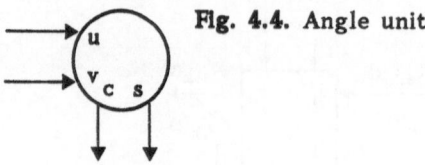

Fig. 4.4. Angle unit

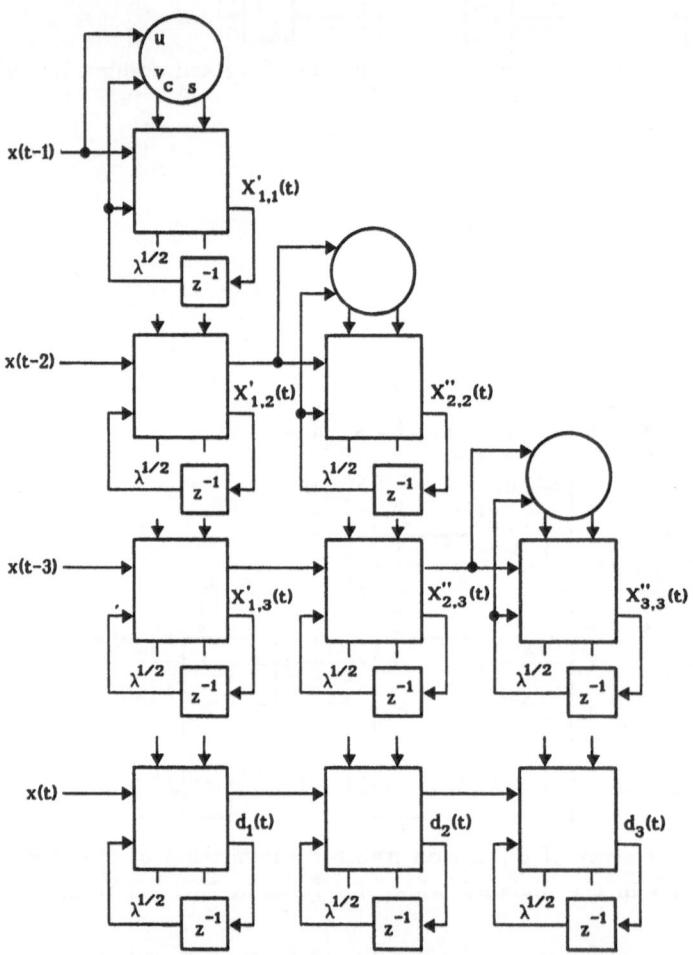

Fig. 4.5. Complete triangular array of order p = 3 with rotational units for rotation of $\mathbf{X}(t)$ (triangular array section) and $\mathbf{d}(t)$ (linear array section) as well as the associated angle units

In general, the back-substitution process is determined by the recursion

$$a_p(t) = d_p(t) / X_{p,p}(t) \tag{4.37a}$$

FOR   m = 1, 2, . . . , p-1

$$a_{p-m}(t) = \frac{d_{p-m}(t) - \sum\limits_{i=0}^{m-1} X_{p-m,\,p-i}(t)\,a_{p-i}(t)}{X_{p-m,\,p-m}(t)} \tag{4.37b}$$

where X is a shorthand notation of X' and X", respectively. This back substitution can be realized by a triangular array as shown in Fig. 4.6. Note that in contrast to Fig. 4.5, the array shown in Fig. 4.6. has different processing cells. The elements of $\mathbf{X}(t)$ and $\mathbf{d}(t)$ are used as entries in the back-substitution array of Fig. 4.6.

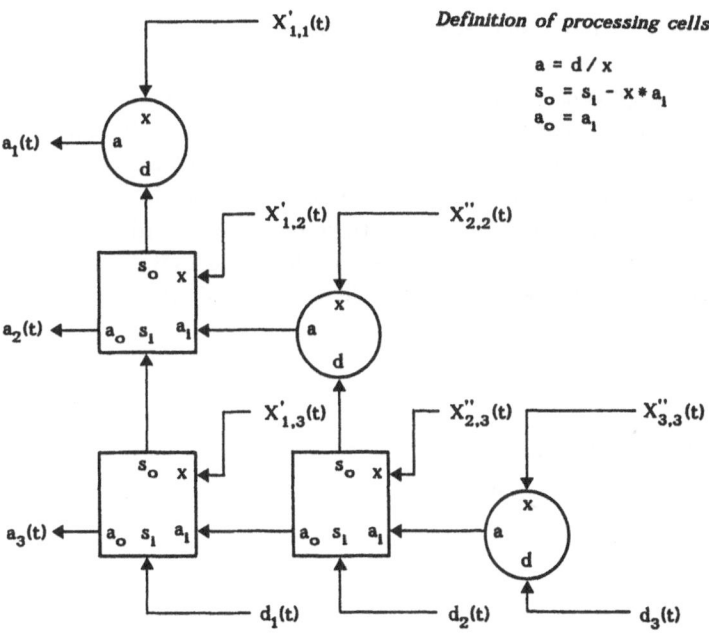

Fig. 4.6. Interleaved back-substitution array with processing cells

4.4 Iterative Vector Rotations – The CORDIC Algorithm

We saw that the use of coordinate rotation operators plays a central role in linear prediction and least-squares theory. The computation of Givens rotations involves such "strange" operations as

$$f(x,y) = \left(x^2 + y^2\right)^{1/2} . \tag{4.38}$$

In multiply-add oriented arithmetic units, operations such as (4.38) can only be obtained by a cumbersome and time-consuming evaluation of truncated series expansions. An interesting alternative to this approach is available in the form of COordinate Rotation DIgital Computer (CORDIC) arithmetic [4.9] . The CORDIC techniques provide an *iterative* rotation of a two-dimensional vector with components x and y along *circular, linear* or even *hyperbolic* trajectories. A given pair {x,y} represents a vector in two-dimensional space, and a rotation by a predetermined angle Δφ can be applied to this vector, e.g., for the purpose of rotating the vector into one of the axis coordinates in order to *annihilate* one of the components of the vector. This is exactly what we do when an upper triangular form is generated by Givens rotations. Indeed, as will become apparent later in this section, CORDIC arithmetic is the most natural form of computing Givens rotations. By using CORDIC arithmetic, Givens rotations can be conveniently implemented by using only shift and add operations. As mentioned earlier, the basis of the CORDIC algorithm is coordinate rotation in a circular, linear, or hyperbolic coordinate system, depending on which function of the pair {x,y} is to be calculated. Such coordinate systems are parametrized by m, in which the radius r and the angle φ of a vector $\mathbf{p} = [x, y]^T$, shown in Fig. 4.7, are defined as

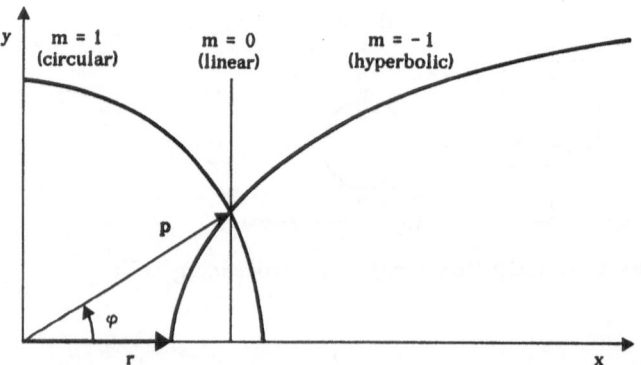

Fig. 4.7. Angle φ and radius r of a vector $\mathbf{p} = [x, y]^T$ in the three coordinate systems

$$r = \left(x^2 + m\ y^2\right)^{1/2} , \tag{4.39}$$

$$\varphi = \left(1/\sqrt{m}\right) \tan^{-1}\left(\sqrt{m}\ y/x\right) . \tag{4.40}$$

Three distinct values of the parameter m correspond to the circular (m=1), linear (m=0), and hyperbolic (m=-1) coordinate systems.

For example, a new vector $\mathbf{p'} = [x', y']^T$ can be obtained from the vector $\mathbf{p} = [x, y]^T$ by clockwise rotation through $\Delta\varphi$, in the circular coordinate system. This operation is expressed as

$$\begin{bmatrix} x' \\ y' \end{bmatrix} = \begin{bmatrix} \cos \Delta\varphi & \sin \Delta\varphi \\ -\sin \Delta\varphi & \cos \Delta\varphi \end{bmatrix} \begin{bmatrix} x \\ y \end{bmatrix} . \tag{4.41}$$

We remember our early discussion about circular rotations in the context of the Givens reduction, as treated in Chap. 3. Now, for small values of $\Delta\varphi$, one can employ the approximation

$$\sin \Delta\varphi = \delta , \tag{4.42a}$$

$$\cos \Delta\varphi = 1 , \tag{4.42b}$$

which, for small values of $\Delta\varphi$, changes (4.41) into

$$\begin{bmatrix} x_{i+1} \\ y_{i+1} \end{bmatrix} = \begin{bmatrix} 1 & \delta_i \\ -\delta_i & 1 \end{bmatrix} \begin{bmatrix} x_i \\ y_i \end{bmatrix} , \tag{4.43}$$

where i is a *recursion index*. With (4.43), we have established an *iterative* rotation of the vector $[x_i, y_i]^T$ along a *circular* trajectory. The extension to the more general linear and hyperbolic cases is easily obtained by incorporation of the parameter m as follows:

$$\begin{bmatrix} x_{i+1} \\ y_{i+1} \end{bmatrix} = \begin{bmatrix} 1 & m\,\delta_i \\ -\delta_i & 1 \end{bmatrix} \begin{bmatrix} x_i \\ y_i \end{bmatrix} . \tag{4.44}$$

The key to a simple implementation of the CORDIC rotation (4.44) lies in the choice of the incremental angle δ_i. When this angle is an integral power of the machine radix, e.g., $\delta_i = 2^{-i}$, scaling by δ_i reduces to mere shift and add operations. For a more detailed discussion of reasonable choices of δ_i, see [4.9]. Details of the convergence behavior of the algorithm, depending on the choice of δ_i may be found in [4.10].

When the δ_i's are chosen as suggested, there are two main goals in CORDIC rotations. First, there may be cases where one applies the

CORDIC rotations (4.44) in order to achieve a total rotation by an angle $\Delta\varphi$. Secondly, one applies the CORDIC rotations for the purpose of annihilating one component of the rotated vector. Note again that both aims are reached *iteratively*, by repeated application of incremental rotations as expressed by (4.44). At subsequent iterations, the angle and radius of the new vector are updated by

$$\varphi_{i+1} = \varphi_i + \alpha_i , \qquad\qquad (4.45a)$$

$$r_{i+1} = k_i r_i , \qquad\qquad (4.45b)$$

where α_i is the incremental angle and k_i is the incremental radius factor. These quantities reflect the influence of the approximations (4.42a,b) as functions of the incremental angle δ_i. The equations for computing α_i and k_i from δ_i are given in Table 4.3 for the three different coordinate systems.

Table 4.3. Equations for incremental angle $\alpha_i^{(m)}$ and incremental radius factor $k_i^{(m)}$ as functions of the increment δ_i for the three coordinate systems

Coordinate system	$\alpha_i^{(m)}$	$k_i^{(m)}$
m = 1 (circular)	$\tan^{-1}\delta_i$	$(1 + \delta_i^2)^{1/2}$
m = 0 (linear)	δ_i	1
m = -1 (hyperbolic)	$\tanh^{-1}\delta_i$	$(1 - \delta_i^2)^{1/2}$

After n iterations, one finds

$$\varphi_n = \varphi_0 - \alpha^{(m)} , \qquad\qquad (4.46a)$$

$$r_n = k^{(m)} r_0 , \qquad\qquad (4.46b)$$

where

$$\alpha^{(m)} = \sum_{i=0}^{n-1} \alpha_i^{(m)} , \qquad\qquad (4.47a)$$

$$k^{(m)} = \prod_{i=0}^{n-1} k_i^{(m)} \quad . \tag{4.47b}$$

The total change in angle is just the total sum over the incremental changes, whereas the total change in radius is the product of the incremental radius changes. A third variable z is introduced for the accumulation of angular variations

$$z_n = z_0 + \alpha^{(m)} \quad . \tag{4.48}$$

The input in the sequence of CORDIC rotations is thus specified by two variables x_0 and y_0 and the desired angular variable z_0 , which can be preset to any desired angle and is subsequently decremented during the CORDIC iteration procedure until the angle z_n remaining after n iterations is zero and the desired rotation of the input vector is obtained. The second important type of CORDIC iterations is the case where CORDIC iterations are successively applied until one of the components of the resulting vector is annihilated. Without loss of generality we consider the case where $y_n = 0$ is achieved by n successive CORDIC iterations. The two types of operations are illustrated in Tables 4.4 and 4.5 for the three different coordinate systems.

Returning to the problem of computing the Givens rotation, we see that the boundary cells of the array shown in Fig. 4.3 perform the radius computation and can therefore be implemented using the CORDIC con-figuration of Table 4.5, circular (m=1) rotation. The rotational units (triangular part of the array) can be conveniently implemented with the CORDIC configuration shown in Table 4.4, and here again with the processor that performs a circular rotation (m=1).

A principal difficulty appears when the CORDIC iterations are completed. Some error stemming from the linear approximation (4.42a,b) must be expected in the result. The corresponding terms $k^{(*)}$ are just these errors. The values of $k^{(*)}$ are quantified in Table 4.3. These errors may be corrected for (or neglected) based on the desired accuracy and the stepsize δ_i . With an arithmetic like CORDIC, quite a number of "unpleasant" operations, mainly trigonometric functions and square roots can be computed in a unified way, where the result is basically obtained through a bit-serial technique. Owing to its generality in use, CORDIC arithmetic has found many applications in calculator design. Besides the case of the Givens reduction, we have a second important area in linear prediction theory where CORDIC can be applied as well, namely, the class of square-root normalized ladder algorithms which will be treated in Chap. 9.

Table 4.4. Case 1: Angular variable z_0 preset to obtain a prescribed rotation angle so that $z_n \longrightarrow 0$ is reached by a sequence of n CORDIC iterations

$x_0 \longrightarrow$ | CORDIC | $\longrightarrow k^{(1)}(x_0 \cos z_0 - y_0 \sin z_0)$

$y_0 \longrightarrow$ | | $\longrightarrow k^{(1)}(x_0 \sin z_0 + y_0 \cos z_0)$

$z_0 \longrightarrow$ | | $\longrightarrow z_n \longrightarrow 0$

circular (m = 1)

$x_0 \longrightarrow$ | CORDIC | $\longrightarrow x_0$

$y_0 \longrightarrow$ | | $\longrightarrow y_0 + x_0 z_0$

$z_0 \longrightarrow$ | | $\longrightarrow z_n \longrightarrow 0$

linear (m = 0)

$x_0 \longrightarrow$ | CORDIC | $\longrightarrow k^{(-1)}(x_0 \cosh z_0 + y_0 \sinh z_0)$

$y_0 \longrightarrow$ | | $\longrightarrow k^{(-1)}(x_0 \sinh z_0 + y_0 \cosh z_0)$

$z_0 \longrightarrow$ | | $\longrightarrow z_n \longrightarrow 0$

hyperbolic (m = -1)

Table 4.5. Case 2: Radius computation by iterative annihilation of y_n using a sequence of CORDIC iterations

---

$x_0 \longrightarrow$
$y_0 \longrightarrow$ [CORDIC]
$z_0 \longrightarrow$

$\longrightarrow k^{(1)}(x_0^2 + y_0^2)^{1/2}$

$\longrightarrow y_n \longrightarrow 0$

$\longrightarrow z_0 + \tan^{-1}(y_0/x_0)$

circular (m = 1)

$x_0 \longrightarrow$
$y_0 \longrightarrow$ [CORDIC]
$z_0 \longrightarrow$

$\longrightarrow x_0$

$\longrightarrow y_n \longrightarrow 0$

$\longrightarrow z_0 + (y_0/x_0)$

linear (m = 0)

$x_0 \longrightarrow$
$y_0 \longrightarrow$ [CORDIC]
$z_0 \longrightarrow$

$\longrightarrow k^{(-1)}(x_0^2 - y_0^2)^{1/2}$

$\longrightarrow y_n \longrightarrow 0$

$\longrightarrow z_0 + \tanh^{-1}(y_0/x_0)$

hyperbolic (m = -1)

4.5 Recursive QR Decomposition Using a Second-Order Window

The derivation of the recursive least-squares algorithm using the Givens reduction, as shown in the previous section, was based on the assumption of an exponential decay of the data. It is well-known, however, that a simple exponential window has both a *poor tracking behavior* and a *poor steady-state behavior*. Intuitively, we can explain the poor steady-state behavior by the fact that, with an exponential window, the actual (incoming) data sample attains the largest weight. Statistical fluctuations in the input data thus cause the estimated parameters $a(t)$ to be very noisy. The exponential weighting factor must be close to 1 to obtain acceptably smoothed parameter estimates. In this case, however, the decay is small and a large number of past data samples can contribute to the actual estimate; hence causing a poor tracking behavior as soon as the statistics of the input data changes.

To cope with this difficulty, one can incorporate *higher-order windows* in the derivation of recursive least-squares algorithms. A simple but effective example is the recursive modified Barnwell window [4.11,12] introduced in Chap. 2. This window gives a small weight to the actual data sample but also provides a fast decay for past data. This way the Barnwell window ensures both a near-optimum tracking and steady-state behavior. The incorporation of higher-order window functions in the derivation of recursive algorithms can sometimes be troublesome and in some schemes, such as the Kalman type algorithms of Chap. 5, recursive algorithms with higher-order windows are obstructed by the unpleasant properties of the Sherman-Morrison identity . The recursive QR decomposition, however, can also be derived for the Barnwell window; although this seems to be a completely new result since, to the author's knowledge, a recursive algorithm of this type has never been proposed in the literature.

Consider the weighting matrix $B(t)$

$$B(t) = diag\left(B_{k,k}(t)\right); \qquad 0 \le k \le t\text{-}1 \qquad (4.49)$$

with

$$B_{k,k}(t) = (1 + k)\beta^k ; \qquad \beta < 1 \qquad (4.50)$$

or, explicitly,

$$B(t) = \begin{bmatrix} 1 & 0 & \cdots\cdots\cdots & 0 \\ 0 & 2\beta & \ddots & \vdots \\ \vdots & \ddots & 3\beta^2 & \ddots & \vdots \\ & & & \ddots & 0 \\ 0 & \cdots\cdots\cdots & 0 & t\beta^{t-1} \end{bmatrix} . \tag{4.51}$$

Note that we wish to apply directly $B(t)$, rather than its square root $B^{1/2}(t)$, to the observation matrix $X(t)$. A necessary condition for the existence of a recursive algorithm based on the decay $B(t)$ is the existence of a partitioning scheme for $B(t)$ similar to the partitioning scheme of $\Lambda(t)$ given in (4.8). Introducing an exponential-like decay matrix $B^*(t)$

$$B^*(t) = \begin{bmatrix} 1 & 0 & \cdots\cdots\cdots & 0 \\ 0 & \beta & \ddots & \vdots \\ \vdots & \ddots & \beta^2 & \ddots & \vdots \\ & & & \ddots & 0 \\ 0 & \cdots\cdots\cdots & 0 & \beta^{t-1} \end{bmatrix} \tag{4.52}$$

and its associated partitioning scheme

$$B^*(t) = \begin{bmatrix} 1 & 0 \cdots\cdots 0 \\ 0 \\ \vdots & \beta B^*(t-1) \\ 0 \end{bmatrix} , \tag{4.53}$$

it is easy to check that the partitioning scheme for $B(t)$ will be of the form

$$B(t) = \begin{bmatrix} 1 & 0 \cdots\cdots\cdots\cdots 0 \\ 0 \\ \vdots & \beta\left(B(t-1) + B^*(t-1)\right) \\ 0 \end{bmatrix} . \tag{4.54}$$

Now we proceed in a similar way as for the case of the simple exponential window. Similarly to (4.5), the incorporation of the weighting matrix $B(t)$ in the QR decomposition gives

$$Q^T(t-1)B(t-1)X(t-1) = B(t-1)\begin{bmatrix} R(t-1) \\ 0 \end{bmatrix} = \begin{bmatrix} R_B(t-1) \\ 0 \end{bmatrix} . \tag{4.55}$$

Similarly, the transformed input data vector might be expressed as

$$Q^T(t-1)B(t-1)x(t-1) = B(t-1)\begin{bmatrix} d_1(t-1) \\ d_2(t-1) \end{bmatrix} = \begin{bmatrix} d_{B1}(t-1) \\ d_{B2}(t-1) \end{bmatrix} . \tag{4.56}$$

Making use of the partitioning scheme of the updated orthogonal matrix (4.7) and the updating property of the observation matrix (4.9 and 10), the updated version of expression (4.55) can be expressed as

$$\bar{Q}^T(t-1)B(t)X(t) = B(t)\begin{bmatrix} z^T(t) \\ R(t-1) \\ 0 \end{bmatrix} \tag{4.57}$$

or, equivalently,

$$\begin{bmatrix} 1 & 0 \ldots . 0 \\ 0 & \\ \vdots & Q^T(t-1) \\ 0 & \end{bmatrix}\begin{bmatrix} 1 & 0 \ldots \ldots \ldots \ldots 0 \\ 0 & \\ \vdots & \beta\left(B(t-1) + B^*(t-1)\right) \\ 0 & \end{bmatrix}\begin{bmatrix} z^T(t) \\ X(t-1) \end{bmatrix}$$

$$= \begin{bmatrix} z^T(t) \\ \beta Q^T(t-1)\left(B(t-1) + B^*(t-1)\right)X(t-1) \end{bmatrix}$$

$$= \begin{bmatrix} 1 & 0 \ldots \ldots \ldots \ldots 0 \\ 0 & \\ \vdots & \beta\left(B(t-1) + B^*(t-1)\right) \\ 0 & \end{bmatrix}\begin{bmatrix} z^T(t) \\ R(t-1) \\ 0 \end{bmatrix}$$

$$= \begin{bmatrix} z^T(t) \\ \beta\left(R_B(t-1) + R_{B^*}(t-1)\right) \\ 0 \end{bmatrix} . \tag{4.58}$$

With an appropriate sequence of Givens rotations $T_B(t)$, the orthogonalization of the right-hand side of (4.58) can be achieved by

$$T_B(t)\bar{Q}^T(t-1)B(t)X(t) = T_B(t)\begin{bmatrix} z^T(t) \\ \beta\left(R_B(t-1) + R_{B^*}(t-1)\right) \\ 0 \end{bmatrix}, \tag{4.59}$$

where

$$Q(t) = T_B(t)\bar{Q}^T(t-1) . \tag{4.60}$$

When the QR decomposition at time step t is expressed as

$$Q^T(t)B(t)X(t) = B(t)\begin{bmatrix} R(t) \\ 0 \end{bmatrix} = \begin{bmatrix} R_B(t) \\ 0 \end{bmatrix}, \tag{4.61}$$

a comparison of (4.61) with (4.59 and 60) reveals that

$$
\begin{bmatrix} R_B(t) \\ 0 \end{bmatrix} = T_B(t) \begin{bmatrix} z^T(t) \\ \beta \left(R_B(t-1) + R_{B*}(t-1) \right) \\ 0 \end{bmatrix} ,
\tag{4.62}
$$

which is the second-order windowed counterpart of (4.17). A similar scheme holds for the data vector $x(t)$ in that

$$
Q^T(t) B(t) x(t)
$$

$$
= T_B(t) \begin{bmatrix} 1 & 0 \ldots \ldots & 0 \\ 0 & & \\ \vdots & Q^T(t-1) & \\ 0 & & \end{bmatrix} \begin{bmatrix} 1 & 0 \ldots \ldots \ldots \ldots & 0 \\ 0 & & \\ \vdots & \beta \left(B(t-1) + B^*(t-1) \right) \\ 0 & & \end{bmatrix} \begin{bmatrix} x(t) \\ \\ x(t-1) \end{bmatrix}
\tag{4.63}
$$

or, equivalently,

$$
Q^T(t) B(t) x(t) = T_B(t) \begin{bmatrix} x(t) \\ \beta Q^T(t-1) \left(B(t-1) + B^*(t-1) \right) x(t-1) \end{bmatrix} .
\tag{4.64}
$$

Comparing (4.64) with the partitioning scheme (4.56), one sees that

$$
\begin{bmatrix} d_{B1}(t) \\ \\ d_{B2}(t) \end{bmatrix} = T_B(t) \begin{bmatrix} x(t) \\ \beta \left(d_{B1}(t-1) + d_{B*1}(t-1) \right) \\ \beta \left(d_{B2}(t-1) + d_{B*2}(t-1) \right) \end{bmatrix} ,
\tag{4.65}
$$

which is the desired equation for updating the transformed data vector in time. Note, however, that with (4.65 and 62), we do not yet have a closed recursion because of the variables $d_{B*1}(t-1)$, $d_{B*2}(t-1)$ and $R_{B*}(t-1)$. A closer look at (4.52) shows that $B^*(t)$ is an exponential decay matrix, and hence, we can use our previously developed algorithm for the exponential decay to update these quantities in time. According to (4.17 and 21), the updating of the remaining quantitites is accomplished as follows:

$$
\begin{bmatrix} R_{B*}(t) \\ 0 \end{bmatrix} = T_{B*}(t) \begin{bmatrix} z^T(t) \\ \beta R_{B*}(t-1) \\ 0 \end{bmatrix} ,
\tag{4.66}
$$

$$
\begin{bmatrix} d_{B*1}(t) \\ \\ d_{B*2}(t) \end{bmatrix} = T_{B*}(t) \begin{bmatrix} x(t) \\ \beta d_{B*1}(t-1) \\ \beta d_{B*2}(t-1) \end{bmatrix} .
\tag{4.67}
$$

From (4.66 and 67), together with (4.62 and 65) one sees that we can first compute the update of (4.66 and 67) by using the algorithm for the exponential decay, as given in Table 4.2. In a similar way, we can compute the update of (4.62 and 65) by using again the exponential decay algorithm of Table 4.2. After these updates have been completed , we establish the element-wise composition of the right sides of (4.62 and 65), which gives the starting point for the next recursion. Interestingly, the updates can be computed *absolutely in parallel*. We simply need two algorithms of the type given in Table 4.2, which are operated in parallel. The computational complexity for the second-order windowed recursive Givens reduction is therefore roughly a factor of two higher than in the case of the simple exponential weighted algorithm.

Figure 4.8 illustrates the principal structure of the simple exponentially weighted recursive Givens reduction as summarized in Table 4.2 and the structure of the more refined second-order windowed counterpart.

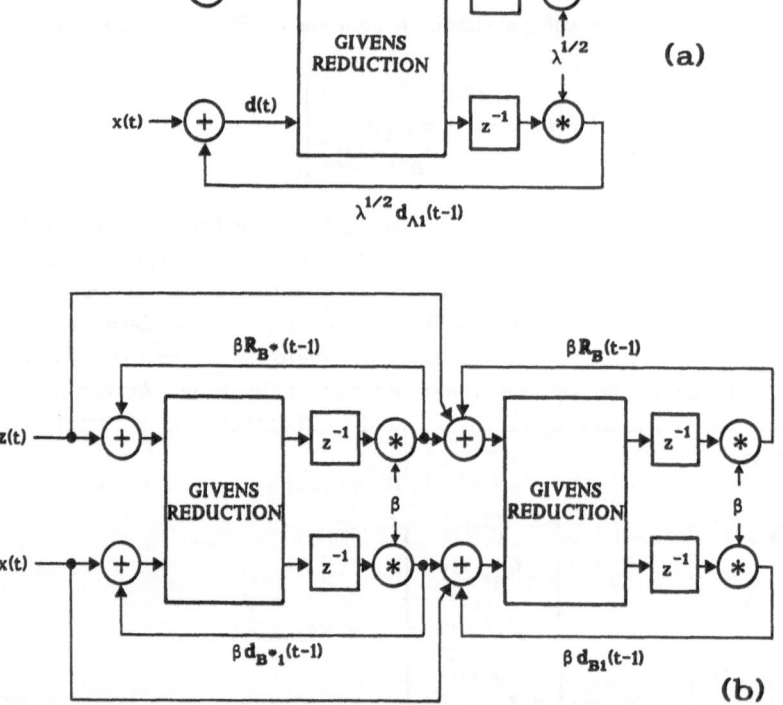

Fig. 4.8. Signal flow-graph illustration of recursive Givens reduction algorithms. (a) Exponentially weighted case (algorithm as listed in Table 4.2). (b) Second-order weighted case (algorithm as listed in Table 4.6)

Table 4.6 provides a summary of this recursive least-squares algorithm based on second-order windowed Givens reductions.

Table 4.6. Recursive least-squares algorithm based on the Givens reduction. The algorithm is based on a second-order windowing of the input data. The parameter β characterizes the second-order window. The quantities $X_{j,k}(t)$, $X^*_{j,k}(t)$, $d_j(t)$, and $d^*_j(t)$ need to be stored twice to avoid overwriting. For a reasonable choice of the parameter β, see Chap. 2. This algorithm involves division. Whenever q is small, set $c=1$ and $s=0$. Whenever q^* is small, set $c^* = 1$ and $s^* = 0$

Initialize: $R_{B^*}(-1) = R_B(-1) = 0$; $d_{B^*_1}(-1) = d_{B1}(-1) = \begin{bmatrix} 0 \ldots 0 \end{bmatrix}^T$

FOR each time step t = 0, 1, 2, 3, . . . do the following:

  available at time t: x(t), z(t), $R_{B^*}(t-1)$, $R_B(t-1)$, $d_{B^*_1}(t-1)$, $d_{B1}(t-1)$, β

  set:

$$X^*(t) = \begin{bmatrix} z^T(t) \\ \beta\, R_{B^*}(t-1) \end{bmatrix} ; \qquad d^*(t) = \begin{bmatrix} x(t) \\ \beta\, d_{B^*_1}(t-1) \end{bmatrix} ;$$

$$X(t) = \begin{bmatrix} z^T(t) \\ \beta\big(R_B(t-1) + R_{B^*}(t-1)\big) \end{bmatrix} ; \quad d(t) = \begin{bmatrix} x(t) \\ \beta\big(d_{B1}(t-1) + d_{B^*_1}(t-1)\big) \end{bmatrix}$$

  FOR j = 1, 2, . . . , p

$$q = \big(X^2_{j,j}(t) + X^2_{j+1,j}(t)\big)^{1/2} \qquad q^* = \big(X^{*2}_{j,j}(t) + X^{*2}_{j+1,j}(t)\big)^{1/2}$$

$$c = X_{j,j}(t)/q \qquad\qquad c^* = X^*_{j,j}(t)/q^*$$

$$s = X_{j+1,j}(t)/q \qquad\qquad s^* = X^*_{j+1,j}(t)/q^*$$

  FOR k = j, j+1, j+2, . . . , p

$$X_{j,k}(t) = c\, X_{j,k}(t) + s\, X_{j+1,k}(t) \qquad\qquad \text{rotate } X(t)$$

$$X_{j+1,k}(t) = -s\, X_{j,k}(t) + c\, X_{j+1,k}(t)$$

$$X^*_{j,k}(t) = c^*\, X^*_{j,k}(t) + s^*\, X^*_{j+1,k}(t) \qquad \text{rotate } X^*(t)$$

$$X^*_{j+1,k}(t) = -s^*\, X^*_{j,k}(t) + c^*\, X^*_{j+1,k}(t)$$

  rotate d(t) and $d^*(t)$:

$$d_j(t) = c\, d_j(t) + s\, d_{j+1}(t) ; \qquad\qquad d^*_j(t) = c^*\, d^*_j(t) + s^*\, d^*_{j+1}(t)$$

$$d_{j+1}(t) = -s\, d_j(t) + c\, d_{j+1}(t) ; \qquad\qquad d^*_{j+1}(t) = -s^*\, d^*_j(t) + c^*\, d^*_{j+1}(t)$$

$$\begin{bmatrix} R_B(t) \\ 0\ldots0 \end{bmatrix} = X(t); \quad \begin{bmatrix} d_{B1}(t) \\ 0 \end{bmatrix} = d(t); \quad \begin{bmatrix} R_{B^*}(t) \\ 0\ldots0 \end{bmatrix} = X^*(t); \quad \begin{bmatrix} d_{B^*_1}(t) \\ 0 \end{bmatrix} = d^*(t)$$

$R_B(t)\, a(t) = d_{B1}(t) \longrightarrow a(t)$ *through back substitution.*

4.6 Alternative Formulations of the QRLS Problem

Recursive least-squares algorithms using the QR decomposition (QRLS algorithms) are based on the idea that the triangular structure of the observation matrix $X(t)$ can be regenerated, after each time update, by annihilation of only p elements, where p denotes the system order. So far, we have only considered the case where this "restoration" was achieved through the annihilation of the down-shifted diagonal elements in the upper part of $X(t)$ in order to obtain an upper-right triangular form in $X(t)$. This is, however, not the only way of deriving a QRLS algorithm. Besides the annihilation of down-shifted diagonal elements in $X(t)$, there exist alternative schemes that directly annihilate the elements in the update vector $z(t)$, thus generating the triangular form at the *bottom* of $X(t)$. Interestingly, a second alternative exists in that the triangular matrix in a QR-decomposition scheme need not necessarily be an *upper-right* triangular form. The idea of decomposing the observation matrix $X(t)$ into an orthogonal part and an associated triangular part works just as well if an *upper-left* triangular form is assumed, as will become apparent in this section. Upper-left triangular forms lead to much simpler initialization schemes, since the observation matrix is originally an upper-left triangular form in the start-up phase of the algorithm. Recall (4.3) for an illustration of this statement. From these considerations, it turns out that we have *four* choices for setting up a QRLS algorithm, as discussed above. These four types of QRLS algorithms, as obtained by combining upper right/left triangular forms and by the type of recursive annihilation (update vector or down-shifted diagonal), are illustrated in Fig. 4.9.

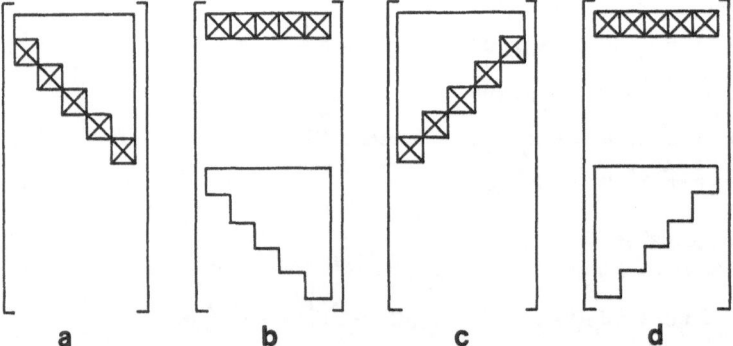

$$\quad a \qquad\qquad b \qquad\qquad c \qquad\qquad d$$

Fig. 4.9. Four types of QRLS algorithms. (a) Upper-right triangular form, annihilation of diagonal elements. (b) Upper-right triangular form, annihilation of update vector. (c) Upper-left triangular form, annihilation of diagonal elements. (d) Upper-left triangular form, annihilation of update vector

So far, our discussion has been based on case (a) in Fig. 4.9. Next, we discuss case (d). In this case, the QRLS scheme is based on an upper–left triangular form. First, it is shown that upper–left triangular forms satisfy the QR-decomposition problem.

Consider again the Normal Equations

$$\mathbf{X}^T\mathbf{X}\mathbf{a} = \mathbf{X}^T\mathbf{x} ; \quad \mathbf{X} \in \mathbb{R}^{L\times P} , \quad \mathbf{a} \in \mathbb{R}^P . \tag{4.68}$$

In contrast to (3.21), we assume an *upper–left triangular form* at the bottom of $\mathbf{R}'$, hence

$$\mathbf{R}' = \begin{bmatrix} \mathbf{0} \\ \mathbf{R} \end{bmatrix} ; \quad \mathbf{R}' \in \mathbb{R}^{L\times P} ; \quad \mathbf{R} \in \mathbb{R}^{P\times P} ; \quad \mathbf{0} \in \mathbb{R}^{(L-p)\times P} , \tag{4.69}$$

where $\mathbf{R}$ is now an *upper-left* triangular form

$$\mathbf{R} = \begin{bmatrix} \diagup 0 \\ \cdots \vdots \\ 0 \cdots 0 \end{bmatrix} \tag{4.70}$$

and $\mathbf{0}$ is again the null matrix of appropriate dimension. Similarly to (3.20), we set

$$\mathbf{X} = \mathbf{Q}\mathbf{R}' , \tag{4.71}$$

which gives

$$\mathbf{X}^T\mathbf{X}\mathbf{a} = \begin{bmatrix} \mathbf{0} \\ \mathbf{R} \end{bmatrix}^T \mathbf{Q}^T\mathbf{Q} \begin{bmatrix} \mathbf{0} \\ \mathbf{R} \end{bmatrix} \mathbf{a} = \mathbf{X}^T\mathbf{x} = \begin{bmatrix} \mathbf{0} \\ \mathbf{R} \end{bmatrix}^T \mathbf{Q}^T\mathbf{x} . \tag{4.72}$$

Partitioning $\mathbf{Q}^T\mathbf{x}$ in the opposite way to in (3.26),

$$\mathbf{Q}^T\mathbf{x} = \begin{bmatrix} \mathbf{d}_2 \\ \mathbf{d}_1 \end{bmatrix} ; \quad \mathbf{d}_1 \in \mathbb{R}^P ; \quad \mathbf{d}_2 \in \mathbb{R}^{L-P} , \tag{4.73}$$

one obtains

$$\mathbf{X}^T\mathbf{x} = \mathbf{R}^T\mathbf{d}_1 + \mathbf{0}^T\mathbf{d}_2 . \tag{4.74}$$

Utilizing the orthogonality relation of $\mathbf{Q}$, equation (4.72) reduces to

$$\mathbf{X}^T\mathbf{X}\mathbf{a} = \mathbf{R}^T\mathbf{R}\mathbf{a} = \mathbf{X}^T\mathbf{x} = \mathbf{R}^T\mathbf{d}_1 , \tag{4.75}$$

yielding

$$R^T d_1 = X^T x \longrightarrow d_1 \, , \qquad (4.76a)$$

$$Ra = d_1 \longrightarrow a \qquad (back \ substitution) \, . \qquad (4.76b)$$

It turns out that we can work with the familiar schemes, even if R is defined as an *upper–left* triangular form. Based on the upper–left triangular form and the direct annihilation of the update vector $z(t)$ (case d in Fig. 4.9), we may develop an alternative QRLS algorithm as follows (exponentially weighted case again). Similarly to (4.5), we may write

$$Q^T(t-1)\Lambda^{1/2}(t-1)X(t-1) = \Lambda^{1/2}(t-1) \begin{bmatrix} 0 \\ R(t-1) \end{bmatrix} = \begin{bmatrix} 0 \\ R_\Lambda(t-1) \end{bmatrix} . \qquad (4.77)$$

As a counterpart to (4.6), one obtains

$$Q^T(t-1)\Lambda^{1/2}(t-1)x(t-1) = \Lambda^{1/2}(t-1) \begin{bmatrix} d_2(t-1) \\ d_1(t-1) \end{bmatrix} = \begin{bmatrix} d_{\Lambda 2}(t-1) \\ d_{\Lambda 1}(t-1) \end{bmatrix} . \qquad (4.78)$$

A time–updated version of (4.77) can be stated as

$$\overline{Q}^T(t-1)\Lambda^{1/2}(t)X(t) = \Lambda^{1/2}(t) \begin{bmatrix} z^T(t) \\ 0 \\ R(t-1) \end{bmatrix} = \begin{bmatrix} z^T(t) \\ 0 \\ \lambda^{1/2}R_\Lambda(t-1) \end{bmatrix} , \qquad (4.79)$$

which, using the partitioning schemes (4.7–9), gives

$$\begin{bmatrix} z^T(t) \\ \lambda^{1/2} Q^T(t-1)\Lambda^{1/2}(t-1)X(t-1) \end{bmatrix} = \begin{bmatrix} z^T(t) \\ 0 \\ \lambda^{1/2}R_\Lambda(t-1) \end{bmatrix} . \qquad (4.80)$$

Accordingly, one can express the updated version of (4.78) as

$$\overline{Q}^T(t-1)\Lambda^{1/2}(t)x(t) = \Lambda^{1/2}(t) \begin{bmatrix} x(t) \\ d_2(t-1) \\ d_1(t-1) \end{bmatrix} = \begin{bmatrix} x(t) \\ \lambda^{1/2}d_{\Lambda 2}(t-1) \\ \lambda^{1/2}d_{\Lambda 1}(t-1) \end{bmatrix} , \qquad (4.81)$$

which [with the left–hand side partitioned as in (4.80)], taking into account (4.18), results in

$$\begin{bmatrix} x(t) \\ \lambda^{1/2} Q^T(t-1)\Lambda^{1/2}(t-1)x(t-1) \end{bmatrix} = \begin{bmatrix} x(t) \\ \lambda^{1/2}d_{\Lambda 2}(t-1) \\ \lambda^{1/2}d_{\Lambda 1}(t-1) \end{bmatrix} . \qquad (4.82)$$

Note that at time step t

$$\mathbf{Q}^T(t)\boldsymbol{\Lambda}^{1/2}(t)\mathbf{X}(t) = \begin{bmatrix} \mathbf{0} \\ \mathbf{R}_\Lambda(t) \end{bmatrix} \quad \text{and} \tag{4.83}$$

$$\mathbf{Q}^T(t)\boldsymbol{\Lambda}^{1/2}(t)\mathbf{x}(t) = \begin{bmatrix} \mathbf{d}_{\Lambda2}(t) \\ \mathbf{d}_{\Lambda1}(t) \end{bmatrix} . \tag{4.84}$$

Recall that an appropriate rotation matrix $\mathbf{T}_p(t)$ is introduced such that (4.15) holds, and hence a comparison of (4.83) with (4.79) yields

$$\begin{bmatrix} \mathbf{0} \\ \mathbf{R}_\Lambda(t) \end{bmatrix} = \mathbf{T}_p(t) \begin{bmatrix} \mathbf{z}^T(t) \\ \mathbf{0} \\ \lambda^{1/2}\mathbf{R}_\Lambda(t-1) \end{bmatrix} . \tag{4.85}$$

Similarly, a comparison of (4.84) with (4.81) reveals

$$\begin{bmatrix} \mathbf{d}_{\Lambda2}(t) \\ \mathbf{d}_{\Lambda1}(t) \end{bmatrix} = \mathbf{T}_p(t) \begin{bmatrix} \mathbf{x}(t) \\ \lambda^{1/2}\mathbf{d}_{\Lambda2}(t-1) \\ \lambda^{1/2}\mathbf{d}_{\Lambda1}(t-1) \end{bmatrix} . \tag{4.86}$$

Clearly, the rotation matrix $\mathbf{T}_p(t)$ must be determined such that the update vector $\mathbf{z}(t)$ in (4.85) is annihilated. Thus it is easy to check that $\mathbf{T}_p(t)$ must represent a sequence of Givens rotations of the type

$$\mathbf{T}_p(t) = \begin{bmatrix} c_p(t) & & & -s_p(t) \\ & 1 & & \\ & & \ddots & \\ & & & 1 \\ s_p(t) & & & c_p(t) \end{bmatrix} \begin{bmatrix} c_{p-1}(t) & & & -s_{p-1}(t) \\ & 1 & & \\ & & \ddots & \\ & & & 1 \\ & s_{p-1}(t) & & c_{p-1}(t) \\ & & & & 1 \end{bmatrix} \cdots$$

$$\cdots \begin{bmatrix} c_1(t) & -s_1(t) & & & \\ & 1 & & & \\ & & \ddots & & \\ & & & 1 & \\ s_1(t) & & & c_1(t) & \\ & & & & 1 & \\ & & & & & \ddots \\ & & & & & & 1 \end{bmatrix}$$

$$= \begin{bmatrix} c_p(t) & & & -s_p(t) \\ & 1 & & \\ & & \ddots & \\ & & & 1 \\ s_p(t) & & & c_p(t) \end{bmatrix} \begin{bmatrix} & & & 0 \\ & & & \vdots \\ & \mathbf{T}_{p-1}(t) & & \\ & & & 0 \\ 0 \cdots \cdots \cdots 0 & 1 \end{bmatrix} , \tag{4.87}$$

where the sines and cosines in (4.87) are determined such that the respective component in the update vector $z(t)$ is annihilated, hence

$$
\begin{bmatrix} z'_{p+1-j}(t) = 0 \\ \\ R_{Aj,\,p+1-j}(t) \end{bmatrix} = \begin{bmatrix} c_j(t) & -s_j(t) \\ \\ s_j(t) & c_j(t) \end{bmatrix} \begin{bmatrix} z_{p+1-j}(t) \\ \\ R_{Aj,\,p+1-j}(t-1) \end{bmatrix} , \tag{4.88}
$$

where

$$
s_j(t) = z_{p+1-j}(t) \big/ q_j(t) \ , \tag{4.89a}
$$

$$
c_j(t) = R_{Aj,\,p+1-j}(t) \big/ q_j(t) \ , \tag{4.89b}
$$

$$
q_j(t) = \left[z^2_{p+1-j}(t) + R^2_{Aj,\,p+1-j}(t) \right]^{1/2} , \tag{4.89c}
$$

$$
1 \le j \le p \ .
$$

More compactly, we may express the multiple rotation matrix $T_p(t)$ as

$$
T_p(t) = \prod_{j=1}^{p} G^T_{t-p+1-j,1}(t) \ , \tag{4.90}
$$

where $G_{t-p+1-j,1}(t)$ is defined as in (3.46). Evaluating the multiple matrix product (4.87), one finds that $T_p(t)$ is structured as

$$
T_p(t) = \begin{bmatrix}
* & 0 \cdots 0 & * \cdots * \\
0 & 1 & 0 \cdots 0 \\
\vdots & \ddots & \vdots \\
0 & 1 & 0 \cdots 0 \\
* & 0 \cdots 0 & * \\
\vdots & \vdots & \ddots \\
* & 0 \cdots 0 & * \cdots *
\end{bmatrix} , \tag{4.91}
$$

where $*$ denotes any nonzero element.

This completes the derivation of the alternative QRLS algorithm based on an upper-left triangular form and on the elimination of the update vector $z(t)$. Some important notes seem to be in order.

(1) The implementation of the alternative QRLS algorithm as stated by the equations (4.85 and 86) leads to what is known as the Gentleman and Kung array. See [4.4, 5] for details. This array implementation is, however, very similar to what we have discussed in the earlier part of this chapter.

(2) The alternative equations possess a favorable partitioning structure which will be particularly useful in further considerations.

(3) The incorporation of an upper-left form leads to a more natural initial phase of the algorithm since the observation matrix is an upper-left form for $t \le p$.

(4) In Chap. 3, we showed how one can obtain the prediction error *energy* directly from the decomposition result. With this alternative QRLS algorithm it is even possible to calculate the actual prediction error sample *implicitly* from intermediate decomposition variables instead of from an explicit computation of the prediction error filter. This will be the topic of the next section.

An explicit summary of this recursive QRLS algorithm is presented in Table 4.7 at the end of the next section. This algorithm summary also contains McWhirter's implicit error computation formulas as derived below.

4.7 Implicit Error Computation

Next we derive an implicit computation method for the prediction error that circumvents the explicit computation of the transversal prediction error filter at each time step. This procedure proposed by McWhirter [4.5] in 1983 is interesting because in some applications one is interested in the prediction error rather than the parameters themselves. Since in such a procedure the parameters are not required, their explicit calculation via the back-substitution part can be omitted. This leads to a considerable reduction in computation. Recall the prediction error filter (2.8) in the exponentially weighted case:

$$\Lambda^{1/2}(t)\, e(t) \;=\; \Lambda^{1/2}(t)\, x(t) \;-\; \Lambda^{1/2}(t)\, X(t)\, a(t) \;. \tag{4.92}$$

Premultiplication of both sides of (4.92) by $Q^T(t)$ gives

$$Q^T(t)\, \Lambda^{1/2}(t)\, e(t) \;=\; Q^T(t)\, \Lambda^{1/2}(t)\, x(t) \;-\; Q^T(t)\, \Lambda^{1/2}(t)\, X(t)\, a(t) \;. \tag{4.93}$$

Comparison of (4.93) with (4.83 and 84) reveals that we can express (4.93) by the triangular forms and "least-squares transformed" data vectors

$$Q^T(t)\, \Lambda^{1/2}(t)\, e(t) \;=\; \begin{bmatrix} d_{\Lambda 2}(t) \\ d_{\Lambda 1}(t) \end{bmatrix} - \begin{bmatrix} 0 \\ R_{\Lambda}(t) \end{bmatrix} a(t) \;. \tag{4.94}$$

But, since

$$R_{\Lambda}(t)\, a(t) \;=\; d_{\Lambda 1}(t) \tag{4.95}$$

[recall (3.29)], we may express the prediction error vector $e(t)$ *implicitly* by means of the transformed data vector $d_{\Lambda 2}(t)$:

$$\mathbf{Q}^T(t)\,\Lambda^{1/2}(t)\,\mathbf{e}(t) = \begin{bmatrix} \mathbf{d}_{\Lambda 2}(t) \\ \\ \mathbf{d}_{\Lambda 1}(t) \end{bmatrix} - \begin{bmatrix} 0 \\ \vdots \\ 0 \\ \mathbf{d}_{\Lambda 1}(t) \end{bmatrix} = \begin{bmatrix} \mathbf{d}_{\Lambda 2}(t) \\ 0 \\ \vdots \\ 0 \end{bmatrix} . \tag{4.96}$$

We continue with a more detailed inspection of the multiple rotation matrix $\mathbf{T}_p(t)$. An evaluation of the product expression (4.87) yields

$$\mathbf{T}_p(t) = \begin{bmatrix} \gamma_p^{1/2}(t) & 0 \cdots 0 & \gamma_p^{1/2}(t)\boldsymbol{\beta}_p^T(t) \\ 0 & 1 & 0 \cdots 0 \\ \vdots & \ddots & \vdots \\ 0 & 1 & 0 \cdots 0 \\ & 0 \cdots 0 & \\ \mathbf{g}_p(t) & \vdots \quad \vdots & \mathbf{Z}_p(t) \\ & 0 \cdots 0 & \end{bmatrix} , \tag{4.97}$$

where

$$\mathbf{g}_p(t) = - \mathbf{Z}_p(t)\boldsymbol{\beta}_p(t) , \tag{4.98a}$$

$$\gamma_p^{1/2}(t) = \prod_{i=1}^{p} c_i(t) , \tag{4.98b}$$

$$\boldsymbol{\beta}_p(t) = \left[\beta_1(t), \beta_2(t), \ldots, \beta_p(t) \right]^T , \tag{4.98c}$$

$$\beta_j(t) = - s_j(t) \big/ \gamma_j^{1/2}(t) ; \qquad 1 \le j \le p . \tag{4.98d}$$

Substitution of (4.96) into (4.86) gives

$$\begin{bmatrix} \mathbf{d}_{\Lambda 2}(t) \\ \\ \mathbf{d}_{\Lambda 1}(t) \end{bmatrix} = \begin{bmatrix} \gamma_p^{1/2}(t)\mathbf{x}(t) + \gamma_p^{1/2}(t)\,\lambda^{1/2}\boldsymbol{\beta}_p^T(t)\mathbf{d}_{\Lambda 1}(t-1) \\ \lambda^{1/2}\mathbf{d}_{\Lambda 2}(t-1) \\ - \mathbf{Z}_p(t)\,\boldsymbol{\beta}_p(t)\mathbf{x}(t) + \lambda^{1/2}\mathbf{Z}_p(t)\,\mathbf{d}_{\Lambda 1}(t-1) \end{bmatrix} . \tag{4.99}$$

From (4.99), we may deduce a partitioning scheme for $\mathbf{d}_{\Lambda 2}(t)$

$$\mathbf{d}_{\Lambda 2}(t) = \begin{bmatrix} \gamma_p^{1/2}(t)\,\alpha_p(t) \\ \\ \lambda^{1/2}\mathbf{d}_{\Lambda 2}(t-1) \end{bmatrix} , \tag{4.100a}$$

where

$$\alpha_p(t) = \mathbf{x}(t) + \lambda^{1/2}\boldsymbol{\beta}_p^T(t)\mathbf{d}_{\Lambda 1}(t-1) . \tag{4.100b}$$

Recalling

$$\mathbf{Q}^T(t) = \mathbf{T}_p(t) \begin{bmatrix} 1 & 0 \ldots 0 \\ 0 & \\ \vdots & \mathbf{Q}^T(t-1) \\ 0 & \end{bmatrix} \tag{4.101}$$

and introducing the partitioning scheme

$$\mathbf{Q}^T(t-1) = \begin{bmatrix} \mathbf{S}^T(t-1) \\[1em] \mathbf{F}^T(t-1) \end{bmatrix} \qquad \begin{array}{l} \mathbf{F}(t-1) \in \mathbb{R}^{(t-1)\times(p)} \\[1em] \mathbf{S}(t-1) \in \mathbb{R}^{(t-1)\times(t-1-p)} \end{array} \quad ; \tag{4.102}$$

one can express $\mathbf{Q}^T(t)$ as

$$\mathbf{Q}^T(t) = \mathbf{T}_p(t) \begin{bmatrix} 1 & 0 \cdots 0 \\ 0 & \\ \vdots & \mathbf{S}^T(t-1) \\ 0 & \\ 0 & \\ \vdots & \mathbf{F}^T(t-1) \\ 0 & \end{bmatrix}$$

$$= \begin{bmatrix} \gamma_p^{1/2}(t) & 0 \cdots 0 & \gamma_p^{1/2}(t)\boldsymbol{\beta}_p^T(t) \\ 0 & 1 & 0 \cdots \cdots 0 \\ \vdots & \ddots & \vdots \cdots \cdots \vdots \\ 0 & 1 & 0 \cdots \cdots 0 \\ & 0 \cdots 0 & \\ \mathbf{g}_p(t) & \vdots \quad \vdots & \mathbf{Z}_p(t) \\ & 0 \cdots 0 & \end{bmatrix} \begin{bmatrix} 1 & 0 \cdots 0 \\ 0 & \\ \vdots & \mathbf{S}^T(t-1) \\ 0 & \\ 0 & \\ \vdots & \mathbf{F}^T(t-1) \\ 0 & \end{bmatrix}$$

$$= \begin{bmatrix} \gamma_p^{1/2}(t) & \gamma_p^{1/2}(t)\boldsymbol{\beta}_p^T(t)\mathbf{F}^T(t-1) \\ 0 & \\ \vdots & \mathbf{S}^T(t-1) \\ 0 & \\ \mathbf{g}_p(t) & \mathbf{Z}_p(t)\mathbf{F}^T(t-1) \end{bmatrix} \tag{4.103}$$

Moreover,

$$\mathbf{Q}^T(t) = \begin{bmatrix} \mathbf{S}^T(t) \\[1em] \mathbf{F}^T(t) \end{bmatrix} . \tag{4.104}$$

Comparison of (4.103) with (4.104) yields

$$\mathbf{S}^T(t) = \begin{bmatrix} \gamma_p^{1/2}(t) & \gamma_p^{1/2}(t)\boldsymbol{\beta}_p^T(t)\mathbf{F}^T(t-1) \\ 0 & \\ \vdots & \mathbf{S}^T(t-1) \\ 0 & \end{bmatrix} . \tag{4.105}$$

Introducing a partitioning scheme for (4.96),

$$\Lambda^{1/2}(t)\,e(t) \;=\; Q(t)\begin{bmatrix} d_{\Lambda 2}(t) \\ 0 \\ \vdots \\ 0 \end{bmatrix} \;=\; \begin{bmatrix} S(t) & F(t) \end{bmatrix}\begin{bmatrix} d_{\Lambda 2}(t) \\ 0 \\ \vdots \\ 0 \end{bmatrix}$$

$$= S(t)\,d_{\Lambda 2}(t) \;, \qquad\qquad (4.106)$$

we find by substitution of (4.105 and 100a) into (4.106) that

$$\Lambda^{1/2}(t)\,e(t) \;=\; S(t)\,d_{\Lambda 2}(t)$$

$$= \begin{bmatrix} \gamma_p(t)\,\alpha_p(t) \\[2mm] \gamma_p(t)\,F(t-1)\,\beta_p(t)\,\alpha_p(t) \;+\; \lambda^{1/2}\,S(t-1)\,d_{\Lambda 2}(t-1) \end{bmatrix}. \qquad (4.107)$$

Clearly, the actual output of the prediction error filter is just the top component of $e(t)$, since the weighting matrix $\Lambda^{1/2}(t)$ does not alter the top component [a fact that can be easily verified by recalling the definition of the weighting matrix (4.4)]. The top component of a vector is easily extracted as the inner product of the respective vector and a *pinning vector* π

$$\pi = \begin{bmatrix} 1, 0 \ldots 0 \end{bmatrix}^T \qquad\qquad (4.108)$$

of appropriate dimension, hence the top component $e_0(t)$ of $e(t)$ is given by

$$e_0(t) \;=\; \pi^T\Lambda^{1/2}(t)\,e(t) \;=\; \pi^T\Lambda^{1/2}(t)\begin{bmatrix} e_0(t), *, \ldots, * \end{bmatrix}^T, \qquad (4.109)$$

where $*$ denotes any nonzero element. Now, simply by comparing (4.107 and 109), it is easy to see that

$$e_0(t) \;=\; \gamma_p(t)\,\alpha_p(t) \;, \qquad\qquad (4.110)$$

which is the desired result. Note that the variables $\gamma_p(t)$ and $\alpha_p(t)$ were defined by (4.98b) and (4.100b), respectively.

Furthermore, a pinning of (4.103) gives

$$\Gamma(t) \;=\; Q^T(t)\,\pi \;=\; T_p(t)\,\pi$$

$$= \begin{bmatrix} \gamma_p^{1/2}(t), 0 \ldots 0, g_p^T(t) \end{bmatrix}^T. \qquad (4.111)$$

The fact that $Q^T(t)\,\pi = T_p(t)\,\pi$ becomes more apparent from (4.101). Table 4.7 is a complete summary of this alternative QRLS algorithm including also the implicit error computation formulas as derived in this section. In Table 4.7, the formulas for the implicit computation of the

Table 4.7. Alternative QRLS algorithm (case d in Fig. 4.9) with implicit error computation (optional) and back substitution (optional). The algorithm uses an exponential decay of past data. The variables $X_{1,k}(t)$ and $d_1(t)$ need to be stored twice to avoid overwriting. This algorithm involves division. Whenever $q_j(t)$ is small, set $c_j(t) = 1$ and $s_j(t) = 0$ (identity rotor). Whenever $\gamma_j^{1/2}(t)$ is small, set $1/\gamma_j^{1/2}(t) = 1$. In the back-substitution part of the algorithm, set $1/X = X$ whenever X is small

---

Initialize: $\quad \mathbf{R}_\Lambda(-1) = \mathbf{0}; \quad \mathbf{d}_{\Lambda 1}(-1) = \begin{bmatrix} 0 \dots 0 \end{bmatrix}^T$

FOR each time step $t = 0, 1, 2, 3, \dots$ do the following:

$\quad$ *available at time t:* $x(t)$, $z(t)$, $\mathbf{R}_\Lambda(t-1)$, $\mathbf{d}_{\Lambda 1}(t-1)$, $0 \leq \lambda \leq 1$

$\quad$ *set:* $\quad \mathbf{X}(t) = \begin{bmatrix} \mathbf{z}^T(t) \\ \lambda^{1/2}\mathbf{R}_\Lambda(t-1) \end{bmatrix}$; $\quad d(t) = \begin{bmatrix} x(t) \\ \lambda^{1/2}\mathbf{d}_{\Lambda 1}(t-1) \end{bmatrix}$

$\quad \gamma_0^{1/2}(t) = 1$

$\quad$ FOR $j = 1, 2, \dots, p$

$\qquad q_j(t) = \begin{bmatrix} X_{1,p+1-j}^2(t) + X_{1+j,p+1-j}^2(t) \end{bmatrix}^{1/2}$

$\qquad s_j(t) = X_{1,p+1-j}(t)/q_j(t)$

$\qquad c_j(t) = X_{1+j,p+1-j}(t)/q_j(t)$

$\qquad$ FOR $k = 1, 2, \dots, p+1-j$ $\hfill$ *rotate* $\mathbf{X}(t)$

$\qquad\quad X_{1,k}(t) \;\; = c_j(t) X_{1,k}(t) - s_j(t) X_{1+j,k}(t)$

$\qquad\quad X_{1+j,k}(t) = s_j(t) X_{1,k}(t) + c_j(t) X_{1+j,k}(t)$

$\qquad d_1(t) \;\; = c_j(t) d_1(t) - s_j(t) d_{1+j}(t)$ $\hfill$ *rotate* $d(t)$

$\qquad d_{1+j}(t) = s_j(t) d_1(t) + c_j(t) d_{1+j}(t)$

$\qquad \gamma_j^{1/2}(t) = \gamma_{j-1}^{1/2}(t) c_j(t)$ $\quad\left.\rule{0pt}{28pt}\right\}$ *optional*

$\qquad \beta_j(t) = -\, s_j(t)/\gamma_j^{1/2}(t)$

$\quad \begin{bmatrix} 0 \dots 0 \\ \mathbf{R}_\Lambda(t) \end{bmatrix} = \mathbf{X}(t)$; $\quad \begin{bmatrix} 0 \\ \mathbf{d}_{\Lambda 1}(t) \end{bmatrix} = d(t)$

$\quad \alpha_p(t) = x(t) + \lambda^{1/2}\begin{bmatrix} \beta_1(t), \beta_2(t), \dots, \beta_p(t) \end{bmatrix} \mathbf{d}_{\Lambda 1}(t-1)$ $\;\left.\rule{0pt}{18pt}\right\}$ *implicit error*
$\quad e_0(t) = \gamma_p(t)\alpha_p(t)$ $\hfill$ *computation*

$\quad a_1(t) = d_{p+1}(t)/X_{p+1,1}(t)$ $\hfill$ *back substitution*

$\quad$ FOR $j = 1, 2, \dots, p$

$\qquad a_j(t) = \begin{bmatrix} d_{p+2-j}(t) - \displaystyle\sum_{i=1}^{j-1} a_i(t) X_{p+2-j,i}(t) \end{bmatrix}/X_{p+2-j,j}(t)$

prediction error are grouped as follows. The quantities $\gamma_j^{1/2}(t)$ and $\beta_j(t)$ are recursively updated according to (4.98b and 98d) where j is the order recursion index. After the order recursion is completed, the quantity $\alpha_p(t)$ is computed according to (4.100b). The final prediction error is then obtained from (4.110).

Clearly, the implicit computation of the prediction error via (4.100b, 98c,d and 110) constitutes a new type of prediction error filter, where only $\beta_p(t)$ is interpreted as a *coefficient vector* and $\lambda^{1/2}d_{\Lambda 1}(t)$ is the "least-squares frequency transformed" *state vector*. The "coefficients" $\{\beta_j(t), 1 \le j \le p\}$ are easily obtained from the *rotors* $\{c_j(t), s_j(t), 1 \le j \le p\}$. The computation of rotors, as provided by the recursive Givens reduction QRLS algorithms, can be interpreted as an alternative *parametrization* of the observed process in terms of sines and cosines of angles instead of transversal predictor parameters (Chap. 5) or ladder reflection coefficients (Chap. 6).

Another comment seems to be in order. From the implicit error computation shown in this section it is already apparent that neither the computation of the prediction error nor the computation of the rotors (sines and cosines) requires explicit knowledge of the elements of the triangular matrix $R_\Lambda(t)$. Therefore, it is possible to derive QRLS algorithms that completely *circumvent* the explicit computation of the triangular matrix $R_\Lambda(t)$. These procedures, recently presented by Cioffi [4.13,14], are "fast" in terms of a linear dependence of the computational complexity on the model order p.

4.8 Chapter Summary

It was the purpose of this chapter to demonstrate the use of a classical scheme such as the QR decomposition to derive computationally attractive recursive least-squares algorithms, the so-called QRLS algorithms. The QRLS problem can be traced back to the problem of triangularization of an *augmented* (updated) observation matrix. The earliest publication on this subject seems to be [4.15]. We derived the QRLS algorithms for the example of a subspace of past observations X where the basis vectors were assumed to be *shifted versions* $x(t-1)$, $x(t-2)$, . . . , $x(t-p)$ of the *same* process. Note, however, that the recursive QR decomposition is more general in use, since the shift property of the data matrix X is not a necessary condition for the existence of a QRLS algorithm. One could use the presented QRLS algorithms alternatively in a problem where a process $x_0(t)$ is to be modeled from a number p of *independent* processes (or measurements) $x_1(t)$, $x_2(t)$, . . . , $x_p(t)$.

We presented *four* types of $O(p^2)$ QRLS algorithms, which are distinguishable by their kind of recursive annihilation and the property of

employing either an upper-left or (more familiarly) upper-right triangular matrix. These algorithms were initially discussed for the simple
pre-windowed case with exponential weighting. Subsequently, it was
shown how a "second-order" windowed algorithm with much better
tracking and steady-state behavior can be obtained. Obviously, all the
different forms of $O(p^2)$ QRLS algorithms exhibit quite a similar triangular array implementation structure. The discussion of a systolic array
implementation was illustrated with a typical example. Moreover, all the
algorithms are entirely based on Givens (plane) rotations and require the
computation of p square roots in each recursion. This suggests the
CORDIC method as another quite natural form of implementation of
QRLS algorithms. We concluded this chapter with a discussion of
McWhirter's implicit computation of prediction errors [4.5]. Interestingly,
the implicit error computation formulas constitute another type of
prediction error filter where rotors act on least-squares frequency transformed data vectors. The recursive QR decomposition is in fact only a
different kind of parametrization of an observed process in terms of
rotors (sines and cosines) instead of the more familiar transversal predictor parameters (Chap. 5) or ladder reflection coefficients (Chap. 6).
The rotors can be converted into transversal predictor parameters or into
ladder reflection coefficients. A simple and straightforward way to convert the rotors into the transversal predictor parameters is back substitution. The major problem with back substitution is that this procedure
is ill-conditioned and destroys the nice accuracy of the rotors. Therefore,
great interest has been aroused in the development of signal analysis
methods based directly on the rotors rather than on the more familiar
transversal or ladder parameters. See the introductory paper by Bellanger
[4.16]. An interesting point with QRLS algorithms is that neither the
implicit error computation of McWhirter nor the updating of the rotors
requires explicit knowledge of the triangularized matrix of past
observations. One can therefore extend McWhirter's implicit error
computation procedure to a complete algorithm that does not require
the explicit updating of the triangular data matrix. Such a procedure was
recently proposed by Cioffi [4.13, 14]. This method is "fast" in terms of
a linear dependence of the total operations count on the model order p.

5. Recursive Least-Squares Transversal Algorithms

In Chap. 2, we discussed the recursive laws of the Normal Equations, and in Chap. 4, we saw how these properties can be used to obtain fast processing schemes for solving the Normal Equations in the recursive case based on the Givens reduction. This chapter is devoted to the recursive least-squares (RLS) algorithms based on a *transversal* predictor structure. Contrary to the order in this book, recursive solutions of the Normal Equations were first investigated for the case of a transversal prediction error filter. The first commonly recognized algorithm of this type was derived by Godard [5.1] in 1974. But the RLS algorithm was apparently found independently by several authors. The earliest reference seems to be Plackett [5.2]. The RLS algorithm exploits the fact that an actual solution of the Normal Equations can be computed *recursively* from the previous (one time-step delayed) solution plus some update information, which depends on the actual (incoming) process sample. Therefore, the RLS algorithm exhibits the structure of a *Kalman filter* [5.3, 4]. See also Chap. 10 for a discussion of the relationships between parameter estimation and Kalman filter theory.

The main goal in RLS transversal algorithms is the recursive updating of the *inverse* covariance matrix of the observed process. Godard's algorithm has a computational complexity of $O(p^2)$. Since the recursive updating of the covariance matrix requires only $O(p)$ operations per recursion, one could assume that there must exist solutions that accomplish the recursive updating of its inverse with the same order of complexity. Such an algorithm, which additionally allows the recursive updating of the predictor parameters, was first introduced by Ljung et al. [5.5] in 1978. This scheme became known as the *fast Kalman algorithm*. There are several ways to derive RLS transversal algorithms. Besides Kalman filter theory (Chap. 10) and geometrical operator theory (Chap. 9), one may exploit a relation in matrix algebra known as the *matrix inversion lemma* [5.6, 7] for deriving RLS transversal algorithms. The matrix inversion lemma is attributed to Sherman and Morrison [5.8] and is henceforth called the *Sherman-Morrison identity*. RLS transversal algorithms can only be derived for the growing exponential decay window, the sliding rectangular window [5.9], or simple combinations thereof. To explain the basic ideas behind the derivation of RLS transversal update equations, we restrict our considerations to the case of a growing window, which is simply a special case of an exponential window with a forgetting factor of unity. Later in this chapter we describe the fast RLS transversal algorithms by utilizing a theorem that is known as the matrix inversion lemma (Sherman-Morrison identity) for *partitioned* matrices. Using this important theorem, we shall derive the update equations of the fast Kalman algorithm and the FAEST algorithm of Carayannis et al. [5.10, 11], which is probably the fastest RLS technique ever developed.

The recursive structure of RLS algorithms can cause serious stability problems in the case of rank deficiencies in the input data covariance matrix and in the case of a nonpersistently exciting input signal, i.e., when the input data contains chains of zeros. We present an analysis of the update structure underlying Kalman type algorithms and an associated strategy for stabilizing the RLS algorithm in the case of nonpersistently exciting input data. Note, however, that all RLS transversal algorithms are based on the initial assumption that the inverse system matrix of the Normal Equations *always exists*. This, in fact, presents the most severe problem in real-time applications where the existence of the system matrix of the Normal Equations is frequently violated.

We conclude this chapter by introducing the least mean squares (LMS) algorithm of Widrow and Hoff [5.12-14] as a special simplified case of an RLS transversal algorithm where the Kalman gain vector is replaced by the state vector of the prediction error filter multiplied by a constant *stepsize*. Besides the interpretation of the LMS algorithm as a simplified case of an RLS transversal algorithm, we show its derivation from the steepest descent algorithm, which, in turn, is deducible from the classical Newton algorithm.

5.1 The Recursive Least-Squares Algorithm

This section presents Godard's algorithm [5.1], which has become known as the *recursive least-squares (RLS) algorithm*. This scheme was the first commonly recognized approach to an algorithm that consistently exploits the properties of *recursively* updated *inverse* covariance matrices to obtain a more efficient solution of the Normal Equations in the recursive case.

For a derivation of the RLS algorithm, we may rewrite the Normal Equations in the following shorthand notation:

$$\mathbf{A}(t)\, \mathbf{a}(t) = \mathbf{h}(t) \; , \tag{5.1}$$

where

$$A(t) = \mathbf{X}^T(t) \, \mathbf{X}(t) \ , \tag{5.2a}$$

$$h(t) = \mathbf{X}^T(t) \, \mathbf{x}(t) \ . \tag{5.2b}$$

The solution of the system (5.1) can be expressed as

$$\mathbf{a}(t) = \mathbf{A}^{-1}(t) \, h(t) \ . \tag{5.3}$$

Note that this approach implies that the inverse of the system matrix of the Normal Equations *exists*. This assumption underlying the derivation of RLS schemes based on the Sherman-Morrison identity is indeed a serious drawback, since in real-time applications with quantized data, there will always be a nonzero probability that $\mathbf{A}(t)$ becomes singular. Since a singular system matrix $\mathbf{A}(t)$ violates the initial assumption (5.3), an RLS scheme based on (5.3) may become *unstable* under such conditions. Recall, at this point, the discussion about the particularly useful concept of the *Penrose pseudoinverse* given in Chap. 3. The pseudoinverse provides a *more general* solution of the Normal Equations when compared to (5.3). Nevertheless, we shall continue with consideration of the special case (5.3). Instead of computing the inverse of $\mathbf{A}(t)$ *explicitly*, we are seeking a method for updating the solution vector $\mathbf{a}(t)$ *recursively* from the *previous* solution $\mathbf{a}(t-1)$, which is assumed to be known at time step t. More precisely, we are interested in a recursion that accomplishes the transition

$$\mathbf{a}(t-1) \longrightarrow \mathbf{a}(t) \ . \tag{5.4}$$

For this purpose, we consider first the possibilities that exist for a recursive updating of the system matrix $\mathbf{A}(t)$ and the right-hand-side vector $h(t)$. According to the definition of the data covariance matrix (2.26) the element (i,j) of the system matrix $\mathbf{A}(t)$ is given by

$$A_{i,j}(t) = \mathbf{x}^T(t-1-i) \, \mathbf{x}(t-1-j) \ ; \qquad 0 \le i,j \le p-1 \ , \tag{5.5}$$

and the ith component of the right-hand-side vector is defined by

$$h_i(t) = \mathbf{x}^T(t-1-i) \, \mathbf{x}(t) \ ; \qquad 0 \le i \le p-1 \ , \tag{5.6}$$

where the signal vectors are defined as

$$\mathbf{x}(t-i) = \left[x(t-i), \, x(t-1-i), \, \ldots \, , \, x(0) \right]^T \ . \tag{5.7}$$

Note that the length of the signal vector $\mathbf{x}(t)$ can grow to infinity. Recall again the formulation of the growing window RLS problem as stated explicitly in Sect. 4.1.

The growing window case (5.7) is the starting point for deriving the RLS algorithm of Godard. The more complex case of a sliding rectangular window can be easily deduced once the growing window RLS algorithm is known. The important steps in the RLS derivation, however, can be explained more easily and without any loss of generality when the growing window case is considered. When using the growing window, we must take into account that the signal energy will also grow to infinity. This makes it necessary to apply an *exponential weighting factor* $\lambda < 1$ to attenuate the influence of past data. Such an exponentially weighted growing window was already considered in Chap. 2, Table 2.1, and in Chap. 4.

Starting from the definitions (5.5 - 7), we can state the recursive forms

$$A_{i,j}(t) = A_{i,j}(t-1) + x(t-1-i)\,x(t-1-j) \quad ; \qquad 0 \le i,j \le p-1 \quad , \qquad (5.8)$$

$$h_i(t) = h_i(t-1) + x(t-1-i)\,x(t) \quad ; \qquad 0 \le i \le p-1 \quad . \qquad (5.9)$$

Recalling the state vector $z(t)$ of the transversal prediction error filter

$$z(t) = \left[x(t-1),\ x(t-2),\ \dots,\ x(t-p+1) \right]^T , \qquad (5.10)$$

we may rewrite (5.8 and 9) in a more convenient matrix-vector notation

$$A(t) = A(t-1) + z(t)\,z^T(t) \quad , \qquad (5.11)$$

$$h(t) = h(t-1) + z(t)x(t) \quad , \qquad (5.12)$$

or, with an exponential weighting of past data,

$$A(t) = \lambda\,A(t-1) + z(t)\,z^T(t) \quad , \qquad (5.13)$$

$$h(t) = \lambda\,h(t-1) + z(t)\,x(t) \quad . \qquad (5.14)$$

The expressions (5.13 and 14) constitute the recursions

$$A(t-1) \longrightarrow A(t) \quad , \qquad (5.15)$$

$$h(t-1) \longrightarrow h(t) \quad . \qquad (5.16)$$

We are now seeking the recursion [assuming that $A^{-1}(t)$ exists]

$$A^{-1}(t-1) \longrightarrow A^{-1}(t) \quad , \qquad (5.17)$$

that is, we have to find

$$\mathbf{A}^{-1}(t) = \left[\lambda \mathbf{A}(t-1) + \mathbf{z}(t)\,\mathbf{z}^T(t) \right]^{-1} . \tag{5.18}$$

This can be solved by applying the matrix inversion lemma (also known as the *Sherman-Morrison identity*) to (5.18).

Sherman-Morrison Identity

Let $\mathbf{A} \in \mathbb{R}^{p \times p}$; $\mathbf{B} \in \mathbb{R}^{p \times n}$; $\mathbf{C} \in \mathbb{R}^{n \times p}$ be three matrices related by

$$\left(\mathbf{A} + \mathbf{B}\mathbf{C} \right) . \tag{5.19a}$$

According to the Sherman-Morrison identity, we may express the inverse of (5.19a) as

$$\left(\mathbf{A} + \mathbf{B}\mathbf{C} \right)^{-1} = \mathbf{A}^{-1} - \mathbf{A}^{-1}\mathbf{B}\left(\mathbf{I} + \mathbf{C}\,\mathbf{A}^{-1}\mathbf{B} \right)^{-1}\mathbf{C}\,\mathbf{A}^{-1} , \tag{5.19b}$$

where $\mathbf{I}$ is the identity matrix.

Proof. Premultiplying both sides of (5.19b) by $\left(\mathbf{A} + \mathbf{B}\mathbf{C} \right)$ gives

$$\mathbf{I} = \left(\mathbf{A} + \mathbf{B}\mathbf{C} \right)\left[\mathbf{A}^{-1} - \mathbf{A}^{-1}\mathbf{B}\left(\mathbf{I} + \mathbf{C}\,\mathbf{A}^{-1}\mathbf{B} \right)^{-1}\mathbf{C}\,\mathbf{A}^{-1} \right] . \tag{5.19c}$$

The objective is now to demonstrate that the right-hand side of (5.19c) can be reduced to the identity matrix. By direct manipulation, we obtain

$$\left(\mathbf{A} + \mathbf{B}\mathbf{C} \right)\left[\mathbf{A}^{-1} - \mathbf{A}^{-1}\mathbf{B}\left(\mathbf{I} + \mathbf{C}\,\mathbf{A}^{-1}\mathbf{B} \right)^{-1}\mathbf{C}\,\mathbf{A}^{-1} \right]$$

$$= \mathbf{A}\mathbf{A}^{-1} + \mathbf{B}\mathbf{C}\mathbf{A}^{-1} - \mathbf{A}\mathbf{A}^{-1}\mathbf{B}\left(\mathbf{I} + \mathbf{C}\,\mathbf{A}^{-1}\mathbf{B} \right)^{-1}\mathbf{C}\,\mathbf{A}^{-1}$$

$$\quad - \mathbf{B}\mathbf{C}\mathbf{A}^{-1}\mathbf{B}\left(\mathbf{I} + \mathbf{C}\,\mathbf{A}^{-1}\mathbf{B} \right)^{-1}\mathbf{C}\,\mathbf{A}^{-1}$$

$$= \mathbf{I} + \mathbf{B}\mathbf{C}\mathbf{A}^{-1} - \mathbf{B}\left(\mathbf{I} + \mathbf{C}\,\mathbf{A}^{-1}\mathbf{B} \right)^{-1}\mathbf{C}\,\mathbf{A}^{-1}$$

$$\quad - \mathbf{B}\mathbf{C}\mathbf{A}^{-1}\mathbf{B}\left(\mathbf{I} + \mathbf{C}\,\mathbf{A}^{-1}\mathbf{B} \right)^{-1}\mathbf{C}\,\mathbf{A}^{-1}$$

$$= \mathbf{I} + \mathbf{B}\mathbf{C}\mathbf{A}^{-1} - \mathbf{B}\left(\mathbf{I} + \mathbf{C}\,\mathbf{A}^{-1}\mathbf{B} \right)\left(\mathbf{I} + \mathbf{C}\,\mathbf{A}^{-1}\mathbf{B} \right)^{-1}\mathbf{C}\,\mathbf{A}^{-1}$$

$$= \mathbf{I} + \mathbf{B}\mathbf{C}\mathbf{A}^{-1} - \mathbf{B}\mathbf{C}\mathbf{A}^{-1}$$

$$= \mathbf{I} .$$

End of proof.

Returning to the original problem of setting up a recursion for updating the inverse system matrix (5.18), we set

$$\mathbf{A} = \lambda \, \mathbf{A}(t-1) \ , \tag{5.20a}$$

$$\mathbf{B} = \mathbf{z}(t) \ , \tag{5.20b}$$

$$\mathbf{C} = \mathbf{z}^T(t) \ . \tag{5.20c}$$

Substitution of (5.20a-c) into the Sherman-Morrison identity (5.19b) gives

$$\mathbf{A}^{-1}(t) = \lambda^{-1}\mathbf{A}^{-1}(t-1)$$
$$- \lambda^{-1}\mathbf{A}^{-1}(t-1)\mathbf{z}(t)\Big[\mathbf{I} + \lambda^{-1}\mathbf{z}^T(t)\mathbf{A}^{-1}(t-1)\mathbf{z}(t)\Big]^{-1}\lambda^{-1}\mathbf{z}^T\mathbf{A}^{-1}(t-1) \ .$$

Note that the expression

$$\Big[\mathbf{I} + \lambda^{-1}\mathbf{z}^T(t)\mathbf{A}^{-1}(t-1)\mathbf{z}(t)\Big]$$

is a *scalar* quantity, and hence we may write

$$\mathbf{A}^{-1}(t) = \lambda^{-1}\left(\mathbf{A}^{-1}(t-1) - \frac{\mathbf{A}^{-1}(t-1)\mathbf{z}(t)\mathbf{z}^T(t)\mathbf{A}^{-1}(t-1)}{\lambda + \mathbf{z}^T(t)\mathbf{A}^{-1}(t-1)\mathbf{z}(t)}\right) \ , \tag{5.21}$$

which is the desired recursive law for updating the inverse system matrix in time. Substitution of expression (5.21) into the solution of the Normal Equations (5.3) yields

$$\mathbf{a}(t) = \lambda^{-1}\left(\mathbf{A}^{-1}(t-1)\mathbf{h}(t) - \frac{\mathbf{A}^{-1}(t-1)\mathbf{z}(t)\mathbf{z}^T(t)\mathbf{A}^{-1}(t-1)\mathbf{h}(t)}{\lambda + \mathbf{z}^T(t)\mathbf{A}^{-1}(t-1)\mathbf{z}(t)}\right) \ . \tag{5.22}$$

Substituting the time recursion (5.14) of $\mathbf{h}(t)$ into (5.22) one sees

$$\mathbf{a}(t) = \lambda^{-1}\left(\mathbf{A}^{-1}(t-1)\big(\lambda\mathbf{h}(t-1) + \mathbf{z}(t)\mathbf{x}(t)\big)\right.$$
$$\left. - \frac{\mathbf{A}^{-1}(t-1)\mathbf{z}(t)\mathbf{z}^T(t)\mathbf{A}^{-1}(t-1)\big(\lambda\mathbf{h}(t-1) + \mathbf{z}(t)\mathbf{x}(t)\big)}{\lambda + \mathbf{z}^T(t)\mathbf{A}^{-1}(t-1)\mathbf{z}(t)}\right) \ . \tag{5.23}$$

With the one time-step delayed solution

$$\mathbf{a}(t-1) = \mathbf{A}^{-1}(t-1)\,\mathbf{h}(t-1) \ , \tag{5.24}$$

(5.23) reduces to

$$\mathbf{a}(t) = \mathbf{a}(t-1) + \frac{\mathbf{A}^{-1}(t-1)\mathbf{z}(t)\big(\mathbf{x}(t) - \mathbf{z}^T(t)\mathbf{a}(t-1)\big)}{\lambda + \mathbf{z}^T(t)\mathbf{A}^{-1}(t-1)\mathbf{z}(t)} \ . \tag{5.25}$$

Expression (5.25) already constitutes the desired recursive law for up-dating the solution vector a(t) of the Normal Equations in time. Intro-ducing some intermediate variables, we may deduce the RLS algorithm directly from (5.25). See Table 5.1 for a summary of the RLS algorithm.

Table 5.1. The recursive least-squares (RLS) algorithm for recursive computation of the parameter vector a(t). The quantity λ is an exponential weighting factor. I is the identity matrix. σ is a large initial constant. This algorithm involves division. Whenever the divisor is small, set $1/x = x$

Initialize: $\mathbf{A}^{-1}(-1) = \sigma \mathbf{I}$; $\sigma \gg 1$

$$\mathbf{a}(-1) = \begin{bmatrix} 0 \ldots 0 \end{bmatrix}^{\mathrm{T}}$$

FOR t = 0, 1, 2, . . .

Input: $\mathbf{z}(t)$, $x(t)$, $0 \le \lambda \le 1$

$$\mathbf{k}(t) = \mathbf{A}^{-1}(t-1)\,\mathbf{z}(t) \tag{5.26a}$$

$$\alpha(t) = \lambda + \mathbf{z}^{\mathrm{T}}(t)\mathbf{k}(t) \tag{5.26b}$$

$$\mathbf{k}^*(t) = \mathbf{k}(t)\big/\alpha(t) \tag{5.26c}$$

$$\mathbf{A}^{-1}(t) = \lambda^{-1}\Big[\,\mathbf{A}^{-1}(t-1) - \mathbf{k}^*(t)\,\mathbf{k}^{\mathrm{T}}(t)\,\Big] \tag{5.26d}$$

$$\varepsilon(t) = x(t) - \mathbf{z}^{\mathrm{T}}(t)\,\mathbf{a}(t-1) \qquad \textit{a priori error} \tag{5.26e}$$

$$\mathbf{a}(t) = \mathbf{a}(t-1) + \mathbf{k}^*(t)\,\varepsilon(t) \tag{5.26f}$$

$$e(t) = x(t) - \mathbf{z}^{\mathrm{T}}(t)\,\mathbf{a}(t) \qquad \textit{a posteriori error} \tag{5.26g}$$

The intermediate variable $\mathbf{k}(t)$ is sometimes termed the *Kalman gain vector*. The variable $\varepsilon(t)$ is called the *a priori error*. From Table 5.1, we recognize that $\varepsilon(t)$ is the prediction error assuming the *"old"* parameter set $\mathbf{a}(t-1)$, *prior* to updating. Hence the terminology "a priori error". Later in this chapter, we will see that we can pose the whole RLS problem also in terms of *a posteriori errors* $e(t)$, that is, in terms of the errors obtained from the *already updated* prediction error filter. Hence the terminology "a posteriori errors". Interestingly, as will become apparent later in this chapter, the a posteriori errors can be determined *before* the prediction error filter produces them. From these considerations, we will see that we can derive *two types* of RLS algorithms: one that is based on the a priori error, which is the classical solution (Table 5.1), and its counterpart based on the a posteriori errors. The same situation will appear in the case of the fast RLS algorithms. The classical fast

Kalman algorithm is related to the a priori formulation of the RLS problem, whereas the FAEST technique of Carayannis, which we shall present later in this chapter, is the a posteriori based counterpart of the fast Kalman algorithm.

In the nonstationary case, one can assume that the a priori errors have a higher variance than the a posteriori errors, which are produced by the already updated prediction error filter.

To derive the similarities between the a priori and a posteriori formulation of the RLS problem, we first consider an interesting alternative to the a priori error based parameter update equation (5.26f). For this purpose, the time recursion (5.13, 14) is substituted into the Normal Equations (5.1) to obtain

$$\left[\lambda A(t-1) + z(t) z^T(t) \right] a(t) = h(t) = \lambda h(t-1) + z(t) x(t) .$$

Utilizing (5.24), one sees that

$$\lambda A(t-1) a(t) + z(t) z^T(t) a(t) = \lambda A(t-1) a(t-1) + z(t) x(t) .$$

Introducing the Kalman gain vector (5.26a), one may deduce

$$\lambda a(t) = \lambda a(t-1) + k(t) \left[x(t) - z^T(t) a(t) \right] .$$

Using (5.26g), we finally obtain

$$a(t) = a(t-1) + \lambda^{-1} k(t) e(t) , \tag{5.27}$$

which is the desired recursive law for updating the parameter vector $a(t)$ using the a posteriori errors. Comparing (5.27) with its a priori error based counterpart (5.26f), we see that the modified Kalman gain vector $k^*(t)$ has been replaced by a weighted Kalman gain vector $\lambda^{-1} k(t)$.

Another interesting relationship that directly emerges from the comparison of the relations (5.26f) and (5.27) is the fact that modified Kalman gains and a priori errors are related to Kalman gains and a posteriori errors as

$$k^*(t) \varepsilon(t) = \lambda^{-1} k(t) e(t) , \tag{5.28}$$

or, taking into account (5.26c),

$$e(t) = \lambda \varepsilon(t) / \alpha(t) . \tag{5.29}$$

With these relations in mind, one could state the alternative a posteriori error based solution to the RLS problem by simply replacing (5.26f) with

(5.29) and by replacing (5.26g) with (5.27). Interestingly, this approach succeeded through the fact that the a posteriori errors can be computed by exploitation of (5.29) before the prediction error filter (5.26g) produces them. This will be a useful result also for the derivation of the FAEST algorithm.

Next, we wish to go deeper into an interpretation of intermediate quantities arising in the RLS algorithm. The quantity

$$\alpha(t) = \lambda + \mathbf{z}^{T}(t)\mathbf{A}^{-1}(t-1)\mathbf{z}(t) \qquad (5.30)$$

has some interesting properties. Since we have assumed $\mathbf{A}(t-1)$ to be a positive definite matrix, it follows that its inverse is again positive definite, which gives rise to the inequality

$$\mathbf{z}^{T}(t)\mathbf{A}^{-1}(t-1)\mathbf{z}(t) > 0 \ ,$$

which, in consideration of (5.30), gives

$$0 \le 1/\alpha(t) \le 1/\lambda \ . \qquad (5.31)$$

Introducing the so-called *likelihood variable* $\gamma(t)$

$$\gamma(t) = \mathbf{z}^{T}(t)\mathbf{A}^{-1}(t)\mathbf{z}(t) \ , \qquad (5.32)$$

we may state the following relationship between $\alpha(t)$ and $\gamma(t)$:

$$\gamma(t) = \big(\alpha(t) - \lambda\big)\big/\alpha(t) \qquad (5.33)$$

or, equivalently,

$$\alpha(t) = \lambda\big/\big(1 - \gamma(t)\big) \ , \qquad (5.34)$$

where

$$0 \le \gamma(t) \le 1 \ . \qquad (5.35)$$

The inequality (5.35) reveals that the likelihood variable $\gamma(t)$ can also be interpreted as the *squared cosine* of an angle. This interpretation plays an important role in the derivation of fast RLS algorithms. We will return to this interpretation several times during the derivation of the fast RLS ladder algorithms in Chap. 9. The derivation of the RLS ladder algorithms will be based on a geometrical approach that allows the interpretation of $\gamma(t)$ as the square of the cosine of the angle between subspaces in subsequent time steps.

To conclude this discussion, we are interested in deriving an alternative expression for the modified Kalman gain vector $\mathbf{k}^*(t)$. Postmultiplication of (5.26d) with the state vector $\mathbf{z}(t)$ gives

$$\mathbf{A}^{-1}(t)\,\mathbf{z}(t) = \lambda^{-1}\mathbf{A}^{-1}(t-1)\mathbf{z}(t) - \lambda^{-1}\mathbf{k}^*(t)\mathbf{k}^T(t)\mathbf{z}(t) \ . \tag{5.36}$$

Recognizing that $\mathbf{A}^{-1}(t-1)\mathbf{z}(t) = \mathbf{k}(t)$ and $\mathbf{k}^T(t)\mathbf{z}(t) = \alpha(t) - \lambda$, in consideration of

$$\mathbf{k}(t) = \mathbf{k}^*(t)\alpha(t) \ , \tag{5.37}$$

one finds

$$\mathbf{A}^{-1}(t)\mathbf{z}(t) = \lambda^{-1}\mathbf{k}^*(t)\alpha(t) - \lambda^{-1}\mathbf{k}^*(t)\alpha(t) + \mathbf{k}^*(t) \ , \tag{5.38}$$

which gives

$$\mathbf{k}^*(t) = \mathbf{A}^{-1}(t)\mathbf{z}(t) \ . \tag{5.39}$$

5.2 Potter's Square-Root Normalized RLS Algorithm

The updating of an inverse covariance matrix, as appearing in the RLS algorithm, is known to be an ill-conditioned problem. Several attempts have been made to improve the numerical behavior of the RLS algorithm. One of the most interesting approaches in this field is *Potter's square-root factorization* [5.15], which we shall present in this section.

We shall start with the idea that the system matrix may be factorized as follows:

$$\mathbf{A}(t) = \mathbf{Q}(t)\mathbf{Q}^T(t) \ . \tag{5.40}$$

Here $\mathbf{Q}(t)$ is a $p \times p$ matrix, which is in general not symmetric. $\mathbf{Q}(t)$ should not be confused with the orthogonal matrix of the QR decomposition. Rewriting the update equation (5.11) in terms of (5.40) gives

$$\mathbf{Q}(t)\mathbf{Q}^T(t) = \lambda\mathbf{Q}(t-1)\mathbf{Q}^T(t-1) + \mathbf{z}(t)\mathbf{z}^T(t) \ . \tag{5.41}$$

Next, we introduce the normalized data vector $\mathbf{z}'(t)$, which is defined as

$$\mathbf{z}(t) = \mathbf{Q}(t-1)\,\mathbf{z}'(t) \ . \tag{5.42}$$

Utilizing (5.42), we may rewrite (5.41) in terms of $\mathbf{z}'(t)$:

$$\mathbf{Q}(t)\mathbf{Q}^T(t) = \mathbf{Q}(t-1)\left[\lambda\mathbf{I} + \mathbf{z}'(t)\,\mathbf{z}'^T(t)\right]\mathbf{Q}^T(t-1) \ . \tag{5.43}$$

An elegant derivation of Potter's recursions can be given if we make use of the following projection operators:

$$\mathbf{P}(t) = \mathbf{z}'(t)\left(\mathbf{z}'^T(t)\mathbf{z}'(t)\right)^{-1}\mathbf{z}'^T(t) \ , \tag{5.44a}$$

$$\mathbf{P}^{\perp}(t) = \mathbf{I} - \mathbf{P}(t) \ . \tag{5.44b}$$

Recall the discussion on projection operators given in Chap. 2. Accordingly, the projection operator $\mathbf{P}(t)$ projects vectors *orthogonally* onto the subspace spanned by the single coordinate vector $\mathbf{z}'(t)$ and $\mathbf{P}^{\perp}(t)$ generates the orthogonal complement. The application of $\mathbf{P}(t)$ to any vector of the p-dimensional space decomposes this vector into an *orthogonal component*, which is orthogonal with respect to $\mathbf{z}'(t)$, and an *in-space component*, which lies in the subspace $\mathbf{z}'(t)$ and is therefore *parallel* to $\mathbf{z}'(t)$.

Rearranging (5.44a),

$$\mathbf{z}'(t)\mathbf{z}'^T(t) = \mathbf{P}(t)\left(\mathbf{z}'^T(t)\mathbf{z}'(t)\right) \ ,$$

we may express the factorized update equation (5.43) in terms of the projection operator $\mathbf{P}(t)$:

$$\mathbf{Q}(t)\mathbf{Q}^T(t) = \mathbf{Q}(t-1)\left[\lambda\mathbf{I} + \mathbf{P}(t)\left(\mathbf{z}'^T(t)\,\mathbf{z}'(t)\right)\right]\mathbf{Q}^T(t-1) \ . \tag{5.45}$$

Using the fact that $\mathbf{I} = \mathbf{P}^{\perp}(t) + \mathbf{P}(t)$, equation (5.45) changes to

$$\mathbf{Q}(t)\mathbf{Q}^T(t) = \mathbf{Q}(t-1)\left[\lambda\mathbf{P}^{\perp}(t) + \left(\lambda + \mathbf{z}'^T(t)\,\mathbf{z}'(t)\right)\mathbf{P}(t)\right]\mathbf{Q}^T(t-1) \ . \tag{5.46}$$

Exploiting the *idempotence property* of projection operators as stated in (2.20), and taking into account the *extinguishing property* of orthogonal projection operators

$$\mathbf{P}^{\perp}(t)\mathbf{P}(t) = \mathbf{0} \ , \tag{5.47}$$

we can easily extract a recursion for the isolated matrix $\mathbf{Q}(t)$ by using the matrix product recursion (5.46) as follows:

$$\mathbf{Q}(t) = \mathbf{Q}(t-1)\left[\lambda^{1/2}\,\mathbf{P}^{\perp}(t) + \left(\lambda + \mathbf{z}'^T(t)\,\mathbf{z}'(t)\right)^{1/2}\mathbf{P}(t)\right] \ . \tag{5.48}$$

Moreover, the nice properties of projection operators allow a simple inversion of (5.48):

$$\mathbf{Q}^{-1}(t) = \left[\lambda^{-1/2}\,\mathbf{P}^{\perp}(t) + \left(\lambda + \mathbf{z}'^T(t)\,\mathbf{z}'(t)\right)^{-1/2}\mathbf{P}(t)\right]\mathbf{Q}^{-1}(t-1) \ . \tag{5.49}$$

Noting that

$$\mathbf{A}^{-1}(t-1) = \mathbf{Q}^{-1}{}^{T}(t-1)\mathbf{Q}^{-1}(t-1) \tag{5.50}$$

and

$$\mathbf{z'}^{T}(t)\,\mathbf{z'}(t) = \mathbf{z}^{T}(t)\,\mathbf{A}^{-1}(t-1)\,\mathbf{z}(t) \ , \tag{5.51}$$

we can express the Kalman gain vector (5.26a) in terms of normalized variables

$$\mathbf{k}(t) = \mathbf{Q}^{-1}{}^{T}(t-1)\,\mathbf{z'}(t) \ . \tag{5.52}$$

Similarly, the modified Kalman gain vector (5.26c) is expressed in terms of normalized variables

$$\mathbf{k}^{*}(t) = \mathbf{k}(t)\big/\!\left(\lambda + \mathbf{z'}^{T}(t)\,\mathbf{z'}(t)\right) \tag{5.53}$$

and $\overset{\bullet}{\alpha}(t)$ has an alternative formulation as

$$\alpha(t) = \lambda + \mathbf{z'}^{T}(t)\,\mathbf{z'}(t) \ . \tag{5.54}$$

In a final step, we are interested in expressing the square-root update (5.49) in terms of the intermediate variables (5.52 – 54). Substituting the definition of the projection operators in terms of $\mathbf{z'}(t)$ into the square-root update (5.49) and taking into account that a rearrangment of (5.54) gives $\mathbf{z'}^{T}(t)\,\mathbf{z'}(t) = \alpha(t) - \lambda$, we find

$$\mathbf{Q}^{-1}(t) = \lambda^{-1/2}\,\mathbf{Q}^{-1}(t-1)$$

$$+ \left(\alpha^{-1/2}(t) - \lambda^{-1/2}\right)\!\left(\alpha(t) - \lambda\right)^{-1}\mathbf{z'}(t)\,\mathbf{k}^{T}(t) \ . \tag{5.55}$$

Equation (5.55) already constitutes a closed recursion for updating of the inverse matrix square-root $\mathbf{Q}^{-1}(t)$, and also fully determines the parameter update. Table 5.2 summarizes this algorithm which has become known as *Potter's square-root normalized RLS algorithm* [5.15].

Table 5.2. Potter's square-root normalized RLS algorithm. This algorithm involves division. σ is a large initial constant. Whenever the divisor is small, set $1/x = x$

Initialize:

$$\mathbf{Q}^{-1}(-1) = \sigma\mathbf{I} \quad ; \qquad \sigma \gg 1$$

$$\mathbf{a}(-1) = \begin{bmatrix} 0 \ldots 0 \end{bmatrix}^T$$

FOR t = 0, 1, 2, . . .

Input: $\mathbf{z}(t)$, $x(t)$, $0 \le \lambda \le 1$

$$\mathbf{z}'(t) = \mathbf{Q}^{-1}(t-1)\,\mathbf{z}(t) \tag{5.56a}$$

$$\mathbf{k}(t) = \mathbf{Q}^{-1^T}(t-1)\,\mathbf{z}'(t) \tag{5.56b}$$

$$\alpha(t) = \lambda + \mathbf{z}'^T(t)\,\mathbf{z}'(t) \tag{5.56c}$$

$$\mathbf{k}^*(t) = \mathbf{k}(t)\big/\alpha(t) \tag{5.56d}$$

$$\mathbf{Q}^{-1}(t) = \lambda^{-1/2}\,\mathbf{Q}^{-1}(t-1)$$

$$\qquad + \left(\alpha^{-1/2}(t) - \lambda^{-1/2}\right)\!\left(\alpha(t) - \lambda\right)^{-1}\mathbf{z}'(t)\,\mathbf{k}^T(t) \tag{5.56e}$$

$$\varepsilon(t) = x(t) - \mathbf{z}^T(t)\,\mathbf{a}(t-1) \qquad \text{\textit{a priori error}} \tag{5.56f}$$

$$\mathbf{a}(t) = \mathbf{a}(t-1) + \mathbf{k}^*(t)\,\varepsilon(t) \tag{5.56g}$$

$$e(t) = x(t) - \mathbf{z}^T(t)\,\mathbf{a}(t) \qquad \text{\textit{a posteriori error}} \tag{5.56h}$$

5.3 Update Properties of the RLS Algorithm

The conventional RLS algorithm of Table 5.1 may show serious stability problems when used in realistic real-time (or on-line) applications. Every time we use quantized data or finite-word-length computations (and this is practically always the case) there will be a nonzero probability that the system matrix of the Normal Equations will become *rank deficient*, that is, the rank of the system matrix may drop to a smaller value than the dimension of the system matrix. In this case, unconditional updating of the inverse $\mathbf{A}^{-1}(t)$ can drive the RLS algorithm into an unstable state of operation from which it can never return to normal operation without the entire algorithm being reset. Recall again our considerations of Chap. 3, where we have discussed the more general approach of solving the Normal Equations via the Penrose pseudoinverse.

A simple method to overcome this difficulty is to add a small portion of white noise to the input, hoping that this action may convert the system matrix of the Normal Equations into a (more or less meaningful) nonsingular form. This sometimes reduces the probability that the algorithm will collapse. This simple heuristic method can also help to circumvent a second (although completely different) problem, which is that exponentially weighted RLS algorithms based on the updating of an inverse system matrix require signals that are *"persistently exciting"* for stable operation. A signal is called persistently exciting when there are no samples in the sequence that have a value of exactly zero [5.6,7]. Clearly, in an unquantized process, this will always be the case except for the zero process. For quantized data, it may happen that the process amplitude drops below the quantization stepsize for a certain period of time, resulting in a sequence of zero samples being fed into the algorithm. The observation is that in those cases the algorithm can also become unstable. To remedy this problem, we have to go deeper into the update behavior of the RLS algorithm. For this purpose, the update equation of Potter's square-root normalized RLS algorithm (5.49) is investigated.

We see from (5.49), that the time-updating of the inverse square-root matrix is accomplished by two feedback loops; the *parallel loop* via $\mathbf{P}(t)$ and the *orthogonal loop* via $\mathbf{P}^{\perp}(t)$. This time update of $\mathbf{Q}^{-1}(t)$ is illustrated in Fig. 5.1. The two projection operators $\mathbf{P}(t)$ and $\mathbf{P}^{\perp}(t)$ are *complementary*, i.e., they add up to the identity matrix. Thus they have no direct influence on the overall norm of the matrix estimate. Rather they act as a "switch" between the two loops. Clearly, regarding the definition of $\mathbf{P}(t)$ (5.44a), we see that as soon as one component in the state vector $\mathbf{z}'(t)$ becomes zero, the parallel loop update is switched off for this row vector of the matrix update in that, for example,

$$\mathbf{z}'(t) = [*, *, *, 0, *, *] , \tag{5.57}$$

where "*" denotes any nonzero element. The corresponding projection

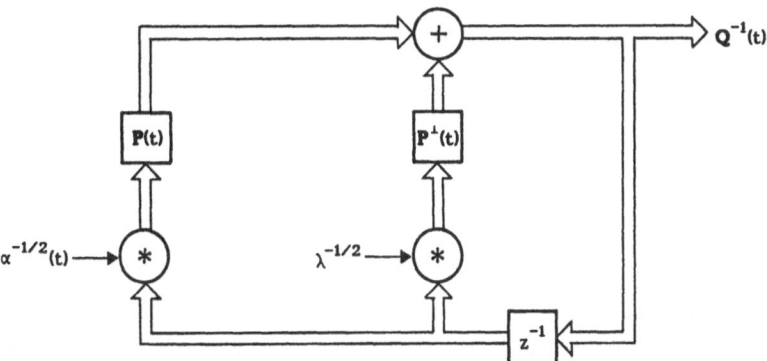

Fig. 5.1. Flow diagram of square-root update in Potter's normalized RLS algorithm

operator $P(t)$ has the structure

$$P(t) = z'(t)\left(z'^T(t) z'(t)\right) z'^T(t) = \begin{bmatrix} * & * & * & 0 & * & * \\ * & * & * & 0 & * & * \\ * & * & * & 0 & * & * \\ 0 & 0 & 0 & 0 & 0 & 0 \\ * & * & * & 0 & * & * \\ * & * & * & 0 & * & * \end{bmatrix} . \tag{5.58}$$

Writing $Q^{-1}(t-1)$ in terms of its column vectors

$$Q^{-1}(t-1) = \left[q_1(t-1), q_2(t-1), \ldots, q_p(t-1)\right]^T , \tag{5.59}$$

we see that the update of $Q^{-1}(t-1)$ through $P(t)$ in this example *annihilates* one row vector in the square-root matrix before the summation point in the flow diagram:

$$P(t) Q^{-1}(t-1)$$

$$= \left[q_1^*(t-1), q_2^*(t-2), q_3^*(t-3), 0, q_5^*(t-1), \ldots, q_p^*(t-1)\right] . \tag{5.60}$$

On the other hand, the flow diagram of Fig. 5.1 reveals that the gain in the orthogonal loop through $P^\perp(t)$ is always greater than 1 if the exponential forgetting factor is less than 1. Therefore, overall stability of the system shown in Fig. 5.1 can only be achieved when the following conditions are satisfied.

(1) The gain in the parallel loop, as represented by the quantity $\alpha^{-1/2}(t)$, must settle on a value less than 1, hence

$$\alpha^{-1/2}(t) < 1 . \tag{5.61}$$

(2) The normalized data vector $z'(t)$ must be persistently exciting, that is, none of the components of this data vector may be equal to zero:

$$z'(t) = \left[z_1'(t), z_2'(t), \ldots, z_p'(t)\right]^T , \tag{5.62}$$

where

$$z_i'(t) \neq 0 \qquad for \qquad 1 \leq i \leq p . \tag{5.63}$$

Condition (2) ensures that each dimension in the p-dimensional column space of $Q^{-1}(t-1)$ passes through the parallel loop at each recursion so as to allow a sufficient "damping" of the otherwise inevitable divergence in this dimension.

In reality, condition (2) is frequently violated. Therefore, the following scenario is likely to appear:

(1) One or more of the p dimensions of the data vector $\mathbf{z}'(t)$ is zero over a certain period of time.

(2) The inverse square-root matrix continues to be fed through the orthogonal loop in the update system shown in Fig. 5.1 and receives considerable amplification through the factor $\lambda^{-1/2}$ since the parallel loop is turned off for the dimensions that lack persistent excitation.

(3) Suddenly, one of these deficient dimensions turns up again in the data and receives a strong damping in the parallel loop due to the feed-forward servomechanism controlled by $\alpha^{-1/2}(t)$.

(4) If the accumulated amplification and the damping are large with respect to the dynamic range of the number representation, the overall result will largely depend on numerical errors. The inverse system matrix is likely to show negative eigenvalues in this case. The square-root normalized algorithm is slightly more robust in that respect, since here the fluctuations of values inside the algorithm generally have a smaller dynamic range. Moreover, one may note that the probability of occurrence of zeros in the normalized data vector $\mathbf{z}'(t)$ is generally smaller than the probability of zeros in $\mathbf{z}(t)$ due to the normalization process (5.56a). This makes the normalized algorithm slightly less sensitive to persistent excitation effects in the input data.

5.4 Kubin's Selective Memory RLS Algorithms

Searching for a way out of the incompatibility of persistent excitation with quantized data and finite-word-length computations, one notes that the update equation (5.49) offers only two options:

(1) Set $\lambda = 1$ to avoid an uncontrolled growth in the orthogonal loop. From Fig. 5.1, it is seen that in this case the inverse of the square-root matrix converges to the zero matrix. Thereby, the algorithm loses its adaptivity as the gain vector (5.56b) is turned off.

(2) Just modify the influence of the weighting factor inside the orthogonal loop by setting

$$\lambda^{-1/2} = 1 \tag{5.64}$$

in the orthogonal loop. Then, the square-root update (5.49) changes to

$$\mathbf{Q}^{-1}(t) = \left[1\, \mathbf{P}^{\perp}(t) + \left(\lambda + \mathbf{z}'^{T}(t)\, \mathbf{z}'(t)\right)^{-1/2} \mathbf{P}(t)\right]\mathbf{Q}^{-1}(t-1) \ . \tag{5.65}$$

From Fig. 5.1, it can be seen that with this choice the uncontrolled growth in the orthogonal feedback loop is avoided. This also means that the matrix estimate is only updated in the single dimension observed through the actual measurement $z'(t)$. This update employs an exponential decay. The other dimensions remain unchanged until they turn up in the data. The influence of past data is not exponentially decayed as in the conventional RLS algorithm.

Equation (5.65) leads directly to Kubin's "selective memory" RLS algorithms [5.16]. As already mentioned, these algorithms are characterized by the property that they do not employ an exponential decay on all past data, but only on the actual component $z(t)$ of the observation matrix $\mathbf{X}(t)$. To obtain closed algorithms, we employ (5.44b and 54), to express (5.65) as

$$\mathbf{Q}^{-1}(t) = \left[\mathbf{I} - \left(1 - \alpha^{-1/2}(t)\right)\mathbf{P}(t)\right]\mathbf{Q}^{-1}(t-1) \ . \tag{5.66}$$

Expressing the projection operator $\mathbf{P}(t)$ in terms of $z'(t)$ (5.44a), and taking into account that $z'^{T}(t)z'(t) = \alpha(t) - \lambda$, we have

$$\mathbf{Q}^{-1}(t) = \left[\mathbf{I} - \left(1 - \alpha^{-1/2}(t)\right)\left(\alpha(t) - \lambda\right)^{-1}z'(t)z'^{T}(t)\right]\mathbf{Q}^{-1}(t-1) \ . \tag{5.67}$$

Noting that $z'(t) = \mathbf{Q}^{-1}(t-1)z(t)$, one ends up with the result

$$\mathbf{Q}^{-1}(t) = \mathbf{Q}^{-1}(t-1) - \left(1 - \alpha^{-1/2}(t)\right)\left(\alpha(t) - \lambda\right)^{-1}z'(t)\mathbf{k}^{T}(t) \ , \tag{5.68}$$

which is the "selective memory" counterpart of the conventional exponentially weighted square-root update given in (5.55).

In order to obtain a closed algorithm, we simply replace (5.56e) in Table 5.2 by (5.68). This then gives Kubin's square-root normalized RLS algorithm with selective memory update [5.16].

To complete these considerations, we may also formulate the selective memory update approach with the original RLS algorithm of Table 5.1., i.e., with the *unnormalized* system matrix $\mathbf{A}(t)$. For this purpose, one basically examines the derivation of Potter's square-root normalized RLS algorithm in the opposite direction. This strategy starts with the square-root product update equation in the selective memory case

$$\mathbf{Q}(t)\mathbf{Q}^{T}(t) = \mathbf{Q}(t-1)\left[\mathbf{I} + \left(\lambda - 1 + z'^{T}(t)z'(t)\right)\mathbf{P}(t)\right]\mathbf{Q}^{T}(t-1) \ . \tag{5.69}$$

One sees that

$$\mathbf{Q}(t)\mathbf{Q}^{T}(t) = \mathbf{Q}(t-1)\mathbf{Q}^{T}(t-1) + \mathbf{Q}(t-1)\left(\lambda - 1 + z'^{T}(t)z'(t)\right)$$
$$\times z'(t)\left(z'^{T}(t)z'(t)\right)^{-1}z'^{T}(t)\mathbf{Q}^{T}(t-1) \ , \tag{5.70}$$

which, by incorporation of (5.40, 42 and 51) reduces to

$$\mathbf{A}(t) = \mathbf{A}(t-1) + \Big[\big(\lambda - 1$$

$$+ \mathbf{z}^T(t)\mathbf{A}^{-1}(t-1)\mathbf{z}(t) \big) \Big/ \big(\mathbf{z}^T(t)\mathbf{A}^{-1}(t-1)\mathbf{z}(t) \big) \Big] \mathbf{z}(t)\,\mathbf{z}^T(t) \qquad (5.71)$$

or equivalently

$$\mathbf{A}(t) = \mathbf{A}(t-1) + s(t)\mathbf{z}(t)\mathbf{z}^T(t) \ , \qquad (5.72a)$$

where

$$s(t) = 1 - \big(1-\lambda\big)\Big/\big(\mathbf{z}^T(t)\mathbf{A}^{-1}(t-1)\mathbf{z}(t)\big) \ . \qquad (5.72b)$$

Taking the inverse of (5.72a), we have

$$\mathbf{A}^{-1}(t) = \mathbf{A}^{-1}(t-1) - s(t) \frac{\mathbf{A}^{-1}(t-1)\,\mathbf{z}(t)\,\mathbf{z}^T(t)\,\mathbf{A}^{-1}(t-1)}{\lambda + \mathbf{z}^T(t)\mathbf{A}^{-1}(t-1)\mathbf{z}(t)} \ . \qquad (5.73)$$

The complete unnormalized RLS algorithm with selective memory update is summarized in Table 5.3. In comparison with its exponentially weighted counterpart of Table 5.1, this algorithm is characterized by the following points:

(1) With $\lambda = 1$, the selective memory RLS algorithm of Table 5.3 is identical with its exponentially weighted counterpart of Table 5.1. The same holds for $\lambda < 1$ when $p = 1$.

(2) While the exponentially weighted estimator forgets a portion $(1-\lambda)$ of all p dimensions of the $\mathbf{A}(t-1)$ matrix during each update, the selective memory algorithm forgets only a portion $(1-\lambda)$ of the innovations matrix

$$\big(\mathbf{z}(t)\mathbf{z}^T(t)\big)\Big/\big(\mathbf{z}^T(t)\mathbf{A}^{-1}(t-1)\mathbf{z}(t)\big) \ . \qquad (5.74)$$

(3) According to Kubin [5.7], the selective memory algorithm always remains stable, independent of the condition of the input data.

(4) For short-word-length computations, the constant $\lambda^{-1/2}$ in (5.49) should be replaced by $1 - \delta$, where δ is a small leakage factor, instead of using the value 1.

In conclusion of this section, we note that methods of stabilizing the RLS algorithm have apparently been developed by different authors. A stabilization of the unnormalized RLS algorithm was given by Bierman [5.17] in 1977. The algorithm is based on a modified update equation for the inverse system matrix of the Normal Equations which is sometimes referred to as the "stabilized Kalman equation". Besides Potter's square-root RLS algorithm, one finds an alternative in Bierman's U-D factorization [5.17] which is slightly more efficient in terms of a smaller number of arithmetic operations than Potter's original algorithm.

Table 5.3. Kubin's unnormalized selective memory RLS algorithm. The value of λ is recommended to be the pth power of the exponential weighting factor in the conventional RLS algorithm for an approximately equivalent tracking behavior. This algorithm involves division. Whenever the divisor is small, set $1/x = x$

Initialize:

$$\mathbf{A}^{-1}(-1) = \sigma\mathbf{I} \; ; \qquad \sigma \gg 1$$

$$\mathbf{a}(-1) = \begin{bmatrix} 0 \ldots 0 \end{bmatrix}^T$$

FOR $t = 0, 1, 2, \ldots$

Input: $\mathbf{z}(t)$, $x(t)$, $0 \le \lambda \le 1$

$$\mathbf{k}(t) = \mathbf{A}^{-1}(t-1)\,\mathbf{z}(t) \tag{5.75a}$$

$$\alpha(t) = \lambda + \mathbf{z}^T(t)\,\mathbf{k}(t) \tag{5.75b}$$

$$\mathbf{k}^*(t) = \mathbf{k}(t)\big/\alpha(t) \tag{5.75c}$$

$$s(t) = 1 - (1-\lambda)/(\alpha(t) - \lambda) \tag{5.75d}$$

$$\mathbf{A}^{-1}(t) = \mathbf{A}^{-1}(t-1) - s(t)\,\mathbf{k}^*(t)\,\mathbf{k}^T(t) \tag{5.75e}$$

$$\varepsilon(t) = x(t) - \mathbf{z}^T(t)\,\mathbf{a}(t-1) \qquad \textit{a priori error} \tag{5.75f}$$

$$\mathbf{a}(t) = \mathbf{a}(t-1) + \mathbf{k}^*(t)\,\varepsilon(t) \tag{5.75g}$$

$$e(t) = x(t) - \mathbf{z}^T(t)\,\mathbf{a}(t) \qquad \textit{a posteriori error} \tag{5.75h}$$

5.5 Fast RLS Transversal Algorithms

The excellent convergence properties of the RLS algorithms, as discussed in the previous section, are achieved at the expense of high computational costs. Particularly, the computation of the gain vector $\mathbf{k}(t)$ requires the updating and storage of a p×p inverse system matrix, where p was the problem order. Accordingly, $O(p^2)$ operations are required for updating the inverse system matrix. This high computational cost is in direct contrast with the recursive updating of the system matrix itself, which – as seen in Chap. 2 – requires only $O(p)$ operations per recursion.

This efficient recursive updating of the data covariance matrix, from which the system matrix is derived via partitioning, was possible due to *shift-invariance properties* arising from the exploitation of *serialized* input data. When the system matrix exhibits these shift invariance properties, then it can be shown that its inverse also obeys a shift invariance law, which, however, is slightly more involved than in the more obvious case of the original matrix. These shift invariance properties of the inverse covariance matrix, as appearing in the case of serialized data, were first seriously studied by Morf [5.18] in 1974 and Morf et al. [5.19] in 1976. The exploration of these properties became possible due to a sophisticated *partitioning scheme* of the covariance matrix and the analytical inversion of the *partitioned* system matrix of the Normal Equations. This "matrix inversion lemma (Sherman-Morrison identity) for partitioned matrices" leads directly to a fast implementation of the RLS algorithm. The first RLS algorithm of this type, which requires only $O(p)$ operations per recursion, was reported by Ljung et al. [5.5] in 1978. This algorithm has become known as the *fast Kalman algorithm*.

Besides the derivation of fast algorithms, the Sherman-Morrison identity for partitioned matrices provides some additional insight into the interrelationships between forward and backward linear prediction. The optimal exploitation of these relationships can be used to derive algorithms that require even fewer arithmetic operations than the original fast Kalman algorithm. In this context, we will discuss the recently introduced *fast a posteriori error sequential technique* (FAEST) presented by Carayannis et al. [5.10, 11] in 1983.

Note that all RLS algorithms based on the Sherman-Morrison identity rely on simple exponential or rectangular windows. The incorporation of higher-order windows, although highly desirable, seems to be impossible owing to the lack of an appropriate inversion lemma for the resulting matrix polynomials in these cases. This limitation is a major drawback when the Sherman-Morrison-based RLS algorithms are compared to their Givens and PORLA-based counterparts where - as seen in Chaps. 4 and 8 and in the Appendix - RLS algorithms with second-order windows and much better tracking and steady-state behavior compared to the exponentially weighted counterparts can be derived.

In Chap. 4, we have seen how exponentially weighted RLS algorithms can be derived using an exponential weighting matrix $\Lambda(t)$ and its partitioned representation. The same basic rules hold for the derivation of the fast RLS algorithms of this chapter. We will therefore omit the notationally burdensome derivation with the general exponential weighting matrix $\Lambda(t)$, and simply replace it by the identity matrix, which corresponds to $\lambda = 1$, for convenience. For this simple pre-windowed case, the derivation can be assimilated much more easily without loss of generality. Later in the algorithm summaries we shall reintroduce the exponential weighting at the appropriate places.

5.5.1 The Sherman-Morrison Identity for Partitioned Matrices

The Sherman-Morrison identity for partitioned matrices constitutes the basis for the derivation of fast RLS algorithms. While the conventional Sherman-Morrison identity deals with the analytical inverse of a matrix polynomial consisting of three matrices, the Sherman-Morrison identity for partitioned marices solves the same problem in the case of *partitioned* matrices.

Recall the classical RLS problem of updating the inverse covariance matrix via the Sherman–Morrison identity, where $\lambda = 1$ is assumed for convenience:

$$
\begin{aligned}
A^{-1}(t) &= \left(A^{-1}(t-1) + z(t)\, z^T(t) \right)^{-1} \\
&= A^{-1}(t-1) - \frac{A^{-1}(t-1)z(t)z^T(t)A^{-1}(t-1)}{1 + z^T(t)A^{-1}(t-1)z(t)} .
\end{aligned}
\tag{5.76}
$$

We wish to discover the corresponding update rule when the matrix $A(t)$ and the observation vector $z(t)$ exhibit the following partitioning schemes:

$$
A(t) = \begin{bmatrix} R & r \\ r^T & r \end{bmatrix} ,
\tag{5.77a}
$$

$$
z(t) = \begin{bmatrix} z_r \\ z \end{bmatrix} .
\tag{5.77b}
$$

This gives rise to the problem

$$
A^{-1}(t) = \begin{bmatrix} R & r \\ r^T & r \end{bmatrix}^{-1} = \quad ? \quad = \begin{bmatrix} U & u \\ u^T & u \end{bmatrix} .
\tag{5.78}
$$

From (5.78), one obtains the relation

$$
\begin{bmatrix} 1 & & 0 \\ & 1 \; \ddots & \\ 0 & & 1 \end{bmatrix} = \begin{bmatrix} R & r \\ r^T & r \end{bmatrix} \begin{bmatrix} U & u \\ u^T & u \end{bmatrix}
$$

$$
= \begin{bmatrix} RU + r u^T & Ru + ru \\ r^T U + r u^T & r^T u + ru \end{bmatrix} .
\tag{5.79}
$$

The partitioning scheme (5.79) can be split into four equations:

$$
RU + r u^T = I ,
\tag{5.80a}
$$

$$
r^T u + ru = 1 ,
\tag{5.80b}
$$

$$
Ru + ru = \begin{bmatrix} 0 \ldots 0 \end{bmatrix}^T ,
\tag{5.80c}
$$

$$\mathbf{r}^T \mathbf{U} + \mathbf{r} \, \mathbf{u}^T = \begin{bmatrix} 0 & \ldots & 0 \end{bmatrix} \, , \tag{5.80d}$$

from which – after some algebra – the quantities $\mathbf{U}$, $\mathbf{u}$ and u can be determined as

$$\mathbf{U} = \left(\mathbf{R} - \frac{1}{r} \mathbf{r} \, \mathbf{r}^T \right)^{-1} \, , \tag{5.81a}$$

$$\mathbf{u} = - \frac{1}{r} \left(\mathbf{R} - \frac{1}{r} \mathbf{r} \, \mathbf{r}^T \right)^{-1} \mathbf{r} \; = \; - \frac{1}{r} \mathbf{U} \, \mathbf{r} \, , \tag{5.81b}$$

$$u = \frac{1}{r} \left[1 + \frac{1}{r} \mathbf{r}^T \left(\mathbf{R} - \frac{1}{r} \mathbf{r} \, \mathbf{r}^T \right)^{-1} \mathbf{r} \right] = \frac{1}{r} \left[1 - \mathbf{r}^T \mathbf{u} \right] \, . \tag{5.81c}$$

With (5.81a-c), a partitioning scheme for the inverse system matrix of the Normal Equations can be stated as

$$\begin{bmatrix} \mathbf{R} & \mathbf{r} \\ \mathbf{r}^T & r \end{bmatrix}^{-1}$$

$$= \begin{bmatrix} \left(\mathbf{R} - \frac{1}{r} \mathbf{r} \, \mathbf{r}^T \right)^{-1} & - \frac{1}{r} \left(\mathbf{R} - \frac{1}{r} \mathbf{r} \, \mathbf{r}^T \right)^{-1} \mathbf{r} \\ - \frac{1}{r} \mathbf{r}^T \left(\mathbf{R} - \frac{1}{r} \mathbf{r} \, \mathbf{r}^T \right)^{-1} & \frac{1}{r} \left[1 + \frac{1}{r} \mathbf{r}^T \left(\mathbf{R} - \frac{1}{r} \mathbf{r} \, \mathbf{r}^T \right)^{-1} \mathbf{r} \right] \end{bmatrix} \, , \tag{5.82a}$$

where, according to the Sherman-Morrison identity,

$$\left(\mathbf{R} - \frac{1}{r} \mathbf{r} \, \mathbf{r}^T \right)^{-1} = \mathbf{R}^{-1} - \frac{\mathbf{R}^{-1} \mathbf{r} \, \mathbf{r}^T \mathbf{R}^{-1}}{r - \mathbf{r}^T \mathbf{R}^{-1} \mathbf{r}} \, . \tag{5.82b}$$

Introducing the "auxiliary" Normal Equations corresponding to the upper-left matrix $\mathbf{R}$

$$\mathbf{R} \, \mathbf{b} = \mathbf{r} \; \longrightarrow \; \mathbf{b} = \mathbf{R}^{-1} \mathbf{r} \tag{5.83}$$

with the property

$$\mathbf{R}^T = \mathbf{R} \tag{5.84}$$

and substituting (5.83) into (5.82b), one finds

$$\left(\mathbf{R} - \frac{1}{r} \mathbf{r} \, \mathbf{r}^T \right)^{-1} = \mathbf{R}^{-1} + \frac{\mathbf{b} \, \mathbf{b}^T}{r - \mathbf{r}^T \mathbf{b}} \, . \tag{5.85}$$

Substitution of (5.85) into (5.82a) finally gives

$$\begin{bmatrix} R & r \\ r^T & r \end{bmatrix}^{-1} = \begin{bmatrix} R^{-1} + \dfrac{b\,b^T}{r - r^T b} & \dfrac{-b}{r - r^T b} \\[2ex] \dfrac{-b^T}{r - r^T b} & \dfrac{1}{r - r^T b} \end{bmatrix} . \qquad (5.86)$$

The vector b (solution of the system (5.83) is frequently termed the *backward predictor*. See also the Appendix for a summary of forward/backward linear prediction relationships.

Relation (5.86) represents the desired law for updating the partitioned inverse system matrix. A similar rule can be stated for the *forward predictor* where the system matrix of the Normal Equations is partitioned in just the *opposite* way as follows:

$$A(t) = \begin{bmatrix} 1 & 1^T \\ 1 & L \end{bmatrix} , \qquad (5.87a)$$

$$z(t) = \begin{bmatrix} z \\ z_1 \end{bmatrix} . \qquad (5.87b)$$

This gives rise to the analogous problem

$$A^{-1}(t) = \begin{bmatrix} 1 & 1^T \\ 1 & L \end{bmatrix}^{-1} = ? = \begin{bmatrix} u & u^T \\ u & U \end{bmatrix} . \qquad (5.88)$$

Similarly to (5.79), we obtain

$$\begin{bmatrix} 1 & & 0 \\ & 1 \ddots & \\ 0 & & 1 \end{bmatrix} = \begin{bmatrix} 1 & 1^T \\ 1 & L \end{bmatrix} \begin{bmatrix} u & u^T \\ u & U \end{bmatrix}$$

$$= \begin{bmatrix} 1u + 1^T u & 1u^T + 1^T U \\ 1u + Lu & 1u^T + LU \end{bmatrix} . \qquad (5.89)$$

Relation (5.89) can be split into four equations:

$$1u + 1^T u = 1 , \qquad (5.90a)$$

$$1u^T + 1^T U = \begin{bmatrix} 0 \ldots 0 \end{bmatrix} , \qquad (5.90b)$$

$$1u + Lu = \begin{bmatrix} 0 \ldots 0 \end{bmatrix}^T , \qquad (5.90c)$$

$$1u^T + LU = I \ , \tag{5.90d}$$

determining the desired variables u, **u** and **U** as follows:

$$U = \left(L - \frac{1}{1}11^T\right)^{-1} , \tag{5.91a}$$

$$u = -\frac{1}{1}\left(L - \frac{1}{1}11^T\right)^{-1}1 = -\frac{1}{1}U1 \ , \tag{5.91b}$$

$$u = \frac{1}{1}\left[1 + \frac{1}{1}1^T\left(L - \frac{1}{1}11^T\right)^{-1}1\right] = \frac{1}{1}\left[1 - 1^Tu\right] . \tag{5.91c}$$

This result (5.91 a-c) has an interesting similarity to (5.81a-c). Indeed, the inversion of a *lower* partitioned matrix is - in some sense - "*dual*" to the problem of inverting an *upper* partitioned matrix. This insight constitutes directly the forward linear predictor, which is determined by the following Normal Equations:

$$L f = 1 \longrightarrow f = L^{-1}1 \ , \tag{5.92}$$

where

$$L^T = L \ . \tag{5.93}$$

As a dual expression to (5.85) one finds

$$\left(L - \frac{1}{1}11^T\right)^{-1} = L^{-1} + \frac{f f^T}{1 - 1^Tf} \ . \tag{5.94}$$

Substitution of (5.94) in the partitioning scheme (5.89) finally gives

$$\begin{bmatrix} 1 & 1^T \\ 1 & L \end{bmatrix}^{-1} = \begin{bmatrix} \dfrac{1}{1 - 1^Tf} & \dfrac{-f^T}{1 - 1^Tf} \\ \dfrac{-f}{1 - 1^Tf} & L^{-1} + \dfrac{f f^T}{1 - 1^Tf} \end{bmatrix} . \tag{5.95}$$

The relationships (5.86 and 95) can be summarized in the following theorem.

Sherman-Morrison Identity for Partitioned Matrices

- Inversion of an *upper-left* partitioned symmetric system matrix:

$$\begin{bmatrix} R & r \\ r^T & r \end{bmatrix}^{-1} = \begin{bmatrix} R^{-1} + bb^T/\alpha^b & -b/\alpha^b \\ -b^T/\alpha^b & 1/\alpha^b \end{bmatrix} , \tag{5.96a}$$

$$\alpha^b = r - \mathbf{r}^T \mathbf{b} \quad , \tag{5.96b}$$

$$\mathbf{R} \mathbf{b} = \mathbf{r} \quad . \tag{5.96c}$$

- Inversion of a *lower-right* partitioned symmetric system matrix:

$$\begin{bmatrix} 1 & \mathbf{1}^T \\ \mathbf{1} & \mathbf{L} \end{bmatrix}^{-1} = \begin{bmatrix} 1/\alpha^f & -\mathbf{f}^T/\alpha^f \\ -\mathbf{f}/\alpha^f & \mathbf{L}^{-1} + \mathbf{f}\,\mathbf{f}^T/\alpha^f \end{bmatrix} \quad , \tag{5.97a}$$

$$\alpha^f = 1 - \mathbf{1}^T \mathbf{f} \quad , \tag{5.97b}$$

$$\mathbf{L}\,\mathbf{f} = \mathbf{1} \quad . \tag{5.97c}$$

Several different versions of the Sherman-Morrison identity have appeared in the literature. A formula for the inversion of a four-term matrix polynomial was given by Ljung and Söderström [5.7]. Other forms of the Sherman-Morrison identity for partitioned matrices may be found in the book by Hsia [5.6].

5.5.2 The Fast Kalman Algorithm

Utilizing the Sherman-Morrison identity for partitioned matrices, we can now derive the fast Kalman algorithm for the pre-windowed case. Again we assume $\lambda=1$ for convenience. The considered linear prediction problem is hence characterized by the actual observation

$$\mathbf{x}(t) = \Big[x(t),\, x(t-1),\, x(t-2),\, \dots,\, x(1),\, x(0),\, 0 \dots 0 \Big]^T , \tag{5.98}$$

the actual matrix of past observations

$$\mathbf{X}_p(t) = \Big[\mathbf{x}(t-1),\, \mathbf{x}(t-2),\, \dots,\, \mathbf{x}(t-p) \Big] \quad , \tag{5.99}$$

the actual least-squares solution (or parameter vector)

$$\mathbf{a}^{(p)}(t) = \Big[a_1(t),\, a_2(t),\, \dots,\, a_p(t) \Big]^T , \tag{5.100}$$

and the state vector of the transversal prediction error filter

$$\mathbf{z}_p(t) = \Big[x(t-1),\, x(t-2),\, \dots,\, x(t-p) \Big]^T . \tag{5.101}$$

Again we use the prediction error filter of order p in a *transversal* structure described by the relation

$$e_p(t) = x(t) - \mathbf{z}_p^T(t)\,\mathbf{a}^{(p)}(t) \quad , \tag{5.102}$$

where $e_p(t)$ is the actual sample of the prediction error after stage p of the prediction error filter. In contrast to this "actual" behavior of the predictor, the prediction error filtering of the *entire* sequence $x(t)$ can be expressed as

$$e_p(t) = x(t) - X_p(t) a^{(p)}(t) \ , \tag{5.103}$$

where

$$e_p(t) = \left[e_{p,t}(t), \ e_{p,t-1}(t), \ \ldots , \ e_{p,1}(t), \ e_{p,0}(t), \ 0 \ldots 0 \right]^T \ . \tag{5.104}$$

Note that (5.103), with (5.104), assumes that we have filtered the *entire* sequence with the parameter set of time step t. Adjusting the parameters $a^{(p)}(t)$ according to the Normal Equations

$$a^{(p)}(t) \ = \ A_p^{-1}(t) h_p(t) \ , \tag{5.105}$$

where

$$A_p(t) = X_p^T(t) X_p(t) \qquad \text{and} \tag{5.106}$$

$$h_p(t) = X_p^T(t) x(t) \tag{5.107}$$

gives, at each time step, a new (updated) parameter set. Operating the prediction error filter (5.102) with this time-varying parameter set, we obtain only the *top component* of the prediction error vector (5.104) at each time step.

We know from the derivation of the conventional RLS algorithm that the recursive computation of the parameter set $a^{(p)}(t)$ at each time step gives rise to the problem of updating an inverse symmetric system matrix (note that we have assumed $\lambda = 1$)

$$A_p^{-1}(t) = \left[A_p(t-1) + z_p(t) z_p^T(t) \right]^{-1} \ , \tag{5.108}$$

$$h_p(t) = h_p(t-1) + z_p(t) x(t) \ . \tag{5.109}$$

As a key step in the derivation of the fast Kalman algorithm, we next consider the state vector and the matrix of past observations of the order p+1 problem:

$$z_{p+1}(t) = \left[x(t-1), \ x(t-2), \ \ldots , \ x(t-p), \ x(t-p-1) \right]^T \ , \tag{5.110}$$

$$X_{p+1}(t) = \left[x(t-1), \ x(t-2), \ \ldots , \ x(t-p), \ x(t-p-1) \right] \ . \tag{5.111}$$

Now it is easy to check that one can introduce the following partitioning schemes for (5.110 and 111):

$$z_{p+1}(t) = \begin{bmatrix} z_p(t) \\ x(t-p-1) \end{bmatrix} = \begin{bmatrix} x(t-1) \\ z_p(t-1) \end{bmatrix} , \qquad (5.112)$$

$$X_{p+1}(t) = \begin{bmatrix} X_p(t) & x(t-p-1) \end{bmatrix} , \qquad (5.113a)$$

$$X_{p+1}(t) = \begin{bmatrix} x(t-1) & X_p(t-1) \end{bmatrix} . \qquad (5.113b)$$

Consider now the system matrix of the order p+1 Normal Equations

$$A_{p+1}(t) = X_{p+1}^T(t)\, X_{p+1}(t) . \qquad (5.114)$$

According to (5.113a,b), we may express $A_{p+1}(t)$ by the following partitioning schemes:

$$A_{p+1}(t) = \begin{bmatrix} X_p^T(t)X_p(t) & X_p^T(t)x(t-p-1) \\ x^T(t-p-1)X_p(t) & x^T(t-p-1)\,x(t-p-1) \end{bmatrix} , \qquad (5.115a)$$

$$A_{p+1}(t) = \begin{bmatrix} x^T(t-1)\,x(t-1) & x^T(t-1)X_p(t-1) \\ X_p^T(t-1)\,x(t-1) & X_p^T(t-1)X_p(t-1) \end{bmatrix} . \qquad (5.115b)$$

With the shorthand notation

$$h_p^f(t) = X_p^T(t-1)x(t-1) , \qquad (5.116a)$$

$$h_p^b(t) = X_p^T(t)x(t-p-1) , \qquad (5.116b)$$

$$h_0^f(t) = x^T(t-1)x(t-1) , \qquad (5.116c)$$

$$h_0^b(t) = x^T(t-p-1)x(t-p-1) , \qquad (5.116d)$$

we may rewrite the partitioning schemes (5.115a,b) as

$$A_{p+1}(t) = \begin{bmatrix} A_p(t) & h_p^b(t) \\ h_p^{bT}(t) & h_0^b(t) \end{bmatrix} , \qquad (5.117a)$$

$$A_{p+1}(t) = \begin{bmatrix} h_0^f(t) & h_p^{f^T}(t) \\ h_p^f(t) & A_p(t-1) \end{bmatrix} . \tag{5.117b}$$

The inversion of these partitioning schemes is a direct application of the Sherman-Morrison identity for partitioned matrices, as summarized in (5.96a-c and 97a-c). Comparing (5.117a) with (5.96a) and (5.117b) with (5.97a), one sees that

$$A_{p+1}^{-1}(t) = \begin{bmatrix} A_p^{-1}(t) + b^{(P)}(t)b^{(P)^T}(t)/\alpha_p^b(t) & -b^{(P)}(t)/\alpha_p^b(t) \\ -b^{(P)^T}(t)/\alpha_p^b(t) & 1/\alpha_p^b(t) \end{bmatrix}, \tag{5.118a}$$

$$\alpha_p^b(t) = h_0^b(t) - h_p^{b^T}(t)\, b^{(P)}(t) \ , \tag{5.118b}$$

$$A_p(t)\, b^{(P)}(t) = h_p^b(t) \ , \tag{5.118c}$$

$$A_{p+1}^{-1}(t) = \begin{bmatrix} 1/\alpha_p^f(t) & -f^{(P)^T}(t)/\alpha_p^f(t) \\ -f^{(P)}(t)/\alpha_p^f(t) & A_p^{-1}(t-1) + f^{(P)}(t)\, f^{(P)^T}(t)/\alpha_p^f(t) \end{bmatrix}, \tag{5.119a}$$

$$\alpha_p^f(t) = h_0^f(t) - h_p^{f^T}(t)\, f^{(P)}(t) \ , \tag{5.119b}$$

$$A_p(t-1)\, f^{(P)}(t) = h_p^f(t) \ . \tag{5.119c}$$

With these partitioning schemes, we have reduced the problem of solving a single, order p+1 Normal Equation to the problem of solving *two*, order p Normal Equations for the backward predictor (5.118c), determined by the parameter vector $b^{(P)}(t)$, and the forward predictor (5.119c), determined by the parameter vector $f^{(P)}(t)$. Note that these are the forward/backward predictors associated with the subsystems of Normal Equations arising through a lower-right and upper-left partitioning. These reduced systems can now be solved using the RLS equations (5.26a-g).

We start with the backward predictor Normal Equations (5.118c). By means of a comparison of (5.2b) with (5.116b), we see that we can use the RLS equations if we simply make the transitions

$$x(t) \longrightarrow x(t-p-1) \quad \text{and} \tag{5.120a}$$

$$X(t) \longrightarrow X_p(t) \ . \tag{5.120b}$$

From (5.26e), one may define the a priori error of the backward predictor as

$$\varepsilon_p^b(t) = x(t-p-1) - \mathbf{z}_p^T(t)\,\mathbf{b}^{(P)}(t-1) \quad, \tag{5.121a}$$

whereas the a posteriori error of the backward predictor derives from (5.26g):

$$e_p^b(t) = x(t-p-1) - \mathbf{z}_p^T(t)\mathbf{b}^{(P)}(t) \quad. \tag{5.121b}$$

Consequently, a closer look at (5.26f) shows that the backward predictor coefficients may be updated via the a priori errors as follows:

$$\mathbf{b}^{(P)}(t) = \mathbf{b}^{(P)}(t-1) + \mathbf{k}_p^*(t)\,\varepsilon_p^b(t) \quad. \tag{5.121c}$$

Equation (5.27) can be used to establish a similar expression for updating the backward predictor from the a posteriori errors:

$$\mathbf{b}^{(P)}(t) = \mathbf{b}^{(P)}(t-1) + \mathbf{k}_p(t)\,e_p^b(t) \quad. \tag{5.121d}$$

Similarly, a comparison of (5.2b) and the right-hand side of the forward predictor Normal Equations (5.116a) reveals that the RLS equations can be exploited for the solution of the system (5.119c) if one introduces the transitions

$$x(t) \longrightarrow x(t-1) \quad, \tag{5.122a}$$

$$\mathbf{X}(t) \longrightarrow \mathbf{X}_p(t-1) \quad. \tag{5.122b}$$

Furthermore, a comparison of (5.1) with (5.119c) relates the system matrix of the forward predictor Normal Equations to the system matrix underlying the RLS equations (5.26a-g) by

$$\mathbf{A}(t) \longrightarrow \mathbf{A}_p(t-1) \quad, \tag{5.122c}$$

which ultimately gives

$$\mathbf{z}(t) \longrightarrow \mathbf{z}_p(t-1) \quad, \tag{5.122d}$$

$$\mathbf{k}(t) \longrightarrow \mathbf{k}_p(t-1) \quad, \tag{5.122e}$$

$$\mathbf{k}^*(t) \longrightarrow \mathbf{k}_p^*(t-1) \quad. \tag{5.122f}$$

Similarly to (5.15a-d), one can now derive the desired expressions for the a priori and a posteriori forward prediction errors

$$\varepsilon_p^f(t) = x(t-1) - z_p^T(t-1)\, f^{(P)}(t-1) \; , \tag{5.123a}$$

$$e_p^f(t) = x(t-1) - z_p^T(t-1)\, f^{(P)}(t) \; , \tag{5.123b}$$

and the corresponding forward predictor coefficient updates

$$f^{(P)}(t) = f^{(P)}(t-1) + k_p^*(t-1)\, \varepsilon_p^f(t) \; , \tag{5.123c}$$

$$f^{(P)}(t) = f^{(P)}(t-1) + k_p(t-1)\, e_p^f(t) \; . \tag{5.123d}$$

From (5.39), we may deduce that

$$k_{p+1}^*(t) = A_{p+1}^{-1}(t)\, z_{p+1}(t) \tag{5.124}$$

holds, and this relation gives us the opportunity of expressing the modified Kalman gain vector $k_{p+1}^*(t)$ in terms of backward/forward prediction partitioning schemes. Thus incorporating (5.118a-c , 119a-c and 112) into (5.124) gives

$$k_{p+1}^*(t)$$

$$= \begin{bmatrix} A_p^{-1}(t) + b^{(P)}(t)b^{(P)T}(t)\big/\alpha_p^b(t) & b^{(P)}(t)\big/\alpha_p^b(t) \\[2mm] - b^{(P)T}(t)\big/\alpha_p^b(t) & 1\big/\alpha_p^b(t) \end{bmatrix} \begin{bmatrix} z_p(t) \\[2mm] x(t-p-1) \end{bmatrix} , \tag{5.125a}$$

$$k_{p+1}^*(t)$$

$$= \begin{bmatrix} 1\big/\alpha_p^f(t) & - f^{(P)T}(t)\big/\alpha_p^f(t) \\[2mm] - f^{(P)}(t)\big/\alpha_p^f(t) & A_p^{-1}(t-1) + f^{(P)}(t)f^{(P)T}(t)\big/\alpha_p^f(t) \end{bmatrix} \begin{bmatrix} x(t-1) \\[2mm] z_p(t-1) \end{bmatrix} . \tag{5.125b}$$

Exploiting expression (5.124) for the order p system together with (5.121b) allows us to express the modified Kalman gain vector of the order p+1 system by the modified Kalman gain vector of the order p system plus some order update that depends on the backward prediction a posteriori errors, the backward predictor coefficients and the quantity $\alpha_p^b(t)$ as

$$k_{p+1}^*(t) = \begin{bmatrix} k_p^*(t) \\[2mm] 0 \end{bmatrix} + \frac{e_p^b(t)}{\alpha_p^b(t)} \begin{bmatrix} - b^{(P)}(t) \\[2mm] 1 \end{bmatrix} . \tag{5.126a}$$

As expected, a similar order recursion of k^* can be established in terms of the one time step delayed modified Kalman gain vector of the order p system plus some update depending on the forward prediction a posteriori errors, the forward predictor coefficients and the quantity $\alpha_p^f(t)$. Viewing again (5.124), together with (5.123b), we can reduce (5.125b) to

$$
k_{p+1}^*(t) = \begin{bmatrix} 0 \\ k_p^*(t-1) \end{bmatrix} + \frac{e_p^f(t)}{\alpha_p^f(t)} \begin{bmatrix} 1 \\ -f^{(P)}(t) \end{bmatrix} . \tag{5.126b}
$$

As a next step, one introduces the following partitioning scheme of the modified Kalman gain vector of the order p+1 system

$$
k_{p+1}^*(t) = \begin{bmatrix} d_p^*(t) \\ \delta_p^*(t) \end{bmatrix} . \tag{5.127}
$$

A comparison of (5.127) with (5.126a) yields the expressions

$$
\delta_p^*(t) = \frac{e_p^b(t)}{\alpha_p^b(t)} , \tag{5.128a}
$$

$$
d_p^*(t) = k_p^*(t) - \delta_p^*(t) b^{(P)}(t) . \tag{5.128b}
$$

Substituting (5.121c) into (5.128b) gives

$$
\begin{aligned}
d_p^*(t) &= k_p^*(t) - \delta_p^*(t) \left[b^{(P)}(t-1) + k_p^*(t) \varepsilon_p^b(t) \right] \\
&= k_p^*(t) \left[1 - \delta_p^*(t) \varepsilon_p^b(t) \right] - \delta_p^*(t) b^{(P)}(t-1) . \tag{5.129}
\end{aligned}
$$

This expression can now be solved for $k_p^*(t)$:

$$
k_p^*(t) = \frac{d_p^*(t) + \delta_p^*(t) b^{(P)}(t-1)}{1 - \delta_p^*(t) \varepsilon_p^b(t)} . \tag{5.130}
$$

In order to obtain a closed algorithm, we are interested in appropriate recursions for updating the variables $\alpha_p^b(t)$ and $\alpha_p^f(t)$ in time.

From (5.118b) one can state the difference between consecutive values of the quantity α_p^b as

$$
\alpha_p^b(t) - \alpha_p^b(t-1) = h_0^b(t) - h_0^b(t-1) + h_p^{bT}(t-1) b^{(P)}(t-1)
$$

$$
- h_p^{bT}(t) b^{(P)}(t) . \tag{5.131}
$$

An inspection of (5.116d) reveals

$$h_0^b(t) - h_0^b(t-1) = x^2(t-p-1) \tag{5.132}$$

and

$$h_p^{b^T}(t) = x^T(t-p-1)X_p(t) \ , \tag{5.133a}$$

$$h_p^{b^T}(t-1) = x^T(t-p-2)X_p(t-1) \ . \tag{5.133b}$$

With the partitioning schemes

$$x^T(t-p-1) = \left[x(t-p-1) \quad x^T(t-p-2) \right] \ , \tag{5.134}$$

$$X_p(t) = \begin{bmatrix} z_p^T(t) \\ X_p(t-1) \end{bmatrix} \ , \tag{5.135}$$

we may express the quantity $h_p^{b^T}(t)$ in terms of its predecessor:

$$h_p^{b^T}(t) = x(t-p-1)z_p^T(t) + h_p^{b^T}(t-1) \ . \tag{5.136}$$

Substituting (5.136) into the differential of α_p^b (5.131) gives

$$\alpha_p^b(t) - \alpha_p^b(t-1) = x(t-p-1)\left[x(t-p-1) - z_p^T(t)b^{(p)}(t) \right]$$
$$+ h_p^{b^T}(t-1) \left[b^{(p)}(t-1) - b^{(p)}(t) \right] \ . \tag{5.137}$$

Continuing this process by substituting (5.121b) into (5.137) results in

$$\alpha_p^b(t) - \alpha_p^b(t-1) = x(t-p-1)e_p^b(t) + h_p^{b^T}(t) \left[b^{(p)}(t-1) - b^{(p)}(t) \right] \ . \tag{5.138}$$

The difference between (5.121b) and (5.121a) may be written as

$$e_p^b(t) - \varepsilon_p^b(t) = z_p^T(t)\left[b^{(p)}(t-1) - b^{(p)}(t) \right] \ . \tag{5.139}$$

With the additional insight that $b^{(p)}(t-1) - b^{(p)}(t) = -k_p^*(t)\varepsilon_p^b(t)$, it follows that

$$\alpha_p^b(t) - \alpha_p^b(t-1) = x(t-p-1)\varepsilon_p^b(t) - h_p^{b^T}(t)k_p^*(t)\varepsilon_p^b(t) \ , \tag{5.140}$$

where one may express the modified Kalman gain vector by the inverse system matrix and the state vector to obtain

$$k_p^*(t) = A_p^{-1}(t)z_p(t) \ , \tag{5.141}$$

which gives

$$\alpha_p^b(t) - \alpha_p^b(t-1) = x(t-p-1)\varepsilon_p^b(t) - \mathbf{h}_p^{b^T}(t)\,\mathbf{A}_p^{-1}(t)\,\mathbf{z}_p(t)\,\varepsilon_p^b(t) \ . \tag{5.142}$$

Noting that $\mathbf{h}_p^{b^T}(t)\,\mathbf{A}_p^{-1}(t) = \mathbf{b}^{(p)^T}(t)$ one sees

$$\alpha_p^b(t) - \alpha_p^b(t-1) = x(t-p-1)\varepsilon_p^b(t) - \mathbf{b}^{(p)^T}(t)\,\mathbf{z}_p(t)\,\varepsilon_p^b(t)$$

$$= \Big[\, x(t-p-1) - \mathbf{b}^{(p)^T}(t)\,\mathbf{z}_p(t)\,\Big]\varepsilon_p^b(t) \ . \tag{5.143}$$

Having in mind that $x(t-p-1) - \mathbf{b}^{(p)^T}(t)\,\mathbf{z}_p(t) = e_p^b(t)$, expression (5.143) reduces to

$$\alpha_p^b(t) = \alpha_p^b(t-1) + e_p^b(t)\,\varepsilon_p^b(t) \ , \tag{5.144}$$

which is the desired recursive law for updating α_p^b in time.

Next, we are interested in the corresponding recursion for the quantity α_p^f. In a similar procedure, we set

$$\alpha_p^f(t) - \alpha_p^f(t-1) = h_0^f(t) - h_0^f(t-1) + \mathbf{h}_p^{f^T}(t-1)\mathbf{f}^{(p)}(t-1)$$

$$- \mathbf{h}_p^{f^T}(t)\mathbf{f}^{(p)}(t) \ . \tag{5.145}$$

Using the relations

$$h_0^f(t) - h_0^f(t-1) = x^2(t-1) \ , \tag{5.146}$$

$$\mathbf{h}_p^{f^T}(t) = \mathbf{x}^T(t-1)\mathbf{X}_p(t-1) \ , \tag{5.147a}$$

$$\mathbf{h}_p^{f^T}(t-1) = \mathbf{x}^T(t-2)\mathbf{X}_p(t-2) \ , \tag{5.147b}$$

along with the partitioning schemes

$$\mathbf{x}^T(t-1) = \Big[x(t-1) \quad \mathbf{x}^T(t-2)\Big] \ , \tag{5.148}$$

$$\mathbf{X}_p(t-1) = \begin{bmatrix} \mathbf{z}_p^T(t-1) \\ \mathbf{X}_p(t-2) \end{bmatrix} \ , \tag{5.149}$$

one obtains the intermediate relation

$$\mathbf{h}_p^{f^T}(t) = \mathbf{h}_p^{f^T}(t-1) + x(t-1)\,\mathbf{z}_p^T(t-1) \ . \tag{5.150}$$

Continuing in this way, we can construct the desired recursive law for updating the quantity α_p^f in time. The fairly complex details of this derivation will be given in the following, for completeness.

Substitution of (5.150) into (5.145), together with (5.146), gives

$$\alpha_p^f(t) - \alpha_p^f(t-1) = x^2(t-1) + h_p^{fT}(t-1) f^{(p)}(t-1) - h_p^{fT}(t-1) f^{(p)}(t)$$

$$- x(t-1) z_p^T(t-1) f^{(p)}(t) . \tag{5.151}$$

Recalling that $x(t-1) - z_p^T(t-1) f^{(p)}(t) = e_p^f(t)$, one can rewrite (5.151) to obtain

$$\alpha_p^f(t) - \alpha_p^f(t-1) = x(t-1) e_p^f(t) + h_p^{fT}(t-1) \left[f^{(p)}(t-1) - f^{(p)}(t) \right] . \tag{5.152}$$

Evaluating the difference between (5.123b) and (5.123a), we have the auxiliary expression

$$e_p^f(t) - \varepsilon_p^f(t) = z_p^T(t-1) \left[f^{(p)}(t-1) - f^{(p)}(t) \right] , \tag{5.153}$$

which, together with the relation

$$h_p^f(t) = h_p^f(t-1) + z_p(t-1) x(t-1) , \tag{5.154}$$

gives

$$\alpha_p^f(t) - \alpha_p^f(t-1) = x(t-1) \varepsilon_p^f(t) + h_p^{fT}(t) \left[f^{(p)}(t-1) - f^{(p)}(t) \right] . \tag{5.155}$$

From (5.123c) one sees that

$$f^{(p)}(t-1) - f^{(p)}(t) = - k_p^*(t-1) \varepsilon_p^f(t) , \tag{5.156}$$

which can be substituted into (5.155) to obtain

$$\alpha_p^f(t) - \alpha_p^f(t-1) = x(t-1) \varepsilon_p^f(t) - h_p^{fT}(t) k_p^*(t-1) \varepsilon_p^f(t) . \tag{5.157}$$

On the other hand, a closer look at (5.124) reveals that (5.157) can alternatively be expressed as

$$\alpha_p^f(t) - \alpha_p^f(t-1) = x(t-1) \varepsilon_p^f(t) - h_p^{fT}(t) A_p^{-1}(t-1) z_p(t-1) \varepsilon_p^f(t) , \tag{5.158}$$

which, subsequently, allows the incorporation of (5.119c) to give

$$\alpha_p^f(t) - \alpha_p^f(t-1) = x(t-1) \varepsilon_p^f(t) - f^{(p)T}(t) z_p(t-1) \varepsilon_p^f(t) . \tag{5.159}$$

Now, in a last step, we check that (5.159) still includes (5.123b), which leads to the final result

$$\alpha_p^f(t) = \alpha_p^f(t-1) + e_p^f(t)\,\varepsilon_p^f(t) \ . \tag{5.160}$$

This recursion completes the derivation. One sees that the updating of the modified Kalman gain vector can be accomplished via the following path of recursions:

$$k_p^*(t-1) \xrightarrow{\ (5.126b)\ } k_{p+1}^*(t) \xrightarrow{\ (5.127, 130)\ } k_p^*(t) \ . \tag{5.161}$$

Table 5.4 summarizes this algorithm, which has become known as the *fast Kalman algorithm*. The recursions of the fast Kalman algorithm are grouped as follows: First, one computes the forward and backward prediction a priori errors ε_p^f and ε_p^b via (5.123a and 121a). Subsequently, the forward predictor parameters $f^{(p)}$ are updated in time using the modified Kalman gain k^* (5.123c). Next, the a posteriori forward prediction error e_p^f is computed according to (5.123b). The a priori and a posteriori forward prediction errors are then used for updating the forward prediction error covariance α_p^f in time. The update of the modified Kalman gain vector k^* is accomplished by the vector equation (5.126b). The result of this vector update is partitioned according to (5.127), and the resulting intermediate vector d_p^* and the scalar quantity δ_p^* are used to compute k^* according to (5.130). The updating of the backward predictor parameter vector $b^{(p)}$ according to (5.121c) concludes the fast update procedure of the modified Kalman gain vector k^* . When one is interested in least-squares FIR filtering, prediction, or parameter tracking, it is additionally required to compute the a priori error of the FIR prediction filter according to (5.26e), the parameter update (5.26f), and the a posteriori error according to (5.26g). These recursions are equivalent to the recursions appearing in the conventional $O(p^2)$ RLS algorithm listed in Table 5.1. As these recursions are not a part of the fast modified Kalman gain update, their computation remains *optional*.

5.5.3 The FAEST Algorithm

A simple inspection of Table 5.4 reveals that the fast Kalman algorithm is mainly based on the *a priori* error formulation. In this section, we will present an alternative formulation which is based, to a great extent, on an *a posteriori* error formulation. In Sect. 5.1, we demonstrated that the conventional RLS algorithm can be expressed in either an a priori or an a posteriori error formulation. It is emphasized that the same situation holds for the fast algorithms. Interestingly, the a posteriori error based counterpart of the fast Kalman algorithm, termed *fast a posteriori error*

Table 5.4. Fast Kalman algorithm. This algorithm requires 8p multiplications or divisions for gain updating and 11p multiplications or divisions for LS FIR filtering. λ is the exponential weighting factor and σ is a small positive constant. This algorithm involves division. Whenever the divisor is small, set $1/x = x$

Initialize:

$$\mathbf{k}_p^*(-1) = \mathbf{f}^{(p)}(-1) = \mathbf{b}^{(p)}(-1) = \mathbf{a}^{(p)}(-1) = \mathbf{z}_p(-1) = \begin{bmatrix} 0 \ldots 0 \end{bmatrix}^T$$

$$\alpha_p^f(-1) = \sigma; \qquad \sigma > 0; \qquad \sigma \ll 1$$

FOR t = 0, 1, 2, . . .

Input: $\mathbf{z}_p(t)$, $x(t)$, $0 \le \lambda \le 1$

$\varepsilon_p^f(t) = x(t-1) - \mathbf{z}_p^T(t-1)\mathbf{f}^{(p)}(t-1)$ *a priori forward prediction error*   (5.162a)

$\varepsilon_p^b(t) = x(t-p-1) - \mathbf{z}_p^T(t)\mathbf{b}^{(p)}(t-1)$ *a priori backward prediction error*   (5.162b)

$\mathbf{f}^{(p)}(t) = \mathbf{f}^{(p)}(t-1) + \mathbf{k}_p^*(t-1)\varepsilon_p^f(t)$ *forward predictor update*   (5.162c)

$e_p^f(t) = x(t-1) - \mathbf{z}_p^T(t-1)\mathbf{f}^{(p)}(t)$ *a posteriori forward prediction error*   (5.162d)

$\alpha_p^f(t) = \lambda\,\alpha_p^f(t-1) + e_p^f(t)\,\varepsilon_p^f(t)$ *forward prediction error covariance*   (5.162e)

$$\begin{bmatrix} \mathbf{d}_p^*(t) \\ \delta_p^*(t) \end{bmatrix} = \begin{bmatrix} 0 \\ \mathbf{k}_p^*(t-1) \end{bmatrix} + \frac{e_p^f(t)}{\alpha_p^f(t)}\begin{bmatrix} 1 \\ -\mathbf{f}^{(p)}(t) \end{bmatrix} \qquad \textit{partitioning} \qquad (5.162f)$$

$$\mathbf{k}_p^*(t) = \frac{\mathbf{d}_p^*(t) + \delta_p^*(t)\,\mathbf{b}^{(p)}(t-1)}{1 - \delta_p^*(t)\,\varepsilon_p^b(t)} \qquad \textit{modified Kalman gain update} \qquad (5.162g)$$

$\mathbf{b}^{(p)}(t) = \mathbf{b}^{(p)}(t-1) + \mathbf{k}_p^*(t)\varepsilon_p^b(t)$ *backward predictor update*   (5.162h)

$\varepsilon_p(t) = x(t) - \mathbf{z}_p^T(t)\mathbf{a}^{(p)}(t-1)$ *a priori error of prediction filter*   (5.162i)

$\mathbf{a}^{(p)}(t) = \mathbf{a}^{(p)}(t-1) + \mathbf{k}_p^*(t)\varepsilon_p(t)$ *prediction filter coefficient update*   (5.162j)

$e_p(t) = x(t) - \mathbf{z}_p^T(t)\mathbf{a}^{(p)}(t)$ *a posteriori error of prediction filter*   (5.162k)

---

sequential technique (FAEST) [5.10], requires even fewer computations. This gain in computational speed is achieved through a better exploitation of relationships between the a priori and a posteriori errors.

The key to the development of the FAEST algorithm is the use of the Kalman gain vector $\mathbf{k}(t)$ instead of the modified Kalman gain vector $\mathbf{k}^*(t)$. Whereas the derivation of the fast Kalman algorithm was based on the expression

$$\mathbf{k}_{p+1}^*(t) = \mathbf{A}_{p+1}^{-1}(t)\,\mathbf{z}_{p+1}(t) \; , \qquad\qquad (5.163)$$

the derivation of the FAEST algorithm relies on

$$\mathbf{k}_{p+1}(t+1) = \mathbf{A}_{p+1}^{-1}(t)\,\mathbf{z}_{p+1}(t+1) \; , \tag{5.164}$$

which is directly obtained from (5.26a). With the partitioning scheme

$$\mathbf{z}_{p+1}(t+1) = \begin{bmatrix} \mathbf{z}_p(t+1) \\ x(t-p) \end{bmatrix} = \begin{bmatrix} x(t) \\ \mathbf{z}_p(t) \end{bmatrix} \; , \tag{5.165}$$

we may establish partitioning schemes of (5.164) that can be viewed as the counterparts of the partitioning schemes (5.125 a,b) as follows:

$$\mathbf{k}_{p+1}(t+1)$$

$$= \begin{bmatrix} \mathbf{A}_p^{-1}(t) + \mathbf{b}^{(p)}(t)\mathbf{b}^{(p)^T}(t)/\alpha_p^b(t) & -\mathbf{b}^{(p)}(t)/\alpha_p^b(t) \\ -\mathbf{b}^{(p)^T}(t)/\alpha_p^b(t) & 1/\alpha_p^b(t) \end{bmatrix} \begin{bmatrix} \mathbf{z}_p(t-1) \\ x(t-p) \end{bmatrix} \; , \tag{5.166a}$$

$$\mathbf{k}_{p+1}(t+1)$$

$$= \begin{bmatrix} 1/\alpha_p^f(t) & -\mathbf{f}^{(p)^T}(t)/\alpha_p^f(t) \\ -\mathbf{f}^{(p)}(t)/\alpha_p^f(t) & \mathbf{A}_p^{-1}(t-1) + \mathbf{f}^{(p)}(t)\mathbf{f}^{(p)^T}(t)/\alpha_p^f(t) \end{bmatrix} \begin{bmatrix} x(t) \\ \mathbf{z}_p(t) \end{bmatrix} \; . \tag{5.166b}$$

Exploiting expression (5.164) for the order p system, together with (5.121a), we may express the Kalman gain vector of the order p+1 system by the Kalman gain vector of the order p system plus an order update that depends on the backward prediction *a priori* errors, the backward predictor coefficients and the quantity $\alpha_p^b(t)$ as follows:

$$\mathbf{k}_{p+1}(t+1) = \begin{bmatrix} \mathbf{k}_p(t+1) \\ 0 \end{bmatrix} + \frac{\varepsilon_p^b(t+1)}{\alpha_p^b(t)} \begin{bmatrix} -\mathbf{b}^{(p)}(t) \\ 1 \end{bmatrix} \; . \tag{5.167a}$$

Analogously, one obtains an expression for updating the Kalman gain vector of the order p+1 system from the Kalman gain vector of the order p system plus an order update that depends on the forward prediction *a priori* errors, the forward predictor coefficients and the quantity $\alpha_p^f(t)$, where we make use of (5.123a):

$$\mathbf{k}_{p+1}(t+1) = \begin{bmatrix} 0 \\ \mathbf{k}_p(t) \end{bmatrix} + \frac{\varepsilon_p^f(t+1)}{\alpha_p^f(t)} \begin{bmatrix} 1 \\ -\mathbf{f}^{(p)}(t) \end{bmatrix} \; . \tag{5.167b}$$

From (5.167a and b), one can deduce the following update scheme of the Kalman gain vector:

$$\mathbf{k}_p(t) \xrightarrow{\text{(5.167b)}} \mathbf{k}_{p+1}(t+1) \xrightarrow{\text{(5.167a)}} \mathbf{k}_p(t+1) \ . \qquad (5.168)$$

We remember the analogous situation in the case of the fast Kalman algorithm (5.161). Utilizing the covariance expression (5.26b) of the RLS algorithm, one sees that [assuming $\lambda = 1$]

$$\alpha_{p+1}(t+1) = 1 + \mathbf{z}_{p+1}^T(t+1)\,\mathbf{k}_{p+1}(t+1) \ . \qquad (5.169)$$

Substituting the partitioning scheme (5.167b) into (5.169), and exploiting the state vector partitioning scheme (5.165), we may express the covariance at stage p+1 and time step t+1 as

$$\alpha_{p+1}(t+1) = 1 + \left[x(t) \ \ \mathbf{z}_p^T(t) \right] \begin{bmatrix} 0 \\ \mathbf{k}_p(t) \end{bmatrix}$$

$$+ \ \frac{\varepsilon_p^f(t+1)}{\alpha_p^f(t)} \left[x(t) \ \ \mathbf{z}_p^T(t) \right] \begin{bmatrix} 1 \\ -\,\mathbf{f}^{(p)}(t) \end{bmatrix} \qquad (5.170)$$

or, equivalently,

$$\alpha_{p+1}(t+1) = 1 + \mathbf{z}_p^T(t)\mathbf{k}_p(t) + \frac{\varepsilon_p^f(t+1)}{\alpha_p^f(t)} \left[x(t) - \mathbf{z}_p^T(t)\,\mathbf{f}^{(p)}(t) \right] \ . \qquad (5.171)$$

Recalling from (5.26b) and (5.123a) that $\alpha_p(t) = 1 + \mathbf{z}_p^T(t)\mathbf{k}_p(t)$ and $\varepsilon_p^f(t+1) = x(t) - \mathbf{z}_p^T(t)\,\mathbf{f}^{(p)}(t)$, we reduce (5.171) to

$$\alpha_{p+1}(t+1) = \alpha_p(t) + \frac{\varepsilon_p^f(t+1)}{\alpha_p^f(t)}\,\varepsilon_p^f(t+1) \ . \qquad (5.172)$$

Similarly, we can start with the partitioning scheme (5.167a) to find

$$\alpha_{p+1}(t+1) = 1 + \left[\mathbf{z}_p^T(t+1) \ \ x(t-p) \right] \begin{bmatrix} \mathbf{k}_p(t+1) \\ 0 \end{bmatrix}$$

$$+ \ \frac{\varepsilon_p^b(t+1)}{\alpha_p^b(t)} \left[\mathbf{z}_p^T(t+1) \ \ x(t-p) \right] \begin{bmatrix} -\,\mathbf{b}^{(p)}(t) \\ 1 \end{bmatrix} \qquad (5.173)$$

or, equivalently,

$$\alpha_{p+1}(t+1) = 1 + z_p^T(t+1) k_p(t+1)$$

$$+ \frac{\varepsilon_p^b(t+1)}{\alpha_p^b(t)} \left[x(t-p) - z_p^T(t+1) b^{(p)}(t) \right] . \tag{5.174}$$

Recalling that $1 + z^T(t+1) k_p(t+1) = \alpha_p(t+1)$ and, according to (5.121a), $x(t-p) - z_p^T(t+1) b^{(p)}(t) = \varepsilon_p^b(t+1)$, we finally obtain a second update equation for the covariance $\alpha_{p+1}(t+1)$ as

$$\alpha_{p+1}(t+1) = \alpha_p(t+1) + \frac{\varepsilon_p^b(t+1)}{\alpha_p^b(t)} \varepsilon_p^b(t+1) . \tag{5.175}$$

From (5.166 and 169) one may deduce the following updating scheme of the error covariance $\alpha_p(t+1)$:

$$\alpha_p(t) \xrightarrow{\quad (5.172) \quad} \alpha_{p+1}(t+1) \xrightarrow{\quad (5.175) \quad} \alpha_p(t+1) . \tag{5.176}$$

Furthermore, one sees that an expression for $\varepsilon_p^b(t+1) / \alpha_p^b(t)$ can be derived from (5.167a) if we set up the partitioning scheme

$$k_{p+1}(t+1) = \begin{bmatrix} d_p(t+1) \\ \\ \delta_p(t+1) \end{bmatrix}$$

$$= \begin{bmatrix} k_p(t+1) - \left(\varepsilon_p^b(t+1) / \alpha_p^b(t) \right) b^{(p)}(t) \\ \\ \left(\varepsilon_p^b(t+1) / \alpha_p^b(t) \right) \end{bmatrix} . \tag{5.177}$$

Expression (5.177) reveals the close relationship between the FAEST algorithm and the fast Kalman algorithm. The partitioning scheme (5.177) is the expression analogous to (5.127 and 128a,b) in the case of the fast Kalman algorithm. The main difference between the two algorithms is mainly that the fast Kalman algorithm is based on the updating of the modified Kalman gain k^*, whereas the FAEST algorithm uses the Kalman gain k.

From (5.177), one may directly deduce the expressions

$$d_p(t+1) = k_p(t+1) - \left(\varepsilon_p^b(t+1) / \alpha_p^b(t) \right) b^{(p)}(t) , \tag{5.178a}$$

$$\delta_p(t+1) = \left(\varepsilon_p^b(t+1) / \alpha_p^b(t) \right) . \tag{5.178b}$$

Substituting (5.178b) into (5.175) gives an alternative formulation of the error covariance update

$$\alpha_{p+1}(t+1) = \alpha_p(t+1) + \delta_p(t+1) \, \varepsilon_p^b(t+1) \; . \tag{5.179}$$

A comparison of (5.121c) and (5.121d) shows that

$$k_p^*(t) \varepsilon_p^b(t) = k_p(t) e_p^b(t) \; , \tag{5.180}$$

which, together with (5.26c), yields

$$e_p^b(t) = \varepsilon_p^b(t) \big/ \alpha_p(t) \; . \tag{5.181}$$

A remarkable consequence of (5.181) is that the a posteriori errors at the actual time instant can be computed *before* the prediction error filter produces them, i.e., before the updating of the prediction error filter parameter set. As a consequence, it is possible that the prediction error filter parameters can be updated via the a posteriori errors as follows:

$$a^{(p)}(t) = a^{(p)}(t-1) + k_p(t) e_p(t) \; . \tag{5.182}$$

Finally, we explore the relationship between the a priori and a posteriori forward prediction errors. A simple inspection of (5.123c and d) reveals that

$$k_p^*(t-1) \varepsilon_p^f(t) = k_p(t-1) e_p^f(t) \; . \tag{5.183}$$

Substituting (5.26c) into (5.183) gives

$$e_p^f(t) = \varepsilon_p^f(t) \big/ \alpha_p(t-1) \; . \tag{5.184}$$

Table 5.5 summarizes this fast RLS algorithm, which has become known as the *fast a posteriori error sequential technique* (FAEST) [5.10]. The FAEST algorithm starts again with the computation of the a priori forward prediction error ε_p^f according to (5.123a). In contrast to the fast Kalman algorithm, which computes the forward prediction error filter *explicitly*, the FAEST method employs the *implicit* computation of the a posteriori forward prediction error e_p^f according to (5.184). The forward prediction parameter vector $f^{(p)}$ can now be updated in time according to (5.123d). The update of the forward prediction error covariance α_p^f is then achieved by (5.160). Subsequently, one computes the intermediate vector d_p and the scalar quantity δ_p from a partitioning (5.177) of the vector update (5.167b). The intermediate quantity δ_p so obtained is used in the computation of the a priori backward prediction error ε_p^b according

Table 5.5. Fast a posteriori error sequential technique (FAEST). The Kalman gain update requires 5p multiplications or divisions per recursion whereas LS FIR filtering requires 7p multiplications or divisions per recursion. This algorithm involves division. Whenever the divisor is small, set $1/x = x$

Initialize:

$$\mathbf{k}_p(-1) = \mathbf{f}^{(p)}(-1) = \mathbf{b}^{(p)}(-1) = \mathbf{a}^{(p)}(-1) = \mathbf{z}_p(-1) = \begin{bmatrix} 0 \ldots 0 \end{bmatrix}^T$$

$$\alpha_p^f(-1) = \alpha_p^b(-1) = \alpha_p(-1) = \sigma; \qquad \sigma > 0; \qquad \sigma \ll 1$$

FOR $t = 0, 1, 2, \ldots$

Input: $\mathbf{z}_p(t), x(t), 0 \le \lambda \le 1$

$$\varepsilon_p^f(t) = x(t-1) - \mathbf{z}_p^T(t-1)\,\mathbf{f}^{(p)}(t-1) \qquad \text{a priori forward prediction error} \qquad (5.185a)$$

$$e_p^f(t) = \varepsilon_p^f(t)/\alpha_p^f(t-1) \qquad \text{a posteriori forward prediction error} \qquad (5.185b)$$

$$\mathbf{f}^{(p)}(t) = \mathbf{f}^{(p)}(t-1) + \mathbf{k}_p(t-1)e_p^f(t) \qquad \text{forward predictor update} \qquad (5.185c)$$

$$\alpha_p^f(t) = \lambda\,\alpha_p^f(t-1) + e_p^f(t)\,\varepsilon_p^f(t) \qquad \text{forward prediction error covariance} \qquad (5.185d)$$

$$\begin{bmatrix} \mathbf{d}_p(t) \\ \delta_p(t) \end{bmatrix} = \begin{bmatrix} 0 \\ \mathbf{k}_p(t-1) \end{bmatrix} + \frac{\varepsilon_p^f(t)}{\alpha_p^f(t-1)} \begin{bmatrix} 1 \\ -\mathbf{f}^{(p)}(t-1) \end{bmatrix} \qquad \text{partitioning} \qquad (5.185e)$$

$$\varepsilon_p^b(t) = \delta_p(t)\,\alpha_p^b(t-1) \qquad \text{a priori backward prediction error} \qquad (5.185f)$$

$$\mathbf{k}_p(t) = \mathbf{d}_p(t) + \delta_p(t)\,\mathbf{b}^{(p)}(t-1) \qquad \text{Kalman gain update} \qquad (5.185g)$$

$$\alpha_{p+1}(t) = \alpha_p(t-1) + \frac{\varepsilon_p^f(t)}{\alpha_p^f(t-1)}\,\varepsilon_p^f(t) \qquad \text{error covariance} \qquad (5.185h)$$

$$\alpha_p(t) = \alpha_{p+1}(t) - \delta_p(t)\varepsilon_p^b(t) \qquad \text{error covariance} \qquad (5.185i)$$

$$e_p^b(t) = \varepsilon_p^b(t)/\alpha_p(t) \qquad \text{a posteriori backward prediction error} \qquad (5.185j)$$

$$\alpha_p^b(t) = \lambda\alpha_p^b(t-1) + \varepsilon_p^b(t)e_p^b(t) \qquad \text{backward prediction error covariance} \qquad (5.185k)$$

$$\mathbf{b}^{(p)}(t) = \mathbf{b}^{(p)}(t-1) + \mathbf{k}_p(t)e_p^b(t) \qquad \text{backward predictor update} \qquad (5.185l)$$

$$\varepsilon_p(t) = x(t) - \mathbf{z}_p^T(t)\mathbf{a}^{(p)}(t-1) \qquad \text{a priori error of prediction filter} \qquad (5.185m)$$

$$e_p(t) = \varepsilon_p(t)/\alpha_p(t) \qquad \text{a posteriori error of prediction filter} \qquad (5.185n)$$

$$\mathbf{a}^{(p)}(t) = \mathbf{a}^{(p)}(t-1) + \mathbf{k}_p(t)e_p(t) \qquad \text{prediction filter coefficient update} \qquad (5.185o)$$

---

to (5.178b). The computation of the Kalman gain vector $\mathbf{k}_p$ is directly obtained from (5.178a). The error covariance α_{p+1} is updated via (5.172) and the recursion (5.179) is finally employed in the computation of α_p. As a consequent step in the algorithm, one computes the a posteriori backward prediction error e_p^b *implicitly* according to (5.181). The time update of the backward prediction error covariance α_p^b is then given by (5.144). The alternative update equation (5.121d) of the backward predictor parameter vector $\mathbf{b}^{(p)}$ using Kalman gains and a posteriori errors concludes the fast algorithm for updating the Kalman gain vector $\mathbf{k}_p$ in time. In the case that least-squares FIR filtering, prediction, or parameter estimation is an issue, one computes additionally the a priori prediction error filter (5.162i). The a posteriori error is again obtained by the implicit computation (5.29). The transversal predictor parameters are finally updated from Kalman gains and a posteriori errors according to (5.27). Note that we have simply cancelled the λ factor in (5.29) with the factor λ^{-1} appearing in (5.27).

5.6 Descent Transversal Algorithms

Besides the closed form solutions of the recursive least-squares problem, *iterative* approaches are still of interest. A very early and obvious iterative procedure is the well-known *Newton algorithm*. This traditional algorithm, which stems from approximation theory, does not have a real application in linear prediction theory, but it can serve as a useful starting point in the derivation of more easily computable iterative algorithms, which are, in fact, all derivable from the classical Newton algorithm by a chain of useful approximations. In this way we will first derive the *steepest descent algorithm* as a special case of the Newton algorithm, where the inverse Hessian matrix is replaced by the identity matrix multiplied by a constant stepsize μ. In the following step we will show how, in turn, the widely used *least mean squares* (LMS) *algorithm* can be conveniently derived as a special case of the steepest descent algorithm, where the system matrix and the right-hand side of the Normal Equations, as appearing in the steepest descent algorithm, are replaced by their *instantaneous estimates*. This way, the LMS algorithm as a very efficient algorithm in terms of a small number of computations is obtained. A detailed analysis of *stability* and *convergence* of the algorithms will accompany the derivation.

Finally, we demonstrate how the LMS algorithm is related to the RLS algorithm simply by setting the modified Kalman gain vector equal to the state vector of the prediction error filter weighted by the stepsize factor μ and thereby omitting the computationally burdensome updating of the Kalman gain vector.

We shall start with a consideration of the squared prediction error as a performance surface, for the discussed iterative algorithms (recall Chap. 2). Reproducing the mean-square error performance surface as originally stated in (2.11),

$$E(t,a(t)) = \mathbf{x}^T(t)\mathbf{x}(t) - \mathbf{a}^T(t)\mathbf{X}^T(t)\mathbf{x}(t) - \mathbf{x}^T(t)\mathbf{X}(t)\mathbf{a}(t)$$

$$+ \mathbf{a}^T(t)\mathbf{X}^T(t)\mathbf{X}(t)\mathbf{a}(t) \quad , \tag{5.186}$$

we see that $E(t,a(t))$ is a function of time and a *quadratic* function of the parameter set $\mathbf{a}(t)$. Recall also Fig. 2.3. There $\mathbf{X}(t)$ was the matrix of past observations and $\mathbf{x}(t)$ was the actual process vector. Using again some convenient abbreviations,

$$\sigma_x^2(t) = \mathbf{x}^T(t)\mathbf{x}(t) \quad , \tag{5.187}$$

$$\mathbf{A}(t) = \mathbf{X}^T(t)\mathbf{X}(t) \quad , \tag{5.188}$$

$$\mathbf{h}(t) = \mathbf{X}^T(t)\mathbf{x}(t) \quad , \tag{5.189}$$

where $\sigma_x^2(t)$ is the actual process variance, $\mathbf{A}(t)$ is the system matrix of the Normal Equations and $\mathbf{h}(t)$ is the right-hand-side vector of the Normal Equations, we may rewrite (5.186) as

$$E(t,a(t)) = \sigma_x^2(t) - 2\mathbf{a}^T(t)\mathbf{h}(t) + \mathbf{a}^T(t)\mathbf{A}(t)\mathbf{a}(t) \tag{5.190}$$

with the underlying Normal Equations

$$\mathbf{A}(t)\,\mathbf{a}(t) = \mathbf{h}(t) \quad . \tag{5.191}$$

Solving (5.191) yields the desired coefficient set of the prediction error filter

$$\hat{\mathbf{a}}(t) = \mathbf{A}^{-1}(t)\mathbf{h}(t) \quad , \tag{5.192}$$

which corresponds to the minimum of the squared prediction error surface

$$E_{min}(t) = E(t,\hat{\mathbf{a}}(t)) \quad . \tag{5.193}$$

Note again that, throughout, we have assumed that $\mathbf{A}^{-1}(t)$ exists. Substituting the solution (5.192) into the expression of the squared prediction error (5.190) yields a relationship between the minimal squared error and the optimal coefficient set:

$$E_{min}(t) = \sigma_x^2(t) - \mathbf{h}^T(t)\hat{\mathbf{a}}(t) \quad . \tag{5.194}$$

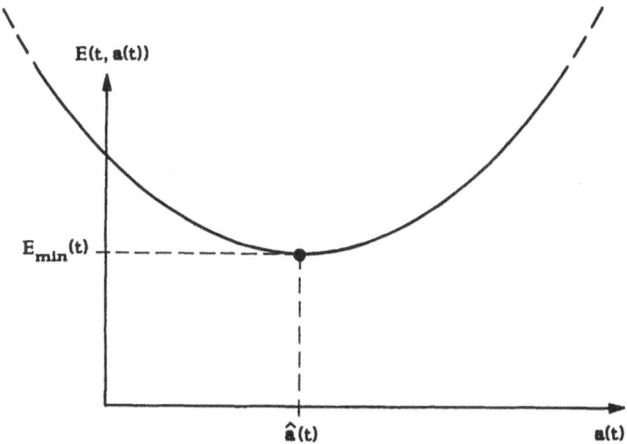

Fig. 5.2. Prediction error energy $E(t,a(t))$ as a convex function of the prediction error filter parameter set $a(t)$

Figure 5.2 illustrates the squared error performance surface as a function of the prediction error filter parameter set.

5.6.1 The Newton Algorithm

The classical Newton algorithm can be thought of as the father of many iterative algorithms. Using the Newton algorithm, the iterative determination of the optimal parameter set $\hat{a}(t)$ can be accomplished via the approach

$$\partial E(t,a_\nu(t))\big/\partial a_\nu(t) + \xi_\nu(t) = \begin{bmatrix} 0 \dots 0 \end{bmatrix}^T ;$$

$$\xi_\nu(t) \longrightarrow \begin{bmatrix} 0 \dots 0 \end{bmatrix}^T : \quad \hat{a}(t) = a_\nu(t) \ . \tag{5.195}$$

Expression (5.195) means that we are seeking the parameter set associated with a gradient of the performance surface that is the zero vector. Here, ν denotes an iteration index. Clearly, a simple inspection of Fig. 5.2 reveals that in a convex performance surface with one unique minimum the location of this minimum corresponds to a zero-crossing of the gradient of the performance surface.

Introducing a shorthand notation for the gradient,

$$\nabla(t,a(t)) = \partial E(t,a(t))\big/\partial a(t) \ , \tag{5.196}$$

we see that the gradient is a function of the parameter set, the system matrix and the right-hand-side vector of the Normal Equations as follows:

$$\nabla(t,a(t)) = -2h(t) + 2A(t)a(t) \ . \tag{5.197}$$

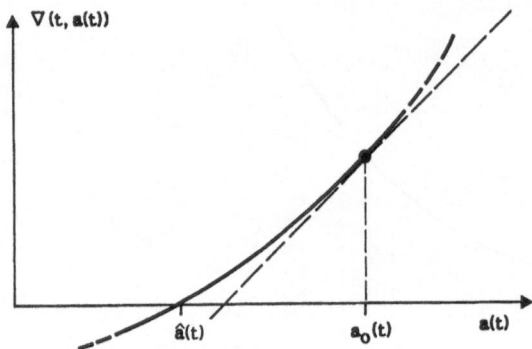

Fig. 5.3. Linearization of the gradient $\nabla(t,a(t))$ around the point $a(t) = a_0(t)$

An iterative procedure for finding the zero-crossing of $\nabla(t,a(t))$ can be established by a successive evaluation of the *linearized gradient* $\overline{\nabla}(t,a(t))$

$$\overline{\nabla}(t,a(t)) = \nabla(t,a_0(t)) + \frac{\partial \nabla(t,a(t))}{\partial a(t)}\bigg|_{a(t) = a_0(t)} \left(a(t) - a_0(t)\right) , \qquad (5.198)$$

where linearization has been carried out at the point $a(t) = a_0(t)$. Figure 5.3 illustrates the linearization of the gradient $\nabla(t,a(t))$ around the point $a(t) = a_0(t)$.

In order to set up an iterative procedure for finding $\hat{a}(t)$, one has to determine the zero-crossing of the linearized gradient by setting (5.198) to zero:

$$\overline{\nabla}(t,a(t)) = \nabla(t,a_0(t)) + \frac{\partial \nabla(t,a(t))}{\partial a(t)}\bigg|_{a(t) = a_0(t)} \left(a(t) - a_0(t)\right)$$

$$= \left[0 \ldots 0\right]^T. \qquad (5.199)$$

Using the abbreviation

$$\Delta(t,a_\nu(t)) = \frac{\partial \nabla(t,a(t))}{\partial a(t)}\bigg|_{a(t) = a_\nu(t)}$$

$$= 2\,A(t) = \Delta(t) \qquad [constant\ over\ a(t)] \qquad (5.200)$$

for the Hessian matrix at position $a(t) = a_\nu(t)$, we can solve (5.199) for the desired next step $a_{\nu+1}(t)$ towards an iterative solution of the Normal Equations:

$$a_{\nu+1}(t) = a_\nu(t) - \Delta^{-1}(t)\,\nabla(t,a_\nu(t)) . \qquad (5.201)$$

In general, with every new iteration, we get closer to the optimum $\hat{a}(t)$. Figure 5.3 illustrates this method of successive approximation of $\hat{a}(t)$ using formula (5.201). This algorithm, which has become known as the *Newton algorithm* , is summarized in Table 5.6.

Table 5.6. Newton algorithm for iterative computation of the parameter vector $\hat{a}(t)$

FOR t = 0, 1, 2, . . .

  Input: $A(t)$, $h(t)$

  Initialize: $a_0(t) = \begin{bmatrix} 0. . .0 \end{bmatrix}^T$ (or other appropriate choice)

  $\Delta^{-1}(t) = (1/2)\,A^{-1}(t)$ (5.202a)

  FOR $\nu = 0, 1, 2, . . . , n - 1$

    $\nabla(t, a_\nu(t)) = -2h(t) + 2A(t)a_\nu(t)$ (5.202b)

    $a_{\nu+1}(t) = a_\nu(t) - \Delta^{-1}(t)\nabla(t, a_\nu(t))$ *iteration*   (5.202c)

    $a_\nu(t) = a_{\nu+1}(t)$ *update of solution vector*   (5.202d)

  $\hat{a}(t) \approx a_{n-1}(t)$ *final solution*   (5.202e)

The Newton algorithm, as listed in Table 5.6 , seems to be a very *unpractical* algorithm since it requires, at each recursion, the inversion of the system matrix of the Normal Equations. Once this inverse is known, one could compute the solution directly via (5.192). Indeed, a simple inspection of the algorithm listed in Table 5.6 reveals that substitution of (5.202a and b) into (5.202c) results in an update term that is zero, that means $a_{\nu+1}(t) = a_\nu(t)$, which indicates that the Newton algorithm converges to the minimum of a quadratic performance surface within only a *single* iteration. Clearly, the gradient (5.197) of the quadratic performance surface (5.190), as appearing in the linear prediction problem, is already a linear function of the parameter set $a(t)$. This indicates that a further linearization of the gradient is unnecessary in this particular problem, and we may be able to derive much faster iterative schemes for solving the Normal Equations.

A situation of practical interest appears as soon as the process correlation is not completely known and in those cases where $A(t)$ changes only smoothly with time. In these cases, the inverse of the system matrix can be set *constant* and need not be updated at each recursion.

5.6.2 The Steepest Descent Algorithm

The analysis of the Newton approach for setting up iterative algorithms for solving the Normal Equations shows that this approach suffers mainly from the requirement of computing the inverse of a Hessian matrix. Loosely interpreted, the Newton algorithm requires the inversion of the Normal Equations *before* solving them iteratively. In order to find a way out of this difficulty, we make the useful approximation

$$\Delta^{-1}(t) = \frac{1}{2}\mu I \quad , \tag{5.203}$$

where μ is a constant stepsize and I is the identity matrix of dimension $p \times p$. This approach simplifies recursion (5.202c) as follows:

$$a_{\nu+1}(t) = a_\nu(t) - \frac{1}{2}\mu \nabla(t, a_\nu(t)) \quad . \tag{5.204}$$

Using (5.197), we can find an alternative formulation of the iteration (5.204) in terms of the system matrix and the right-hand-side vector of the Normal Equations:

$$a_{\nu+1}(t) = \left(I - \mu A(t)\right) a_\nu(t) + \mu h(t) \quad . \tag{5.205}$$

In many applications, there is only time for a single iteration per recursion. In this case, (5.205) may be rewritten as

$$a(t) = \left(I - \mu A(t)\right) a(t-1) + \mu h(t) \quad . \tag{5.206}$$

This algorithm, as represented by expression (5.206), has become known as the *steepest descent algorithm*. Figure 5.4 is a signal flow diagram representation of the steepest descent algorithm determined by expression (5.206). Table 5.7 is a summary of the steepest descent algorithm.

Table 5.7. Steepest descent algorithm for iterative solution of the Normal Equations

---

Initialize: $a(-1) = \begin{bmatrix} 0 \ldots 0 \end{bmatrix}^T$ (or other appropriate choice)

FOR $t = 0, 1, 2, \ldots$

> *Input:* $x(t)$
>
> *update of :* $A(t)$; $h(t)$
>
> $a(t) = \left(I - \mu A(t)\right) a(t-1) + \mu h(t)$ *parameter update*   (5.207a)
>
> $e(t) = x(t) - z^T(t)a(t)$ *prediction error filter*   (5.207b)

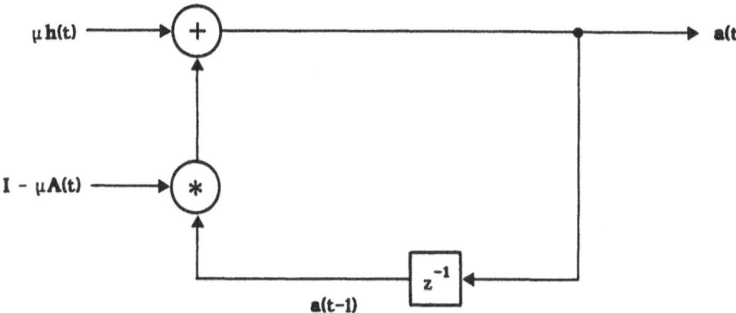

Fig. 5.4. Signal flow diagram representation of the steepest descent algorithm

The steepest descent algorithm, as listed in Table 5.7, is still a computationally burdensome procedure since it requires, at each recursion, the updating of the system matrix and right-hand-side vector of the Normal Equations along with the computation of the $O(p^2)$ parameter update (5.207a).

5.6.3 Stability of the Steepest Descent Algorithm

A simple analysis of the signal flow diagram representation of the steepest descent algorithm in Fig. 5.4 shows that the algorithm is a first-order recursive system that is – of course – not unconditionally stable. Stability depends mainly on the loop gain, which is itself a function of the stepsize μ and the system matrix $\mathbf{A}(t)$. As long as the system matrix is assumed to be time-varying, it is difficult to carry out a stability analysis.

To make the problem analytically tractable, we assume a time-invariant system matrix and right-hand-side vector of the Normal Equations. As a consequence of this assumption, the optimal solution will also be time invariant, hence

$$\mathbf{A} = \mathbf{A}(t) \ , \tag{5.208a}$$

$$\mathbf{h} = \mathbf{h}(t) \ , \tag{5.208b}$$

$$\mathbf{A}\hat{\mathbf{a}} = \mathbf{h} \ . \tag{5.209}$$

Under this assumption of a time-invariant system matrix and right-hand-side vector of the Normal Equations, the steepest descent algorithm of (5.206) reduces to

$$\mathbf{a}(t) = \left(\mathbf{I} - \mu\mathbf{A}\right)\mathbf{a}(t-1) + \mu\mathbf{h} \ . \tag{5.210}$$

Introducing now the *tap error vector* $\mathbf{c}(t)$

$$\mathbf{c}(t) = \mathbf{a}(t) - \hat{\mathbf{a}} \ , \tag{5.211}$$

we note that

$$\mathbf{c}(t-1) = \mathbf{a}(t-1) - \hat{\mathbf{a}} \tag{5.212}$$

because of the time-invariance assumption. Substitution of (5.211 and 212) into (5.210) yields

$$\mathbf{c}(t) + \hat{\mathbf{a}} = \left(\mathbf{I} - \mu\mathbf{A}\right)\left(\mathbf{c}(t-1) + \hat{\mathbf{a}}\right) + \mu\mathbf{h} \ , \tag{5.213}$$

which reduces to an expression for the propagation of the tap error vector $\mathbf{c}(t)$ inside the recursive loop of the steepest descent algorithm:

$$\mathbf{c}(t) = \left(\mathbf{I} - \mu\mathbf{A}\right)\mathbf{c}(t-1) \ . \tag{5.214}$$

Clearly, the problem is to show that the norm of $\mathbf{c}(t)$ converges to zero, which indicates that the algorithm converges stably to the optimal solution. Obviously, the stability of the recursive system described by (5.214) depends solely on the values of μ and $\mathbf{A}$.

Using the unitary similarity transform of $\mathbf{A}$, we may express the system matrix $\mathbf{A}$ as [recall (3.94)]

$$\mathbf{A} = \mathbf{V}\Omega\mathbf{V}^{\mathrm{T}} \ . \tag{5.215}$$

The columns of $\mathbf{V}$ are an orthogonal set of *eigenvectors* and $\Omega = \mathbf{S}^{\mathrm{T}}\mathbf{S}$ is a diagonal matrix containing the associated *eigenvalues* of the system matrix $\mathbf{A}$, hence

$$\Omega = \begin{bmatrix} \omega_1 & & & \\ & \omega_2 & & \\ & & \ddots & \\ & & & \omega_p \end{bmatrix} \ , \tag{5.216}$$

where $\omega_1, \omega_2, \ldots, \omega_p$ is the sequence of eigenvalues of the system matrix $\mathbf{A}$. These eigenvalues are assumed to be all positive and real. Using the *eigenvalue decomposition* (5.215), we may express the update equation of the tap error vector (5.214) as

$$\mathbf{c}(t) = \left(\mathbf{I} - \mu\mathbf{V}\Omega\mathbf{V}^{\mathrm{T}}\right)\mathbf{c}(t-1) \ . \tag{5.217}$$

A left-side multiplication of (5.217) by $\mathbf{V}^{\mathrm{T}}$ along with the orthogonality relation

$$\mathbf{V}\,\mathbf{V}^T = \mathbf{V}^T\mathbf{V} = \mathbf{I} \qquad\qquad (5.218)$$

gives

$$\mathbf{V}^T\mathbf{c}(t) = \left(\mathbf{I} - \mu\Omega\right)\mathbf{V}^T\mathbf{c}(t-1) \ . \qquad\qquad (5.219)$$

Defining the *transformed* tap error vector

$$\mathbf{d}(t) = \mathbf{V}^T\mathbf{c}(t) \ , \qquad\qquad (5.220)$$

we note that a *Parseval's relation* holds in that

$$\mathbf{d}^T(t)\mathbf{d}(t) = \mathbf{c}^T(t)\,\mathbf{V}\,\mathbf{V}^T\mathbf{c}(t) = \mathbf{c}^T(t)\,\mathbf{c}(t) \ . \qquad\qquad (5.221)$$

Equation (5.221) leads to the important conclusion that it is sufficient to investigate the convergence of the transformed tap error vector $\mathbf{d}(t)$, since the convergence of $\mathbf{d}(t)$ automatically yields the convergence of $\mathbf{c}(t)$.

Rewriting (5.219) in terms of the transformed tap error vector $\mathbf{d}(t)$ gives

$$\mathbf{d}(t) = \left(\mathbf{I} - \mu\Omega\right)\mathbf{d}(t-1) \ . \qquad\qquad (5.222)$$

Since Ω is a diagonal matrix, (5.222) yields a set of p *decoupled* first order recursive scalar feedback models of the type

$$d_k(t) = \left(1 - \mu\omega_k\right)d_k(t-1) \ ; \qquad 1 \le k \le p \qquad\qquad (5.223)$$

where ω_k is the kth eigenvalue of $\mathbf{A}$. Figure 5.5 illustrates the scalar feedback model determined by (5.223). The variable $d_k(t)$ represents the *kth natural mode* of the steepest descent algorithm.

Clearly, for stability it is required that all p natural modes of the steepest descent algorithm have sufficient *"damping"* in order to

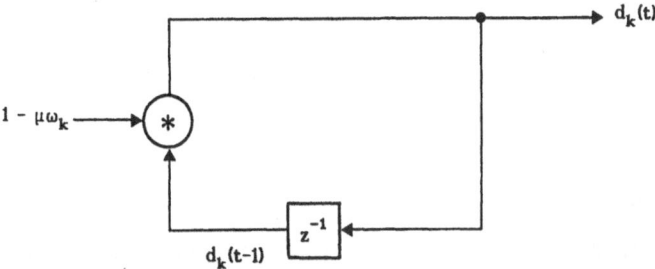

Fig. 5.5. Scalar feedback model of the kth component $d_k(t)$ of the transformed tap error vector $\mathbf{d}(t)$

guarantee the *convergence* of the norm of the transformed tap error vector $\mathbf{d}(t)$ in the loop of the steepest descent algorithm. For all natural modes, stability can be guaranteed if the loop gain for each mode has a magnitude less than unity, and therefore

$$| 1 - \mu\omega_k | < 1 , \qquad \omega_k > 0 ; \qquad 1 \leq k \leq p , \tag{5.224}$$

which gives the condition

$$0 < \mu < 2/\omega_k ; \qquad\qquad 1 \leq k \leq p , \tag{5.225}$$

for stability of mode k. Since we have only a single stepsize that acts on all modes, the value of μ will be bounded on the *largest* eigenvalue ω_{max} as

$$0 < \mu < 2/\omega_{max} , \tag{5.226}$$

which is the final stability condition of the steepest descent algorithm. Note again that this stability criterion has been de ived under the assumption of time-invariant data. In nonstationary ϵ onments, the value of μ might be chosen considerably smaller than   cated by the bound (5.226).

5.6.4 Convergence of the Steepest Descent Algorithm

Referring to Fig. 5.5, we see that, in the case of sta   , the feedback system shown in Fig. 5.5 performs an *exponential* deca_   the kth natural mode $d_k(t)$ of the steepest descent algorithm. Hence the magnitude of $d_k(t+n)$ can be calculated by

$$d_k(t+n) = (1 - \mu\omega_k)^n d_k(t) . \tag{5.227}$$

On the other hand, we can define a *time constant* of the kth natural mode of the steepest descent algorithm as

$$1 - \mu\omega_k = \exp\left(- \frac{1}{\tau_k} \right) . \tag{5.228}$$

The time constant τ_k can be expressed in terms of the stepsize parameter μ and the kth eigenvalue ω_k as

$$\tau_k = \frac{-1}{\ln(1 - \mu\omega_k)} . \tag{5.229}$$

The time constant τ_k defines the time required for the amplitude of the kth natural mode $d_k(t)$ to decay to $1/e$ of its initial value. For a very small stepsize μ, we may approximate the time constant τ_k as

$$\tau_k = 1/(\mu \omega_k) \; ; \qquad \mu \ll 1 \; . \tag{5.230}$$

We are finally interested in the transient behavior of the original parameter vector $a(t)$. Premultiplying both sides of (5.220) by V and taking into account (5.211) we find

$$V\,d(t) = V\,V^T c(t) = c(t) = a(t) - \hat{a} \; . \tag{5.231}$$

Solving (5.231) for $a(t)$, we obtain the desired result

$$a(t) = \hat{a} + V\,d(t) \; . \tag{5.232}$$

With

$$V = \left[v_1, v_2, \ldots , v_p \right] \tag{5.233}$$

we may rewrite (5.232) to obtain

$$a(t) = \hat{a} + \sum_{k=1}^{p} v_k \, d_k(t) \; , \tag{5.234}$$

where $v_1, v_2, \ldots , v_p$ is the sequence of *eigenvectors* associated with the sequence of *eigenvalues* $\omega_1, \omega_2, \ldots , \omega_p$ of the system matrix A. Substituting the decay law (5.227) into (5.234), one obtains the parameter vector $a(t+n)$ at time $t+n$, after n iterations, as

$$a(t+n) = \hat{a} + \sum_{k=1}^{p} v_k (1 - \mu \omega_k)^n d_k(t) \; . \tag{5.235}$$

Equation (5.235) shows that the parameter vector $a(t)$ converges as the *weighted sum* of exponentials of the form $(1 - \mu \omega_k)^n$. Although the time constant of each of these components is given by (5.229), the summation term in (5.235) cannot be expressed in a simple closed form like (5.229). Nevertheless, the slowest rate of convergence is attained when $d_k(t)$ is zero for all k, except for one k associated with the smallest eigenvalue ω_{min} of A; so the upper bound on τ_a is defined by $-1/\ln(1 - \mu \omega_{min})$. The fastest rate of convergence is attained when all $d_k(t)$ are zero except the one that corresponds to the largest eigenvalue ω_{max}, and so the lower bound on τ_a is defined by $-1/\ln(1 - \mu \omega_{max})$. Accordingly, the overall time constant τ_a for any parameter vector of the steepest descent algorithm is bounded as follows:

$$-1/\ln(1 - \mu \omega_{max}) \le \tau_a \le -1/\ln(1 - \mu \omega_{min}) \; . \tag{5.236}$$

When the eigenvalues of the system matrix A are *widely spread*, i.e., in the case of an ill-conditioned system matrix, the settling time of the steepest descent algorithm is limited by the *smallest* eigenvalue or the *slowest natural mode*.

5.6.5 The Least Mean Squares Algorithm

The steepest descent algorithm, as defined by (5.206), requires at each recursion the updating of the system matrix and the right-hand-side vector of the Normal Equations and an additional multiplication of the system matrix $A(t)$ with the actual parameter vector $a(t-1)$. This involves a computational complexity of $O(p^2)$ operations per recursion.

A much simpler algorithm for updating the parameter vector from the incoming data can be developed if we just replace the true system matrix $A(t)$ and the true right-hand-side vector $h(t)$ by their *instantaneous estimates*, which are based on the sample values of the state vector $z(t)$ and the actual process sample $x(t)$ as

$$A(t) \approx z(t)\, z^T(t) \ , \tag{5.237}$$

$$h(t) \approx z(t)\, x(t) \ . \tag{5.238}$$

Substituting these approximations into the steepest descent algorithm (5.206), we obtain

$$a(t) = a(t-1) + \mu\, z(t)\left[x(t) - z^T(t)\, a(t-1) \right] \ . \tag{5.239}$$

We see that the expression $x(t) - z^T(t)\, a(t-1)$ is just the *a priori error* of the transversal prediction error filter. See also (5.26e). Hence, (5.239) may be rewritten as a two-term recurrence relation

$$\varepsilon(t) = x(t) - z^T(t)\, a(t-1) \ , \tag{5.240a}$$

$$a(t) = a(t-1) + \mu\, z(t)\, \varepsilon(t) \ . \tag{5.240b}$$

This algorithm has become known as the *least mean squares* (LMS) algorithm [5.12-14]. Sometimes, this algorithm is alternatively referred as the *"stochastic" gradient* (SG) algorithm. A simple inspection of the recurrence relation (5.240a,b) reveals that the overall computational complexity of the LMS algorithm for adaptation and prediction error filtering is only 2p operations per recursion. This striking simplicity, rather than its performance, seems to be the explanation of why the LMS algorithm is still the most widely accepted algorithm for recursive adaptive filtering, prediction and parameter tracking. Table 5.8 is a summary of the LMS algorithm.

We now wish to explore the relationships between the LMS algorithm of Table 5.8 and the RLS algorithm of Table 5.1. A quick inspection of the RLS algorithm of Table 5.1 gives rise to the statement that the LMS algorithm of Table 5.8 is just a simplified version of the RLS algorithm, since it can be directly derived from the RLS algorithm by means of the approximation

Table 5.8. Least mean squares (LMS) algorithm. μ is a constant stepsize

---

initialize: $\mathbf{a}(-1) = \begin{bmatrix} 0 \ldots 0 \end{bmatrix}^T$

FOR $t = 0, 1, 2, \ldots$

$\qquad\begin{bmatrix}\; Input:\; x(t) \\[4pt] \;\varepsilon(t) = x(t) - \mathbf{z}^T(t)\mathbf{a}(t-1) \qquad prediction\ error\ filter \\[4pt] \;\mathbf{a}(t) = \mathbf{a}(t-1) + \mu\,\mathbf{z}(t)\,\varepsilon(t) \qquad parameter\ update \end{bmatrix}$

$\qquad\qquad\qquad\qquad\qquad\qquad\qquad\qquad\qquad\qquad\qquad\qquad\qquad$ (5.241a)

$\qquad\qquad\qquad\qquad\qquad\qquad\qquad\qquad\qquad\qquad\qquad\qquad\qquad$ (5.241b)

$$\mathbf{k}^*(t) = \mu\,\mathbf{z}(t) \quad . \tag{5.242}$$

Since the modified Kalman gain $\mathbf{k}^*(t)$ is no longer required, the computationally burdensome gain-updating can be completely *omitted*.

More generally, we observe the following chain of approximations by which the algorithms of this chapter are related to each other:

$$\text{Newton} \xrightarrow{(5.203)} \begin{array}{c}\text{Steepest}\\ \text{descent}\end{array} \xrightarrow{(5.237-238)} \text{LMS} \xleftarrow{(5.242)} \text{RLS} \; . \tag{5.243}$$

A final question concerns the choice of the stepsize μ in the case of the LMS algorithm. Clearly, the exact stability criterion (5.226) of the steepest descent algorithm may not apply since the system matrix and the right-hand-side vector have been approximated by their instantaneous estimates. However, Widrow et al. [5.13] have suggested the following sufficient condition for stability of the LMS algorithm:

$$0 < \mu < 2 \Big/ \sum_{j=1}^{P} A_{j,j} \quad . \tag{5.244}$$

On the other hand, the mean-squared prediction error produced by the LMS transversal prediction error filter (5.241a) converges exponentially with the dominant (i.e., largest) time constant τ_{LMS} being given by

$$\tau_{\text{LMS}} = 1 \Big/ (2\mu\omega_{\text{min}}) \quad . \tag{5.245}$$

When μ is chosen to ensure stability according to (5.226) we have

$$\tau_{\text{LMS}} \geq \omega_{\text{max}} \Big/ (4\omega_{\text{min}}) \quad . \tag{5.246}$$

Thus the larger the eigenvalue ratio, the slower will be the convergence of the LMS algorithm.

5.7 Chapter Summary

This chapter has shed some light on the linear prediction algorithms that are based on the *transversal* predictor structure. Both *direct solution* and *iterative solution* algorithms have been discussed. The direct solution of the recursive least-squares problem was solely based on the Sherman-Morrison identity, which is a basis for deriving recursive update equations for the *inverse* system matrix of the Normal Equations, thereby reducing the required computations to $O(p^2)$. An inherent difficulty is present in this approach since in many applications we cannot generally ensure that the inverse of the system matrix of the Normal Equations exists. Godard's RLS algorithm was presented as the first commonly recognized representative of this type of recursive least-squares algorithms. This forerunner position is also the why Godard's algorithm is exclusively referred to as *"The RLS algorithm"*.

Another drawback of algorithms based on the Sherman-Morrison identity is rooted in the fact that only a simple exponential decay, or a (more complex) rectangular window can be used. When the updating of exponentially weighted Normal Equations can be interpreted as a first-order recursive system with the pole inside the unit circle for ($\lambda < 1$), the updating of the *inverse* system matrix of the Normal Equations corresponds to a first-order recursive system with the pole located *outside* the unit circle. In fact, the pole locations for the two cases are related as *reflection points* with respect to the unit circle in the z-plane. These simple considerations ultimately lead to the insight that, although the recursive updating of the system matrix of the Normal Equations is an unconditionally stable process, the recursive updating of its inverse is an intrinsically *unstable* process.

Several methods have been discussed to reduce the problems arising from this fact [5.17, 20]. Among the techniques presented were the "selective memory" RLS algorithms recently introduced by Kubin. Strictly speaking, these algorithms circumvent the problem that the inverse system matrix diverges when the elements of the system matrix itself tend to zero in the case of, for example, the occurrence of a chain of zero samples in the observed process. Later in this chapter, we showed how "fast" $O(p)$ RLS algorithms can be derived by using the Sherman-Morrison identity for *partitioned* matrices. Two important algorithms, namely, the fast Kalman algorithm and the fast a posteriori error sequential technique (FAEST), were discussed and their derivation was presented with great care. It is hoped that this detailed discussion will remove some of the "mystery of the fast Kalman algorithm". In fact, it is interesting to see how the fast algorithms arise from clever partitioning schemes of the system matrix of the Normal Equations and to notice the interesting dualities between forward and backward linear prediction thereby revealed. Fast Kalman type algorithms have been treated extensively in the literature [5.21-23].

It seems in order to mention that - besides the pure algebraic derivation of fast transversal RLS algorithms - there exists an alternative derivation that relies solely on *geometrical* considerations. This way of treating the fast RLS transversal algorithms can be attributed to Cioffi and Kailath [5.9]. A detailed discussion of geometrical methods for deriving fast RLS transversal algorithms can also be found in the books by Honig and Messerschmitt [5.24] and Alexander [5.25]. These techniques have been adopted from the pioneer work of Lee [5.26], who first introduced geometrical and projection operator methods to derive the fast RLS ladder algorithms. Geometrical techniques are the most natural way of deriving the fast RLS ladder algorithms. This will be the topic of Chap. 9 of this book. It is emphasized, however, that algebraic and geometrical approaches apply for both the fast RLS transversal algorithms and for the fast RLS ladder algorithms. It is felt, however, that it is more natural to demonstrate the algebraic approach on the fast RLS transversal algorithms, while the demonstration of the geometrical approach is saved for the fast RLS ladder algorithms.

To conclude this chapter, we discussed the class of *iterative* algorithms for solving the Normal Equations. We started our considerations with the (unpractical) Newton algorithm for the purpose of better understanding the subsequent derivations. The well-known steepest descent algorithm and the widely used LMS algorithm have been introduced as members of a chain of successive approximations. The properties of the LMS algorithm have been studied most in detail. The gain parameter was considered by Bershad [5. 27]. A normalized LMS algorithm was introduced in [5.28]. Frequency domain LMS adaptive filters were discussed by Bershad and Feintuch [5.29] and Cowan and Grant [5.30]. We even showed how the LMS algorithm is related to the RLS algorithm of Godard. This brought us back to the subject of the beginning of the chapter and closed the circle of considerations about transversal algorithms.

6. The Ladder Form

So far, we have worked with the "fixed-order" formulation of the Normal Equations, i.e., we have set the order of the Normal Equations to a *fixed* value before solving them. The solution algorithms which required such a fixed-order description of the linear least-squares prediction problem were merely based on *algebraic* concepts , employing orthogonal rotations (as in the case of the QRLS algorithms based on the recursive Givens reduction), or on the Sherman-Morrison identity in connection with a *transversal* prediction error filter.

The name "normal", however, derives from *geometrical* considerations. Indeed in Sect. 2.2, we explored some of the geometrical background of the Normal Equations. Employing a geometrical interpretation, the problem of predicting an observed process vector $\mathbf{x}(t)$ by a linear combination of past observations $\mathbf{x}(t-1)$, $\mathbf{x}(t-2)$, . . . , $\mathbf{x}(t-m)$ was shown to be equivalent to the problem of projecting $\mathbf{x}(t)$ onto a subspace spanned by (oblique) basis vectors, which are just these vectors of past observations $\mathbf{x}(t-1)$, $\mathbf{x}(t-2)$, . . . , $\mathbf{x}(t-m)$. The predicted process $\hat{\mathbf{x}}^f(t)$ arises as a *weighted* linear combination of the oblique basis vectors and is therefore always an element of the subspace of past observations. The process vector $\mathbf{x}(t)$ is, in general, not an element of the subspace of past observations, and hence $\mathbf{x}(t)$ is generally *not perfectly predictable*. A prediction error $\mathbf{e}(t)$ will occur. The Normal Equations determine the *weighting coefficients* (predictor coefficients) associated with each basis vector of the subspace of past observations such that the prediction error $\mathbf{e}(t)$ is *orthogonal* with respect to the subspace of past observations, since an orthogonal error vector is of least Euclidean length and therefore satisfies the least-squares error criterion. See also Fig. 2.4 for an illustration of the geometrical character of the linear least-squares prediction problem.

These considerations give rise to the question whether one can develop an alternative way of solving the Normal Equations that makes a better use of the geometrical relationships that we have discussed. Indeed, an alternative way of constructing the orthogonal prediction error vector exists as follows. First, construct an *orthogonal basis* of the subspace of past observations, and second, *project* the vector $\mathbf{x}(t)$ successively onto the orthogonal basis vectors. This procedure is termed the

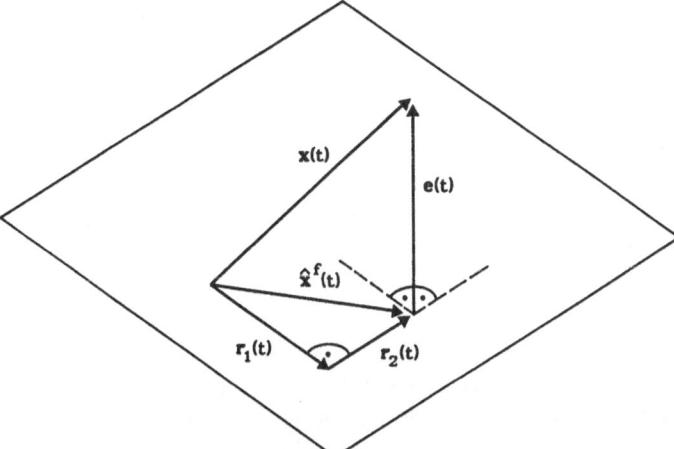

Fig. 6.1. Least-squares projection of a process $\mathbf{x}(t)$ onto a subspace of past observations spanned by two orthogonal basis vectors $\mathbf{r}_1(t)$ and $\mathbf{r}_2(t)$. Here $\hat{\mathbf{x}}^f(t)$ is the predicted process and $\mathbf{e}(t)$ is the prediction error of least Euclidean length, which is therefore orthogonal with respect to the subspace of past observations

"growing-order" or *ladder* formulation of the linear least-squares prediction problem. See Fig. 6.1 for an illustration of this alternative way of projecting the input process successively onto *orthogonalized* basis vectors. This approach leads to an alternative structure of the prediction error filter termed the *ladder form*. As a consequence of the fact that projections are performed onto orthogonal basis vectors, successive stages of the ladder form are *decoupled*, and we can always increase the filter order by adding some more stages while the "old" part of the predictor still remains *optimal*. This means in addition that there is no longer any necessity to work with Normal Equations of a fixed (predetermined) order. When working with ladder forms, we can vary the predictor order *dynamically* with any desired fidelity criterion guiding the decision whether an additional stage of the predictor results in a significant improvement of prediction or not.

   In the context of the structure of linear systems, the ladder forms are the most promising canonical forms for many classical linear systems problems. For instance, their states are *orthogonalized* or *decoupled* so that they become the "natural" canonical forms for many physical modeling problems. We just mention the acoustic tube model in speech processing [6.1], for which the ladder form turns out to be the most natural structural equivalent. This might be why the coefficients of the ladder form are frequently termed *"reflection coefficients"*. Besides these important properties, the ladder forms have many advantages, such as *modularity, robustness,* and *better numerical conditioning* compared to the conventional transversal prediction error filter.

We derive the ladder form by employing the *recursion formula for orthogonal projections,* a convenient formulation of the Gram-Schmidt orthogonalization procedure [6.2]. The most general nonstationary case is treated first. From these results, the *PARCOR ladder forms* [6.3] for stationary processes are derived in a unified manner. We discuss all important time and frequency domain aspects of the ladder forms, including the relationships between transversal forward/backward prediction and the ladder prediction error filter parameters. Moreover, the analysis/synthesis problem of stationary processes using ladder forms and a stability analysis of the feed-back (all-pole) PARCOR ladder form is described. Burg's harmonic mean PARCOR ladder algorithm [6.4] is introduced as an estimator that meets the stability requirements of the feed-back ladder form synthesis filter.

6.1 The Recursion Formula for Orthogonal Projections

The recursion formula for orthogonal projections will be the mathematical tool we use to derive the ladder form. This formula is constituted by the following theorem.

Theorem. (Recursion formula for orthogonal projections.) Let e_m be the complement of the orthogonal projection of a vector x onto a subspace B_m denoted by

$$e_m = x \langle B_m \rangle = x \langle B_{m-1}, b_m \rangle \tag{6.1}$$

with the property

$$e_m^T b_i = 0 ; \qquad 1 \le i \le m , \tag{6.2}$$

where

$$B_m = \left[b_1, b_2, \ldots, b_{m-1}, b_m \right] = \left[B_{m-1}, b_m \right] , \tag{6.3}$$

then the orthogonal complement e_m of the mth-order projection can be constructed *order-recursively* from the orthogonal complement e_{m-1} of the (m-1)th-order projection where

$$e_{m-1} = x \langle B_{m-1} \rangle \tag{6.4}$$

via the following *recursion formula for orthogonal projections:*

$$x\langle B_{m-1}, b_m\rangle = x\langle B_{m-1}\rangle + K_m b_m \langle B_{m-1}\rangle \tag{6.5}$$

or, equivalently,

$$e_m = e_{m-1} + K_m b_m \langle B_{m-1}\rangle \ , \tag{6.6}$$

where

$$K_m = -\ \frac{x^T\langle B_{m-1}\rangle\ b_m\langle B_{m-1}\rangle}{b_m^T\langle B_{m-1}\rangle\ b_m\langle B_{m-1}\rangle} \tag{6.7}$$

and K_m is a scalar parameter.

Proof. Let $e_m(K_m)$ be a vector constructed by the linear combination

$$e_m(K_m) = x\langle B_{m-1}\rangle + K_m b_m \langle B_{m-1}\rangle \ , \tag{6.8}$$

where K_m is a scalar parameter. Then, e_m is orthogonal with respect to the subspace B_{m-1} extended by the vector b_m and equivalently is orthogonal to B_{m-1} and $b_m\langle B_{m-1}\rangle$ if and only if the scalar parameter K_m is adjusted such that the Euclidean norm of $e_m(K_m)$ attains a minimum. This statement follows directly from our geometrical considerations of the Normal Equations, as provided in Sect. 2.2. This statement leads directly to a least-squares determination of K_m via the approach

$$E_m(K_m) = e_m^T e_m \overset{!}{=} min \longrightarrow K_m \ . \tag{6.9}$$

Substituting (6.8) into (6.9) gives

$$E_m(K_m) = \left[x^T\langle B_{m-1}\rangle + K_m b_m^T \langle B_{m-1}\rangle \right]$$

$$\times \left[x\langle B_{m-1}\rangle + K_m b_m\langle B_{m-1}\rangle \right]$$

$$= x^T\langle B_{m-1}\rangle\ x\langle B_{m-1}\rangle + K_m b_m^T \langle B_{m-1}\rangle\ x\langle B_{m-1}\rangle$$

$$+ K_m x^T\langle B_{m-1}\rangle\ b_m\langle B_{m-1}\rangle$$

$$+ K_m^2 b_m^T \langle B_{m-1}\rangle\ b_m\langle B_{m-1}\rangle \ . \tag{6.10}$$

Stating the gradient of $E_m(K_m)$ and setting it equal to zero

$$\frac{\partial E_m(K_m)}{\partial K_m} = 2\,x^T\langle B_{m-1}\rangle\ b_m\langle B_{m-1}\rangle + 2\,K_m b_m^T \langle B_{m-1}\rangle\ b_m\langle B_{m-1}\rangle$$

$$= \left[0 \ldots 0\right]^T \tag{6.11}$$

yields expression (6.7) for K_m, which determines K_m such that $\mathbf{e}_m$ is orthogonal with respect to the *extended* subspace spanned by $\mathbf{B}_{m-1}$ and $\mathbf{b}_m \langle \mathbf{B}_{m-1} \rangle$ or, equivalently, is orthogonal with respect to the subspace $\mathbf{B}_m$.

End of proof.

This recursion formula for orthogonal projections, as stated by (6.5-7), may be illustrated in the three-dimensional space as follows. Consider the vectors $\mathbf{x}$, $\mathbf{b}_1$ and $\mathbf{e}_1$. Then, $\mathbf{e}_1$ is the complement of the orthogonal projection of $\mathbf{x}$ onto $\mathbf{b}_1$ as illustrated in Fig. 6.2.

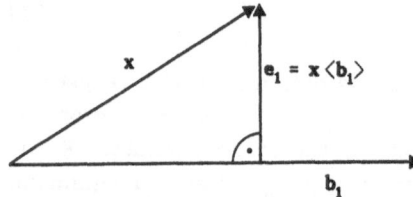

Fig. 6.2. Interpretation of $\mathbf{e}_1$ as the complement of the orthogonal projection of $\mathbf{x}$ onto the subspace $\mathbf{b}_1$ where $\mathbf{e}_1 = \mathbf{x} \langle \mathbf{b}_1 \rangle$

Next, we consider the order-increased subspace spanned by the vectors $\mathbf{b}_1$ and $\mathbf{b}_2$. According to the recursion formula for orthogonal projections (6.5-7), we may construct the complement of the orthogonal projection of $\mathbf{x}$ onto the subspace spanned by $\mathbf{b}_1$ and $\mathbf{b}_2$, namely, $\mathbf{e}_2 = \mathbf{x} \langle \mathbf{b}_1 , \mathbf{b}_2 \rangle$, as

$$\mathbf{e}_2 = \mathbf{x}\langle \mathbf{b}_1 \rangle + K_2 \mathbf{b}_2 \langle \mathbf{b}_1 \rangle = \mathbf{e}_1 + K_2 \mathbf{b}_2 \langle \mathbf{b}_1 \rangle \;,$$

where

$$K_2 = - \frac{\mathbf{e}_1^T \, \mathbf{b}_2 \langle \mathbf{b}_1 \rangle}{\mathbf{b}_2^T \langle \mathbf{b}_1 \rangle \, \mathbf{b}_2 \langle \mathbf{b}_1 \rangle} \;.$$

Figure 6.3 illustrates this construction of $\mathbf{e}_2$ from $\mathbf{e}_1$ plus a weighted portion of the *orthogonal* basis vector $\mathbf{b}_2 \langle \mathbf{b}_1 \rangle$.

6.1.1 Solving the Normal Equations with the Recursion Formula for Orthogonal Projections

From the considerations that we have made so far, it is only a small step to the insight that the recursion formula for orthogonal projections can be a useful mathematical tool for solving the Normal Equations. Indeed, we may project the actual process vector $\mathbf{x}(t)$ successively onto the components of the subspace of past observations spanned by the vectors of $\mathbf{X}(t)$ as follows:

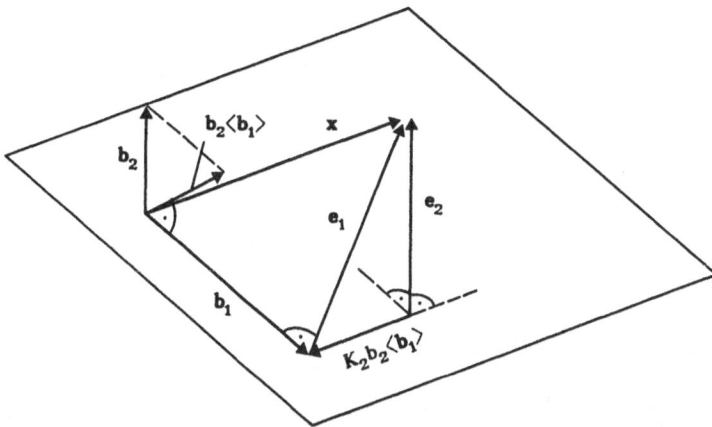

Fig. 6.3. Recursive construction of $\mathbf{e_2}$ from $\mathbf{e_1}$ and the *weighted* orthogonal basis vector $\mathbf{b_2} \langle \mathbf{b_1} \rangle$ with the final result that $\mathbf{e_2} = \mathbf{x} \langle \mathbf{b_1}, \mathbf{b_2} \rangle$

$$\mathbf{e_0}(t) \quad = \mathbf{x}(t) \qquad\qquad\qquad\qquad\qquad\qquad\qquad\qquad\qquad (6.12a)$$

$$\mathbf{e_1}(t) \quad = \mathbf{x}(t) \langle \mathbf{x}(t-1) \rangle \qquad\qquad\qquad\qquad\qquad\qquad (6.12b)$$

$$\mathbf{e_2}(t) \quad = \mathbf{x}(t) \langle \mathbf{x}(t-1), \mathbf{x}(t-2) \rangle \qquad\qquad\qquad\qquad (6.12c)$$

$$\vdots$$

$$\mathbf{e_{m-1}}(t) = \mathbf{x}(t) \langle \mathbf{x}(t-1), \mathbf{x}(t-2), \ldots, \mathbf{x}(t-m+2), \mathbf{x}(t-m+1) \rangle \qquad (6.12d)$$

$$\mathbf{e_m}(t) \quad = \mathbf{x}(t) \langle \mathbf{x}(t-1), \mathbf{x}(t-2), \ldots, \mathbf{x}(t-m+1), \mathbf{x}(t-m) \rangle \; . \qquad (6.12e)$$

Similarly, we may successively construct an orthogonal basis of the subspace of past observations as follows:

$$\mathbf{r_0}(t) \quad = \mathbf{x}(t) \qquad\qquad\qquad\qquad\qquad\qquad\qquad\qquad\qquad (6.13a)$$

$$\mathbf{r_1}(t) \quad = \mathbf{x}(t-1) \langle \mathbf{x}(t) \rangle \qquad\qquad = \mathbf{x}(t-1) \langle \mathbf{r_0}(t) \rangle \qquad (6.13b)$$

$$\mathbf{r_2}(t) \quad = \mathbf{x}(t-2) \langle \mathbf{x}(t), \mathbf{x}(t-1) \rangle \qquad = \mathbf{x}(t-2) \langle \mathbf{r_0}(t), \mathbf{r_1}(t) \rangle \qquad (6.13c)$$

$$\vdots$$

$$\mathbf{r_{m-1}}(t) = \mathbf{x}(t-m+1) \langle \mathbf{x}(t), \mathbf{x}(t-1), \ldots, \mathbf{x}(t-m+3), \mathbf{x}(t-m+2) \rangle \qquad (6.13d)$$

$$\mathbf{r_m}(t) \quad = \mathbf{x}(t-m) \langle \mathbf{x}(t), \mathbf{x}(t-1), \ldots, \mathbf{x}(t-m+2), \mathbf{x}(t-m+1) \rangle \; . \qquad (6.13e)$$

These successive decompositions are still *nonrecursive*. They can be brought into a recursive formulation by applying the recursion formula for orthogonal projections (6.5-7) to the decomposition terms (6.12e and 13e). According to (6.5), the following relations hold:

$$\mathbf{e}_m(t) = \mathbf{x}(t) \langle \mathbf{x}(t-1), \mathbf{x}(t-2), \dots, \mathbf{x}(t-m+2), \mathbf{x}(t-m+1) \rangle$$
$$+ K_m^f(t) \, \mathbf{x}(t-m) \langle \mathbf{x}(t-1), \dots, \mathbf{x}(t-m+2), \mathbf{x}(t-m+1) \rangle , \quad (6.14a)$$

$$\mathbf{r}_m(t) = \mathbf{x}(t-m) \langle \mathbf{x}(t-1), \dots, \mathbf{x}(t-m+3), \mathbf{x}(t-m+2), \mathbf{x}(t-m+1) \rangle$$
$$+ K_m^b(t) \, \mathbf{x}(t) \langle \mathbf{x}(t-1), \dots, \mathbf{x}(t-m+2), \mathbf{x}(t-m+1) \rangle . \quad (6.14b)$$

Expressing (6.14a,b) in terms of the orthogonalized vectors $\mathbf{e}_{m-1}(t)$ and $\mathbf{r}_{m-1}(t)$ of stage m-1, we may establish the recursion laws

$$\mathbf{e}_m(t) = \mathbf{e}_{m-1}(t) + K_m^f(t) \mathbf{r}_{m-1}(t-1) , \quad (6.15a)$$

$$\mathbf{r}_m(t) = \mathbf{r}_{m-1}(t-1) + K_m^b(t) \mathbf{e}_{m-1}(t) , \quad (6.15b)$$

with the initial condition

$$\mathbf{e}_0(t) = \mathbf{r}_0(t) = \mathbf{x}(t) . \quad (6.16)$$

Clearly, the scalar parameters $K_m^f(t)$ and $K_m^b(t)$ can be determined according to the recursion formula for orthogonal projections (6.7) as

$$K_m^f(t) = - \frac{\mathbf{e}_{m-1}^T(t) \, \mathbf{r}_{m-1}(t-1)}{\mathbf{r}_{m-1}^T(t-1) \, \mathbf{r}_{m-1}(t-1)} , \quad (6.17a)$$

$$K_m^b(t) = - \frac{\mathbf{e}_{m-1}^T(t) \, \mathbf{r}_{m-1}(t-1)}{\mathbf{e}_{m-1}^T(t) \, \mathbf{e}_{m-1}(t)} , \quad (6.17b)$$

where the inner products appearing in the numerator and denominator expressions of (6.17a,b) are frequently termed the *"residual energies"* of stage m-1:

$$E_{m-1}(t) = \mathbf{e}_{m-1}^T(t) \, \mathbf{e}_{m-1}(t) , \quad (6.18a)$$

$$R_{m-1}(t) = \mathbf{r}_{m-1}^T(t) \, \mathbf{r}_{m-1}(t) , \quad (6.18b)$$

$$C_{m-1}(t) = \mathbf{e}_{m-1}^T(t) \, \mathbf{r}_{m-1}(t-1) . \quad (6.18c)$$

The scalar parameters $K_m^f(t)$ and $K_m^b(t)$ are termed the *forward reflection coefficient* and the *backward reflection coefficient*, respectively. They can be expressed in terms of the residual energies of stage m-1 as

$$K_m^f(t) = - C_{m-1}(t) \big/ R_{m-1}(t-1) , \quad (6.19a)$$

$$K_m^b(t) = - C_{m-1}(t) \big/ E_{m-1}(t) . \quad (6.19b)$$

The vector order recursions (6.15a,b), their initialization from the observed process (6.16), and the computation of the required reflection coefficients (6.17a,b) already establish a closed recursion for the order recursive solution of the Normal Equations. Interestingly, since this procedure is *order recursive*, it requires *no pre-assumption about the model order* or, equivalently, the order of the Normal Equations. This was achieved by the *decoupled* nature of orthogonal projections, which use the solution of the order m-1 problem directly to obtain the solution of the order m problem. This nice feature of recursive orthogonal projections offers the possibility of increasing the model order depending on an appropriate fidelity criterion that reflects the *accuracy of approximation* of the process $\mathbf{x}(t)$ by $\hat{\mathbf{x}}^f(t)$ at a given order m. Such fidelity criteria will be discussed later in this chapter.

6.1.2 The Feed-Forward Ladder Form

Another important consequence that ultimately emerges from these considerations is that the solution of the Normal Equations appears as a set of *reflection coefficients* $\{K_m^f(t), K_m^b(t), 1 \le m \le p\}$ rather than a parameter set $\{a_m(t), b_m(t), 1 \le m \le p\}$. Later, we will see, however, that both representations of the desired least-squares solution of the linear prediction problem are *equivalent*, in that the a's and b's can be computed from the K's and vice versa. But the properties of the two coefficient sets may *differ* significantly. For example, the K's are known to have a *smaller dynamic range* or variance than the a's and b's.

Since the a's and b's define the transversal forward/backward predictor underlying the fixed-order formulation of the linear least-squares prediction problem, it is now obvious that the K's define an *alternative structure* of the linear prediction error filter emerging from the growing-order formulation of the linear least-squares prediction problem. Consequently, p and m are the fixed-order and the growing-order indices, respectively. Proceeding along this line of thought, we may visualize the vector order recursions (6.15a,b) as an all-zero prediction error filter structure, the so-called *feed-forward ladder form* shown in Fig. 6.4. The coefficients of the feed-forward ladder form can be computed via the *explicit ladder algorithm* listed in Table 6.1.

A quick inspection of the algorithm listed in Table 6.1 shows that the terminology "explicit" ladder algorithm is justified, since the residual vectors $\mathbf{e}_m(t)$ and $\mathbf{r}_m(t)$ are *explicitly* computed. Therefore, the computational complexity grows with O(Lp), making this procedure a fairly *unpractical* method when the length L of the input data record (window length) is relatively long. We note that many estimation problems involve windows of L > 50, or even larger windows. Moreover, the explicit computation of residual signals is an ill-conditioned problem, as became apparent much earlier in this book. Round-off noise of considerable

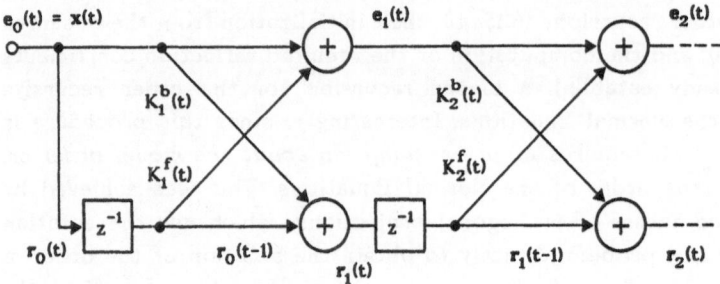

Fig. 6.4. Feed-forward ladder form prediction error filter

Table 6.1. Time recursive least-squares ladder estimation algorithm with *explicit* computation of residual vectors

---

FOR   t = 0, 1, 2, . . .

   Input: $\mathbf{x}(t)$

   Initialize: $\mathbf{e}_0(t) = \mathbf{r}_0(t) = \mathbf{x}(t)$; $\mathbf{r}_0(t-1) = \mathbf{x}(t-1)$ (6.20a)

   FOR   m = 1, 2, 3, . . .

$$E_{m-1}(t) = \mathbf{e}_{m-1}^T(t)\, \mathbf{e}_{m-1}(t) \tag{6.20b}$$

$$R_{m-1}(t) = \mathbf{r}_{m-1}^T(t)\, \mathbf{r}_{m-1}(t) \tag{6.20c}$$

$$C_{m-1}(t) = \mathbf{e}_{m-1}^T(t)\, \mathbf{r}_{m-1}(t-1) \tag{6.20d}$$

$$K_m^f(t) = -\, C_{m-1}(t)\,/\,R_{m-1}(t-1) \tag{6.20e}$$

$$K_m^b(t) = -\, C_{m-1}(t)\,/\,E_{m-1}(t) \tag{6.20f}$$

$$\mathbf{e}_m(t) = \mathbf{e}_{m-1}(t) + K_m^f(t)\,\mathbf{r}_{m-1}(t-1) \tag{6.20g}$$

$$\mathbf{r}_m(t) = \mathbf{r}_{m-1}(t-1) + K_m^b(t)\,\mathbf{e}_{m-1}(t) \tag{6.20h}$$

---

variance can accumulate in the residual vectors during the order recursions (6.20g,h), causing *biased* residual energies and biased reflection coefficients.

    This gives rise to the question whether it is possible to derive more *direct* order recursions for the computation of residual energies without requiring the explicit computation of prediction errors. Proceeding along this line of thought, we may substitute the residual signal vector recursions (6.20g,h) into the expressions for the forward/backward residual energies (6.20b,c) to obtain

$$E_m(t) = e_m^T(t) e_m(t)$$

$$= \left(e_{m-1}^T(t) + K_m^f(t) r_{m-1}^T(t-1) \right) \left(e_{m-1}(t) + K_m^f(t) r_{m-1}(t-1) \right)$$

$$= E_{m-1}(t) + 2 K_m^f(t) C_{m-1}(t) + K_m^f(t) K_m^f(t) R_{m-1}(t-1) \; , \qquad (6.21a)$$

$$R_m(t) = r_m^T(t) r_m(t)$$

$$= R_{m-1}(t-1) + 2 K_m^b(t) C_{m-1}(t) + K_m^b(t) K_m^b(t) E_{m-1}(t) \; . \qquad (6.21b)$$

According to (6.20e,f), the reflection coefficients can be expressed in terms of *residual energies*. In this way, the expressions (6.21a,b) reduce to

$$E_m(t) = E_{m-1}(t) - C_{m-1}^2(t) \big/ R_{m-1}(t-1) \; , \qquad (6.22a)$$

$$R_m(t) = R_{m-1}(t-1) - C_{m-1}^2(t) \big/ E_{m-1}(t) \; . \qquad (6.22b)$$

Alternatively, one may express the recursions (6.22a,b) in terms of the reflection coefficients in the following "condensed" form:

$$E_m(t) = E_{m-1}(t) + K_m^f(t) C_{m-1}(t) \; , \qquad (6.23a)$$

$$R_m(t) = R_{m-1}(t-1) + K_m^b(t) C_{m-1}(t) \; . \qquad (6.23b)$$

Next it would be of great interest to derive a similar order recursion for the prediction error covariance $C_m(t)$. Substitution of the residual signal vector order recursions (6.20g,h) into the expression of $C_m(t)$ (6.20d) gives

$$C_m(t) = e_m^T(t) r_m(t-1)$$

$$= \left(e_{m-1}^T(t) + K_m^f(t) r_{m-1}^T(t-1) \right) \left(r_{m-1}(t-2) + K_m^b(t-1) e_{m-1}(t-1) \right)$$

$$= e_{m-1}^T(t) r_{m-1}(t-2) + K_m^f(t) r_{m-1}^T(t-1) r_{m-1}(t-2)$$

$$+ K_m^b(t-1) e_{m-1}^T(t) e_{m-1}(t-1)$$

$$+ K_m^f(t) K_m^b(t-1) r_{m-1}^T(t-1) e_{m-1}(t-1) \; . \qquad (6.24)$$

Expression (6.24) contains *unknown* residual energies, such as the inner product $e_{m-1}^T(t) r_{m-1}(t-2)$. Furthermore, we note the appearance of a backward reflection coefficent of time instant t-1. It is seen that therefore a simple order recursive law similar to the order recursions of $E_m(t)$ and $R_m(t)$ does *not* hold in the case of the prediction error covariance $C_m(t)$.

Indeed, the order recursive updating of $C_m(t)$ presents a serious problem. Its solution is postponed to Chaps. 7 and 8.

6.2 Computing Time-Varying Transversal Predictor Parameters from the Ladder Reflection Coefficients

As mentioned earlier, the ladder approach to solving the Normal Equations leads to a solution in terms of a set of forward/backward ladder re-flection coefficients $\{K_m^f(t), K_m^b(t), 1 \le m \le p\}$. Sometimes, however, we are interested in reconstructing the equivalent time-varying transver-sal forward/backward predictor parameters (AR parameters) $\{a_m(t), b_m(t), 1 \le m \le p\}$ from the ladder reflection coefficients at each time step. In the stationary case, this task is easily accomplished just by measuring the impulse response of the ladder form after stage m = p. In the more general nonstationary case, such a simple procedure does not hold. Note that in the nonstationary case, the reflection coefficients of the ladder-form prediction error filter are updated at *each* time step. However, we can derive a *Levinson-type recursion* that allows the desired recon-struction of the equivalent time-varying transversal predictor parameters. Such a procedure will be derived in the following. For this purpose, we assume that we relate a forward and a backward prediction error filter with transversal structure and order m to a ladder prediction error filter of the same order m. When the order m is updated by one, another ladder stage is added, while the "old" coefficients remain *unchanged*. In the associated transversal predictors, however, *all* m coefficients will change and one more coefficient is added. Let us therefore denote the transversal forward predictor coefficients that are the equivalents to the order m ladder prediction error filter coefficients by $\{a_j^{(m)}(t), 0 \le j \le m\}$ and the corresponding transversal backward predictor coefficients by $\{b_j^{(m)}(t), 0 \le j \le m\}$. True least-squares prediction is assumed throughout all our considerations.

The coefficients of the transversal predictors are *time varying*, just as the ladder reflection coefficients. Now, the method of finding a relating recursion is to assume that an impulse is applied at time t=0 at the predictor inputs and the states of the predictors are compared while the impulses propagate through the predictors. Here it is very important to check that the predictor coefficients can *change* during the propagation of the impulse. The evaluation of this effect can be accomplished by a *time-order table* that reflects the relationships of propagating impulses in different predictor structures. The appropriate time-order table for comparing a ladder form with associated transversal forward/backward predictors is shown in Table 6.2. It is seen that by using the time-order table we can easily express the coefficients of the transversal predictors

Table 6.2. Time-order table for fitting the transversal predictors $\{a_j^{(m)}(t),$ $b_j^{(m)}(t), 0 \le j \le m\}$ to a least-squares ladder form of order m. The transversal predictor coefficients are expressed in terms of the ladder reflection coefficients. Example shown for the order $p = 2$ problem

		t = 0	t = 1	t = 2
m = 0	$e_0(t)$	$a_0^{(0)}(t) = -1$	0	0
	$r_0(t)$	$b_0^{(0)}(t) = -1$	0	0
m = 1	$e_1(t)$	$a_0^{(1)}(t) = -1$	$a_1^{(1)}(t) = -K_1^f(t)$	0
	$r_1(t)$	$b_0^{(1)}(t) = -K_1^b(t)$	$b_1^{(1)}(t) = -1$	0
m = 2	$e_2(t)$	$a_0^{(2)}(t) = -1$	$a_1^{(2)}(t) = -K_1^f(t) - K_2^f(t)K_1^b(t-1)$	$a_2^{(2)}(t) = -K_2^f(t)$
	$r_2(t)$	$b_0^{(2)}(t) = -K_2^b(t)$	$b_1^{(2)}(t) = -K_1^b(t-1) - K_2^b(t)K_1^f(t)$	$b_2^{(2)}(t) = -1$

as functions of the ladder reflection coefficients, as desired. Note that successively delayed ladder reflection coefficients toggle through the delay elements in the backward path of the ladder form. Hence it is not surprising that the transversal predictor coefficients appear as functions of *delayed* ladder reflection coefficients.

Table 6.2 additionally reveals that there exists an *order recursive law* for constructing the transversal predictor coefficients of an mth order predictor from the coefficients of the predictor of order m-1. Careful inspection of Table 6.2, along with a consideration of the ladder recursions (6.15a,b), reveals that this order recursion can be directly obtained from the ladder recursions in a scalar form,

$$e_m(t) = e_{m-1}(t) + K_m^f(t) r_{m-1}(t-1) , \qquad (6.25a)$$

which leads to the forward predictor coefficient order recursion

$$a_j^{(m)}(t) = a_j^{(m-1)}(t) + K_m^f(t) b_{j-1}^{(m-1)}(t-1) . \qquad (6.25b)$$

Similarly, one may recognize from an evaluation of Table 6.2 that

$$r_m(t) = r_{m-1}(t-1) + K_m^b(t) e_{m-1}(t) \qquad (6.26a)$$

yields the corresponding backward predictor coefficient order recursion

$$b_j^{(m)}(t) = b_{j-1}^{(m-1)}(t-1) + K_m^b(t) a_j^{(m-1)}(t) . \qquad (6.26b)$$

From the expressions (6.25b and 26b), we may establish a *Levinson type recursion* that uniquely relates the time-varying ladder reflection coefficients to the corresponding forward and backward prediction error filters with transversal structure. This important procedure is listed in Table 6.3. Note again that the transversal prediction error filters obtained from this recursion behave *absolutely identically* to the ladder prediction error filter, even in the case of transient signals and time-varying ladder reflection coefficients, since we have assumed the most general non-stationary case throughout the derivations. This statement holds as long as we assume unquantized data and infinite precision arithmetic. When finite precision arithmetic is used, one can expect a better accuracy and a more robust behavior from the ladder realization than from the transversal realization of the forward/backward prediction error filters.

Table 6.3. Levinson type recursion for computing the time-varying least-squares transversal forward/backward predictor coefficients $\{ a_j(t), b_j(t), 0 \leq j \leq p \}$ from the ladder reflection coefficients $\{ K_m^f(t), K_m^b(t), 1 \leq m \leq p \}$

FOR $t = 0, 1, 2, \ldots$

    Input: $K_m^f(t), K_m^b(t), 1 \leq m \leq p$

    Initialize: $a_0^{(1)}(t) = -1$ $a_1^{(1)}(t) = -K_1^f(t)$

                      $b_0^{(1)}(t) = -K_1^b(t)$ $b_1^{(1)}(t) = -1$

    FOR $m = 2, 3, \ldots, p$

        FOR $j = 1, 2, \ldots, m-1$

            $a_j^{(m)}(t) = a_j^{(m-1)}(t) + K_m^f(t) b_{j-1}^{(m-1)}(t-1)$

            $b_j^{(m)}(t) = b_{j-1}^{(m-1)}(t-1) + K_m^b(t) a_j^{(m-1)}(t)$

        $a_0^{(m)}(t) = -1$ $a_m^{(m)}(t) = -K_m^f(t)$

        $b_0^{(m)}(t) = -K_m^b(t)$ $b_m^{(m)}(t) = -1$

    FOR $j = 0, 1, 2, \ldots, p$

        $a_j(t) = a_j^{(p)}(t)$

        $b_j(t) = b_j^{(p)}(t)$

6.3 Stationary Case – The PARCOR Ladder Form

A special situation of interest appears when the observed process $x(t)$ is *stationary*. In Sect. 2.6, we defined a stationary process as a process with *time-invariant autocorrelation coefficients* (2.45). Since the covariance matrix of the observed process can be viewed as a matrix built up from successively delayed autocorrelation vectors, the system matrix of the Normal Equations was shown to be a symmetric Toeplitz matrix in the case of stationary data.

In this chapter, we are interested in finding to what extent the ladder form is affected by the stationarity assumption. For this purpose, we investigate the general (time-varying) feed-forward ladder form as shown in Fig. 6.4 and the associated explicit least-squares ladder algorithm for estimating the ladder reflection coefficients (Table 6.1). From the definition of a stationary process (2.45), it follows that

$$\mathbf{x}^T(t-1)\mathbf{x}(t-1) = \mathbf{x}^T(t)\mathbf{x}(t) \qquad \textit{energy invariance} \; . \tag{6.27}$$

Consequently

$$R_0(t-1) = R_0(t) = E_0(t) \tag{6.28}$$

and therefore

$$K_1^f(t) = K_1^b(t) = K_1 \; . \tag{6.29}$$

Considering the forward/backward energy propagation law (6.23a,b), one sees that

$$E_1 = E_0 + K_1 \, C_0 \quad \text{and} \tag{6.30a}$$

$$R_1 = R_0 + K_1 \, C_0 \; , \tag{6.30b}$$

which, together with (6.28), give

$$E_1 = R_1 \; . \tag{6.31}$$

This law holds for all stages from $m = 0$ to $m = p$:

$$E_m = R_m \; , \tag{6.32}$$

$$K_m^f(t) = K_m^b(t) = K_m \; . \tag{6.33}$$

Proof by induction.

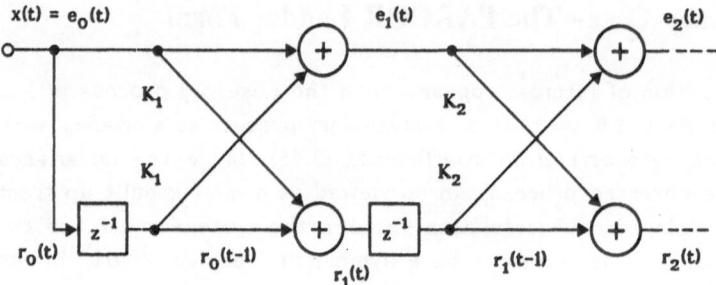

Fig. 6.5. Feed-forward PARCOR ladder form

Relation (6.33) can be substituted in the vector order recursions (6.15a,b) to obtain the PARtial CORrelation (PARCOR) ladder recursions

$$e_m(t) = e_{m-1}(t) + K_m r_{m-1}(t-1) \; , \tag{6.34a}$$

$$r_m(t) = r_{m-1}(t-1) + K_m e_{m-1}(t) \; . \tag{6.34b}$$

Interestingly, the stationary case of linear prediction is associated with a ladder form with *identical* reflection coefficients in the forward and backward predictors. Additionally, the forward and backward predictor residual energies E_m and R_m attain *identical* values. Both the reflection coefficients and the residual energies are *time invariant*. This was a direct consequence of the stationarity assumption underlying the derivation of the PARCOR ladder form. Figure 6.5 shows the PARCOR ladder form. The time-invariant K's are frequently termed the *"PARCOR coefficients"*.

One more remark about the forward/backward residual energies and residual signals in a PARCOR ladder form seems to be in order. Although the residual energies have identical values in the forward and backward predictors, it is emphasized that the associated prediction error vectors are, in general, *not identical* so that

$$e_m(t) \neq r_m(t) \quad for \quad 1 \le m \le p \; . \tag{6.35}$$

The PARCOR ladder form exhibits quite a number of interesting properties. These will be discussed in the following sections.

6.4 Relationships Between PARCOR Ladder Form and Transversal Predictor

In an earlier part of this chapter, we introduced a Levinson type recursion for reconstruction of the transversal predictor parameters from the (time-varying) ladder reflection coefficients in the general nonstationary case. This algorithm was listed in Table 6.3. Since the PARCOR ladder form is a special case of the general time-varying ladder form, it is intuitively expected that a recursion for computing the transversal predictor parameters will be a special case of the procedure listed in Table 6.3. Indeed, through the time-invariance assumption underlying the PARCOR ladder form, it is clear that the associated transversal predictor parameters will also be time-invariant, hence the recursions (6.25b and 26b) will reduce to

$$a_j^{(m)} = a_j^{(m-1)} + K_m b_{j-1}^{(m-1)} \ , \tag{6.36a}$$

$$b_j^{(m)} = b_{j-1}^{(m-1)} + K_m a_j^{(m-1)} \ . \tag{6.36b}$$

The correctness of these recursions can easily be verified by evaluating a time-order table of the PARCOR ladder form. Such a time-order table is listed in Table 6.4 for the case of a maximum order of $p = 3$.

Table 6.4. Time-order table for evaluating the impulse response of the PARCOR ladder form. The table shows the relationships between the PARCOR coefficients $\{K_m, 1 \le m \le p\}$ and the associated transversal forward/backward predictors $\{a_j^{(m)}, b_j^{(m)}, 0 \le j \le m \}$ of stage m. Example shown for a maximum order of $p = 3$

	$t = 0$	$t = 1$	$t = 2$	$t = 3$
$e_0(t)$	$a_0^{(0)} = -1$			
$r_0(t)$	$b_0^{(0)} = -1$			
$e_1(t)$	$a_0^{(1)} = -1$	$a_1^{(1)} = -K_1$		
$r_1(t)$	$b_0^{(1)} = -K_1$	$b_1^{(1)} = -1$		
$e_2(t)$	$a_0^{(2)} = -1$	$a_1^{(2)} = -K_1 - K_1 K_2$	$a_2^{(2)} = -K_2$	
$r_2(t)$	$b_0^{(2)} = -K_2$	$b_1^{(2)} = -K_1 K_2 - K_1$	$b_2^{(2)} = -1$	
$e_3(t)$	$a_0^{(3)} = -1$	$a_1^{(3)} = -K_1 - K_1 K_2 - K_2 K_3$	$a_2^{(3)} = -K_2 - K_3 K_1 K_2 - K_3 K_1$	$a_3^{(3)} = -K_3$
$r_3(t)$	$b_0^{(3)} = -K_3$	$b_1^{(3)} = -K_2 - K_3 K_1 K_2 - K_3 K_1$	$b_2^{(3)} = -K_1 - K_1 K_2 - K_2 K_3$	$b_3^{(3)} = -1$

A closer look at Table 6.4 can be instructive. Clearly, we may deduce from this table that the coefficients of the transversal forward/backward predictors correspond to each other via the relation

$$a_j^{(m)} = b_{m-j}^{(m)} \ ; \qquad\qquad 0 \le j \le m \tag{6.37a}$$

or, equivalently,

$$b_j^{(m)} = a_{m-j}^{(m)} \; ; \qquad\qquad 0 \le j \le m \; . \qquad\qquad (6.37b)$$

In other words, the transversal forward/backward prediction error filters associated with the PARCOR ladder form have *identical* coefficients, which just appear in a *reversed* order. For example, in the case m = 3, the relationship between the coefficients in the forward and backward transversal predictors may be visualized as follows:

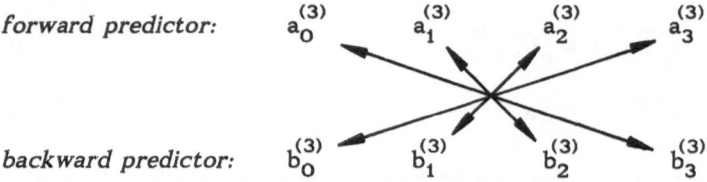

forward predictor: $\quad a_0^{(3)} \qquad a_1^{(3)} \qquad a_2^{(3)} \qquad a_3^{(3)}$

backward predictor: $\quad b_0^{(3)} \qquad b_1^{(3)} \qquad b_2^{(3)} \qquad b_3^{(3)}$

This *reverse ordering property* of coefficients can be exploited to express the mth order transversal forward and backward prediction filters in terms of the reverse-ordered data sequence as

$$e_m(t) = x(t) - \sum_{j=1}^{m} a_j^{(m)} \, x(t-j)$$

$$= - \sum_{j=0}^{m} a_j^{(m)} \, x(t-j) \; , \qquad \textit{forward prediction} \; , \qquad (6.38a)$$

$$r_m(t) = x(t-m) - \sum_{j=1}^{m} a_j^{(m)} \, x(t-m+j)$$

$$= - \sum_{j=0}^{m} a_j^{(m)} \, x(t-m+j) \; , \qquad \textit{backward prediction} \; . \qquad (6.38b)$$

The signal $r_m(t)$ is the prediction error of the process $x(t)$ with *reversed* time direction (backward linear prediction).

6.4.1 Computing Transversal Predictor Parameters from PARCOR Coefficients – The Levinson Recursion

Clearly, in the PARCOR case, only *one* transversal filter parameter set is required to describe the prediction of the stationary time series, in contrast to the more general nonstationary case, where we required *two* (generally different) parameter sets in the forward and backward predictors associated with the different forward/backward ladder reflection coefficients.

Substituting (6.37b) into (6.36a) yields the well-known *Levinson recursion* for computing the transversal predictor parameters from the PARCOR coefficients:

$$a_j^{(m)} = a_j^{(m-1)} + K_m a_{m-j}^{(m-1)} \,. \tag{6.39}$$

Obviously, (6.36b) can be omitted in the stationary case, since the transversal backward predictor is uniquely detemined as soon as the transversal forward predictor coefficients are known. Table 6.5 summarizes the complete Levinson recursion, which appears as the *stationary counterpart* of the more general Levinson type recursion of Table 6.3.

Table 6.5. Levinson recursion for computation of the transversal forward/backward predictor parameters $\{a_j , b_j , 0 \le j \le p\}$ from the PARCOR coefficients $\{K_m, 1 \le m \le p\}$

Input: $K_m, 1 \le m \le p$

Initialize: $a_0^{(1)} = -1$

$a_1^{(1)} = -K_1$

FOR   m = 2, 3, . . . , p

$\quad$ FOR   j = 1, 2, . . . , m-1

$\qquad a_j^{(m)} = a_j^{(m-1)} + K_m a_{m-j}^{(m-1)}$

$\quad a_0^{(m)} = -1 \qquad a_m^{(m)} = -K_m$

FOR   j = 0, 1, 2, . . . , p

$\quad a_j = a_j^{(p)}$

$\quad b_j = a_{p-j}^{(p)}$

6.4.2 Computing PARCOR Coefficients from Transversal Predictor Parameters – The Inverse Levinson Recursion

We continue by deriving the *inverse* Levinson recursion, i.e., a recursion that accomplishes the reconstruction of the PARCOR coefficients from a given set of transversal predictor parameters. For this purpose, one needs to establish a *decomposition* of a given transversal filter parameter set $\{a_j^{(m)}, 0 \le j \le m\}$ into all parameter sets of the lower order transversal predictors as follows:

$$
\begin{array}{lllll}
a_0^{(m)} & a_1^{(m)} & a_2^{(m)} & \cdots\cdots\cdots\cdots\cdots & a_m^{(m)} = -K_m \\
a_0^{(m-1)} & a_1^{(m-1)} & a_2^{(m-1)} & \cdots\cdots\cdots & a_{m-1}^{(m-1)} = -K_{m-1} \\
\vdots \\
a_0^{(2)} & a_1^{(2)} & a_2^{(2)} = -K_2 \\
a_0^{(1)} & a_1^{(1)} = -K_1
\end{array}
\tag{6.40}
$$

In order to derive a general recursion for the K's, we first take the Levinson recursion (6.39) which is rearranged to express the jth coefficient of the (m-1)th-order filter in terms of the coefficients of the reverse (m-1)th-order filter, the coefficients of the mth-order filter and the PARCOR coefficient of the mth-order PARCOR ladder form:

$$
a_j^{(m-1)} = a_j^{(m)} - K_m a_{m-j}^{(m-1)} .
\tag{6.41}
$$

Evaluating this recursion on an m = 3 example gives

$$
a_2^{(2)} = a_2^{(3)} - K_3 a_1^{(2)} ,
\tag{6.42a}
$$

$$
a_1^{(2)} = a_1^{(3)} - K_3 a_2^{(2)} ,
\tag{6.42b}
$$

where $a_0^{(2)} = -1$ and $K_3 = -a_3^{(3)}$.

Proceeding along this line of thought, one substitutes (6.42b) into (6.42a) to obtain

$$
a_2^{(2)} = a_2^{(3)} - K_3 \left[a_1^{(3)} - K_3 a_2^{(2)} \right] .
\tag{6.43a}
$$

Similarly, substitution of (6.42a) into (6.42b) yields

$$
a_1^{(2)} = a_1^{(3)} - K_3 \left[a_2^{(3)} - K_3 a_1^{(2)} \right]
\tag{6.43b}
$$

or, equivalently,

$$
a_2^{(2)} = \left(a_2^{(3)} - K_3 a_1^{(3)} \right) / \left(1 - K_3^2 \right) ,
\tag{6.44a}
$$

$$
a_1^{(2)} = \left(a_1^{(3)} - K_3 a_2^{(3)} \right) / \left(1 - K_3^2 \right) .
\tag{6.44b}
$$

From this result, we may deduce a general recursion for computing the $(m-1)$th-order transversal predictor from the mth-order predictor and the PARCOR coefficient of order m as

$$a_j^{(m-1)} = \left(a_j^{(m)} - K_m a_{m-j}^{(m)}\right) / \left(1 - K_m^2\right) . \qquad (6.45)$$

The recursion (6.45) may be termed the *"inverse"* Levinson recursion, since it starts at a given maximum order transversal predictor and, recursively, computes all the filters of lower orders down to the order $m = 1$ filter, thereby providing the desired decomposition (6.40).

Clearly, a closer look at the decomposition (6.40) reveals that the inverse Levinson recursion can be exploited for computation of the PARCOR coefficients, since

$$K_m = - a_m^{(m)} ; \qquad 1 \leq m \leq p . \qquad (6.46)$$

Table 6.6 provides a closed form of this inverse Levinson recursion in its use for reconstruction of the PARCOR coefficients from the transversal predictor parameters.

Table 6.6. "Inverse" Levinson recursion for computing the PARCOR coefficients $\{K_m, 1 \leq m \leq p\}$ from a given set of transversal predictor parameters $\{a_m^{(p)}, 0 \leq m \leq p\}$, thereby obtaining all the lower-order transversal predictors $\{a_j^{(m)}, 1 \leq m \leq p; 0 \leq j \leq m \}$ as by-products. This algorithm involves division. Whenever the divisor is small, set $1/x = x$

Input: $a_m^{(p)}, 0 \leq m \leq p$

Initialize: $K_p = - a_p^{(p)}$

FOR m = p, p-1, . . . , 3, 2

$\quad a_0^{(m-1)} = 1$

$\quad$ FOR j = 1, 2, . . . , m - 1

$\qquad a_j^{(m-1)} = \left(a_j^{(m)} - K_m a_{m-j}^{(m)}\right) / \left(1 - K_m^2\right)$

$\quad K_{m-1} = - a_{m-1}^{(m-1)}$

6.5 The Feed-Back PARCOR Ladder Form

The application of PARCOR ladder forms has a long history in speech analysis and coding. We refer to the pioneer work of Atal [6.5], Atal and Hanauer [6.6], Itakura and Saito [6.7], and Atal and Schroeder [6.8]. In these applications, the feed-forward PARCOR ladder form shown in Fig. 6.5 is mainly used as an *analysis filter*, where it is the goal to operate the feed-forward ladder form as an *inverse filter* to the under-lying process model, which is assumed to be an *all-pole* (AR) model. Since the feed-forward ladder form represents an *all-zero* (FIR) filter, it can be *matched* to the parameters of the observed process in the least-squares sense. In this way the model parameters can be extracted from the observed process.

Conversely, the obtained parameter set can be used to *synthesize* the process when only the *inverse* structure of the analysis filter is supplied with an appropriate excitation. This leads to the *feed-back PARCOR ladder form* as the inverse representation of the feed-forward PARCOR ladder–form analysis filter. This inverse structure is con-veniently obtained from the feed-forward PARCOR ladder form shown in Fig. 6.5 by applying *Mason's rule* [6.9]:

Mason's Rule

Given a canonical filter structure with a transfer function $H(z)$, one may obtain the structure of the inverse filter characterized by $H^{-1}(z)$ by applying the following rules:

(1) *Reverse the forward signal flow direction.*

(2) *Invert the signs of the gains in the forward path.*

With this useful rule, the canonical all-pole ladder form synthesis filter can be easily deduced from the feed-forward PARCOR ladder form. This structure is termed the feed-back PARCOR ladder form and is shown in Fig. 6.6.

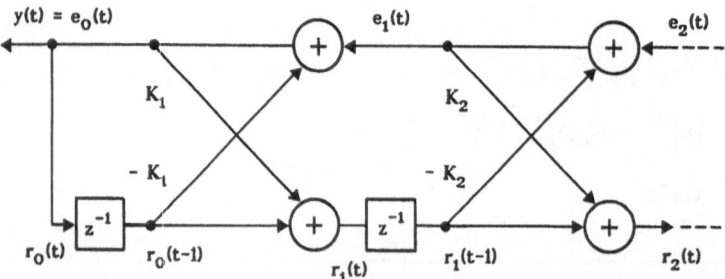

Fig. 6.6. Feed-back PARCOR ladder form (all-pole synthesis filter)

6.6 Frequency Domain Description of PARCOR Ladder Forms

We have introduced the PARCOR ladder forms as a natural consequence of the stationarity assumption of an observed process. Since the PARCOR ladder forms are *time-invariant* filters, they can be described by *transfer functions* in the complex z-plane.

6.6.1 Transfer Function of the Feed-Forward PARCOR Ladder Form

Consider one section of the feed-forward PARCOR ladder form as shown in Fig. 6.7. An isolated section of the feed-forward PARCOR ladder form obeys the following input/output relationships:

$$e_m(t) = e_{m-1}(t) + K_m r_{m-1}(t-1) \; , \tag{6.47a}$$

$$r_m(t) = K_m e_{m-1}(t) + r_{m-1}(t-1) \; . \tag{6.47b}$$

Taking the z-transforms of (6.47a,b) and rewriting the result in a convenient matrix/vector notation, we find the transfer function of an isolated section of a feed-forward PARCOR ladder form becomes

$$\begin{bmatrix} e_m(z) \\ r_m(z) \end{bmatrix} = \begin{bmatrix} 1 & z^{-1}K_m \\ K_m & z^{-1} \end{bmatrix} \begin{bmatrix} e_{m-1}(z) \\ r_{m-1}(z) \end{bmatrix} \; . \tag{6.48}$$

The overall transfer function of an order m filter of cascaded ladder sections can be composed by simply chain multiplying the associated transfer matrices of the cascaded ladder sections

$$\begin{bmatrix} e_m(z) \\ r_m(z) \end{bmatrix} = \begin{bmatrix} 1 & z^{-1}K_m \\ K_m & z^{-1} \end{bmatrix} \begin{bmatrix} 1 & z^{-1}K_{m-1} \\ K_{m-1} & z^{-1} \end{bmatrix} \cdots$$

$$\cdots \begin{bmatrix} 1 & z^{-1}K_1 \\ K_1 & z^{-1} \end{bmatrix} \begin{bmatrix} e_0(z) = x(z) \\ r_0(z) = x(z) \end{bmatrix} \; , \tag{6.49}$$

where $x(z)$ is the z-transform of the input sequence, $e_m(z)$ is the z-transform of the forward prediction error sequence and $r_m(z)$ is the z-transform of the backward prediction error sequence.

We may define a *forward predictor transfer function*

$$H_m^f(z) = e_m(z)/x(z) \; , \tag{6.50a}$$

and a *backward predictor transfer function*

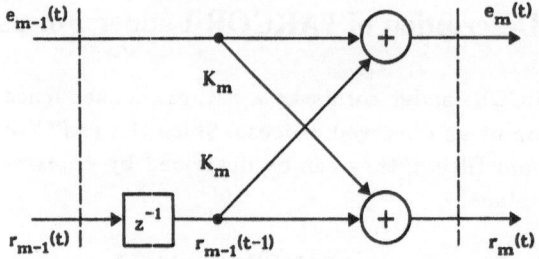

Fig. 6.7. Isolated section of a feed-forward PARCOR ladder form

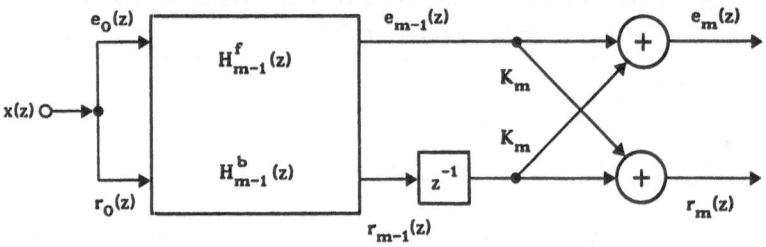

Fig. 6.8. Adding an extra feed-forward PARCOR ladder section to a given filter

$$H_m^b(z) = r_m(z) \big/ x(z) , \qquad\qquad (6.50b)$$

both related to the order m feed-forward ladder form.

With the definition of these forward/backward transfer functions, we may establish an order recursive law of transfer functions, i.e., we may express the transfer functions of the mth order filter in terms of the transfer functions of the filter of order m-1 and some update relationships that describe the effect of the additional ladder stage. See also Fig. 6.8.

Employing the approach

$$\begin{bmatrix} e_m(z) \\ r_m(z) \end{bmatrix} = \begin{bmatrix} 1 & z^{-1}K_m \\ K_m & z^{-1} \end{bmatrix} \begin{bmatrix} H_{m-1}^f(z) \\ H_{m-1}^b(z) \end{bmatrix} x(z) = \begin{bmatrix} H_m^f(z) \\ H_m^b(z) \end{bmatrix} x(z) , \quad (6.51)$$

we see that

$$H_m^f(z) = H_{m-1}^f(z) + z^{-1}K_m H_{m-1}^b(z) , \qquad\qquad (6.52a)$$

$$H_m^b(z) = K_m H_{m-1}^f(z) + z^{-1}H_{m-1}^b(z) . \qquad\qquad (6.52b)$$

6.6.2 Transfer Function of the Feed-Back PARCOR Ladder Form

A set of similar relationships can be derived for the inverse structure. Consider a single (isolated) section of the feed-back PARCOR ladder form as shown in Fig. 6.9. This isolated section is described by the input/output relationships

$$e_{m-1}(t) = e_m(t) - K_m r_{m-1}(t-1) \; , \tag{6.53a}$$

$$r_m(t) = r_{m-1}(t-1) + K_m e_{m-1}(t) \; . \tag{6.53b}$$

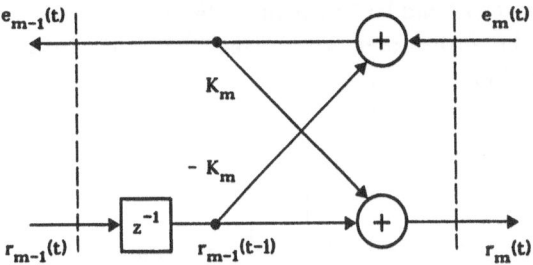

Fig. 6.9. Isolated section of a feed-back PARCOR ladder form

Substituting (6.53a) into (6.53b) gives

$$r_m(t) = K_m e_m(t) + (1 - K_m^2) r_{m-1}(t-1) \; . \tag{6.54}$$

Taking the z-transforms of (6.53a and 54) and rewriting the result in a convenient matrix/vector notation yields the transfer function of an isolated section of the feed-back PARCOR ladder form:

$$\begin{bmatrix} e_{m-1}(z) \\ r_m(z) \end{bmatrix} = \begin{bmatrix} 1 & -z^{-1}K_m \\ K_m & z^{-1}(1 - K_m^2) \end{bmatrix} \begin{bmatrix} e_m(z) \\ r_{m-1}(z) \end{bmatrix} \; . \tag{6.55}$$

Continuing along this line of thought, one may establish forward and backward predictor transfer functions associated with the feed-back PARCOR ladder form as

$$H_m^f(z) = e_0(z) / e_m(z) \; , \tag{6.56a}$$

$$H_m^b(z) = r_m(z) / r_0(z) \; , \tag{6.56b}$$

where $e_0(z) = r_0(z)$ and $e_m(z) = x(z)$. Again, we wish to derive an order recursive law of transfer functions, in order to express the transfer

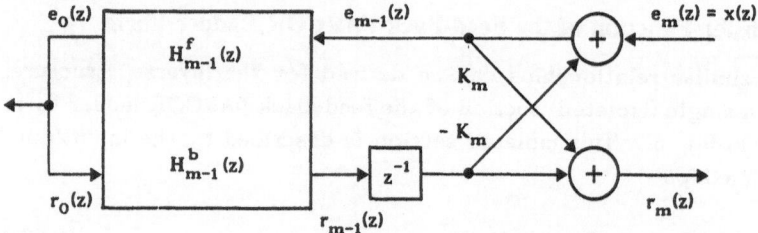

Fig. 6.10. Adding an extra feed-back PARCOR ladder section to a given filter

functions of a feed-back PARCOR ladder form of order m in terms of the transfer functions of the order m-1 filter. See Fig. 6.10 for an illustration of this configuration.

Substitution of (6.56a,b) into (6.55) gives

$$\begin{bmatrix} H_{m-1}^{f^{-1}}(z)\, e_0(z) \\ \\ r_m(z) \end{bmatrix} = \begin{bmatrix} 1 & -z^{-1}K_m \\ \\ K_m & z^{-1}(1 - K_m^2) \end{bmatrix} \begin{bmatrix} e_m(z) \\ \\ H_{m-1}^b(z)\, r_0(z) \end{bmatrix} \tag{6.57}$$

or, explicitly,

$$H_{m-1}^{f^{-1}}(z)\, e_0(z) = e_m(z) - z^{-1}K_m H_{m-1}^b(z)\, r_0(z) \ , \tag{6.58a}$$

$$r_m(z) = K_m e_m(z) + z^{-1}(1 - K_m^2)\, H_{m-1}^b(z)\, r_0(z) \ . \tag{6.58b}$$

Since we know that $e_0(z) = r_0(z) = y(z)$ is the z-transform of the filtered output sequence, we may rewrite (6.58a) to obtain

$$e_0(z)\left[1 + z^{-1}K_m\, H_{m-1}^f(z)\, H_{m-1}^b(z)\right] = H_{m-1}^f(z)\, e_m(z) \ . \tag{6.59}$$

Comparison of (6.59) with (6.56a) finally results in the desired order recursive law of the forward predictor transfer function of the feed-back PARCOR ladder form:

$$H_m^f(z) = \frac{H_{m-1}^f(z)}{1 + z^{-1}K_m\, H_{m-1}^f(z)\, H_{m-1}^b(z)} \ . \tag{6.60}$$

As a next step, one can substitute (6.56a) into (6.58b) to find

$$r_m(z) = \left[K_m H_m^{f^{-1}}(z) + z^{-1}(1 - K_m^2)\, H_{m-1}^b(z)\right] r_0(z) \ . \tag{6.61}$$

Comparison of this expression with the definition of the backward predictor transfer function of the feed-back PARCOR ladder form (6.56b) finally gives the desired recursive law of the backward predictor transfer function:

$$H_m^b(z) = K_m H_m^{f^{-1}}(z) + z^{-1}(1 - K_m^2) H_{m-1}^b(z)$$

$$= \frac{K_m + z^{-1} H_{m-1}^f(z) H_{m-1}^b(z)}{H_{m-1}^f(z)} \quad . \tag{6.62}$$

6.6.3 Relationships Between Forward and Backward Predictor Transfer Functions

In this part of the discussion, we wish to explore further the very important relationships between the forward and backward predictors in a feed-forward PARCOR ladder form. The expressions (6.38a,b) have already revealed that the forward/backward prediction error filters are identical but with reverse-ordered coefficients or, alternatively, an inverse signal flow direction in the backward predictor. This fact leads to a nice relationship between forward and backward predictor transfer functions in the feed-forward PARCOR ladder form as described below.

Taking the z-transforms of (6.38a,b) we have

$$e_m(z) = - \sum_{j=0}^m a_j^{(m)} z^{-j} x(z) \quad , \tag{6.63a}$$

$$r_m(z) = - \sum_{j=0}^m a_j^{(m)} z^{-m} z^j x(z) \quad . \tag{6.63b}$$

Recalling the definitions of forward/backward predictor transfer functions in the case of the feed-forward PARCOR ladder form, we may readily see that

$$H_m^f(z) = e_m(z) / x(z) = - \sum_{j=0}^m a_j^{(m)} z^{-j} \quad , \tag{6.64a}$$

$$H_m^b(z) = r_m(z) / x(z) = - z^{-m} \sum_{j=0}^m a_j^{(m)} z^j \quad . \tag{6.64b}$$

The inverse signal flow direction in the backward predictor is associated with a transfer function that depends on z^{-1} rather than on z alone:

$$- \sum_{j=0}^m a_j^{(m)} z^j = H_m^f(z^{-1}) \quad . \tag{6.65}$$

This gives rise to the following important relationship between forward and backward predictor transfer functions in a feed-forward PARCOR ladder form:

$$H_m^b(z) = z^{-m} H_m^f(z^{-1}) \quad . \tag{6.66}$$

6.7 Stability of the Feed-Back PARCOR Ladder Form

The feed-back PARCOR ladder form is an *all-pole* filter; hence it is *not* unconditionally stable. Stability of the feed-back PARCOR ladder form is guaranteed if and only if all roots of the forward predictor transfer function fall *inside* the unit circle of the z-plane. An equivalent condition is that all zeros of the corresponding feed-forward PARCOR ladder form (inverse filter) fall inside the unit circle. We are interested in deriving a relationship between the PARCOR ladder coefficients and the location of the zeros in the z-plane. Thereby, we obtain the bounds on the PARCOR coefficients that correspond to a transfer function with all zeros located inside the unit circle. The derivation will take advantage of the order recursive properties of the PARCOR ladder transfer functions derived earlier in this chapter. In this way, we assume first that we have given the transfer function of a stable filter (all zeros inside the unit circle) of order m-1. Subsequently, we add one more ladder stage and derive the bounds on this additional PARCOR coefficient.

Consider the forward predictor transfer function of the first-order filter

$$H_1^f(z) = 1 + K_1 z^{-1} \ . \tag{6.67}$$

The single zero of this filter is located at $z_{1,1} = - K_1$ and falls inside the unit circle when the condition

$$|K_1| < 1 \tag{6.68}$$

is satisfied.

Next we investigate the transfer function of a forward predictor of order m in its *order recursive* representation as the weighted sum of the forward and backward transfer functions of order $m-1$. Rewriting (6.52a), we have

$$H_m^f(z) = H_{m-1}^f(z) + z^{-1} K_m H_{m-1}^b(z) \ . \tag{6.69}$$

The forward and backward predictor transfer functions of the order m-1 predictor can be expressed in terms of their zeros $\{z_{m-1,\,i} \ , \ 1 \le i \le m-1\}$

$$H_{m-1}^f(z) = \prod_{i=1}^{m-1} \left(1 - z_{m-1,\,i}\, z^{-1}\right) \ . \tag{6.70}$$

Recalling the relationship between forward and backward predictor transfer functions in a feed-forward PARCOR ladder form (6.66),

$$H_{m-1}^b(z) = z^{-m+1} H_{m-1}^f(z^{-1}) \ , \tag{6.71}$$

one may express the backward predictor transfer function of order $m-1$ by the zeros of the forward predictor (6.70) as

$$H_{m-1}^{b}(z) = z^{-m+1} \prod_{i=1}^{m-1} \left(1 - z_{m-1,i} \, z \right) = \prod_{i=1}^{m-1} \left(z^{-1} - z_{m-1,i} \right) . \quad (6.72)$$

Exploiting the expressions (6.71 and 72) allows the transfer function of the mth order forward predictor to be expressed in terms of the zeros of the order $m-1$ forward predictor and the PARCOR coefficient K_m of the additional ladder section:

$$H_{m}^{f}(z) = \prod_{i=1}^{m-1} \left(1 - z_{m-1,i} \, z^{-1} \right) + z^{-1} K_m \prod_{i=1}^{m-1} \left(z^{-1} - z_{m-1,i} \right) . \quad (6.73)$$

By setting the transfer function $H_{m}^{f}(z)$ to zero, one can determine the location of the zeros of the order m predictor in the z-plane and how these locations depend on K_m and on the zeros of the "old" order $m-1$ predictor. This gives the functional that is required for developing a stability condition, as desired.

Introducing the zeros of the order m predictor $\{z_{m,j}, 1 \le j \le m\}$, we have

$$H_{m}^{f}(z_{m,j}) = \prod_{i=1}^{m-1} \left(1 - z_{m-1,i} \, z_{m,j}^{-1} \right)$$

$$+ z_{m,j}^{-1} K_m \prod_{i=1}^{m-1} \left(z_{m,j}^{-1} - z_{m-1,i} \right) \overset{!}{=} 0 \quad (6.74)$$

or, equivalently,

$$K_m = - z_{m,j} \prod_{i=1}^{m-1} \frac{\left(1 - z_{m-1,i} \, z_{m,j}^{-1} \right)}{\left(z_{m,j}^{-1} - z_{m-1,i} \right)} . \quad (6.75)$$

Expression (6.75) must be evaluated for all zeros $\{z_{m,j}, 1 \le j \le m\}$.

Since the coefficients of $H_{m}^{f}(z)$ are all real-valued, the zeros will be real or will appear as complex conjugate pairs. Hence, it is clear that we can *reorder* the product terms in the denominator of (6.75) to obtain

$$K_m = - z_{m,j} \prod_{i=1}^{m-1} A_i(z_{m,j}) ; \qquad 1 \le j \le m , \quad (6.76a)$$

where

$$A_i(z_{m,j}) = \frac{\left(1 - z_{m-1,i} \, z_{m,j}^{-1} \right)}{\left(z_{m,j}^{-1} - z_{m-1,i}^{*} \right)} ; \qquad 1 \le j \le m . \quad (6.76b)$$

Clearly, the product terms (6.76b) are functions of the type

$$A(z) = \frac{1 - z_0 z^{-1}}{z^{-1} - z_0^*} \ , \tag{6.77}$$

where z_0^* is the *complex conjugate* of z_0 and

$$0 < |z_0| < 1 \tag{6.78}$$

because we have assumed the overall stability of the order m-1 feed-back ladder form through the condition

$$0 < |z_{m-1,i}| < 1 \ ; \qquad\qquad 1 \le i \le m-1 \ . \tag{6.79}$$

Taking the squared absolute value of (6.76a)

$$|K_m|^2 = |z_{m,j}|^2 \prod_{i=1}^{m-1} |A_i(z_{m,j})|^2 \ ; \qquad 1 \le j \le m \tag{6.80}$$

makes it necessary to evaluate functions of the type

$$|A(z)|^2 = \left| \frac{1 - z_0 z^{-1}}{z^{-1} - z_0^*} \right|^2 \ . \tag{6.81}$$

After some calculations, the squared modulus of A(z) may be expressed as

$$|A(z)|^2 = \frac{|z|^2 - 2 \, Re\{z_0 z^*\} + |z_0|^2}{1 - 2 \, Re\{z_0 z^*\} + |z_0|^2 |z|^2} = \frac{|N(z)|^2}{|D(z)|^2} \ . \tag{6.82}$$

Taking the difference of the numerator and the denominator moduli, we obtain

$$|N(z)|^2 - |D(z)|^2 = |z|^2 + |z_0|^2 - 1 - |z_0|^2 |z|^2$$

$$= - \left(1 - |z|^2 \right) \left(1 - |z_0|^2 \right) \ . \tag{6.83}$$

From (6.78), it is easy to check that the second term of (6.83) is always positive:

$$0 < 1 - |z_0|^2 < 1 \ . \tag{6.84}$$

The *difference* between the numerator and denominator moduli (6.83) is zero if $|z|^2 = 1$. In this case, the value of the squared numerator modulus is *equal* to the value of the squared denominator modulus. For $|z|^2 < 1$, the difference (6.83) is *negative*, indicating that the squared denominator modulus is *greater* than the squared numerator modulus. Finally, for $|z|^2 > 1$, the difference (6.83) is *positive* and therefore the squared numerator modulus is *greater* than the squared denominator modulus.

The result of these considerations can be summarized as follows:

$$|N(z)|^2 - |D(z)|^2 \begin{cases} < 0 & \text{for } |z|^2 < 1 \longrightarrow |A(z)|^2 < 1 \\ = 0 & \text{for } |z|^2 = 1 \longrightarrow |A(z)|^2 = 1 \\ > 0 & \text{for } |z|^2 > 1 \longrightarrow |A(z)|^2 > 1 \end{cases} \quad (6.85)$$

This result (6.85) can be directly used for evaluation of our original problem (6.80). Clearly, it follows that for all $\{z_{m,j}, 1 \le j \le m\}$ the following condition holds:

$$|K_m|^2 \begin{cases} < 1 & \text{for } |z_{m,j}|^2 < 1 \\ = 1 & \text{for } |z_{m,j}|^2 = 1 \\ > 1 & \text{for } |z_{m,j}|^2 > 1 \end{cases} \quad (6.86)$$

More precisely, we can interpret (6.86) as follows. Only the strict condition

$$|K_m|^2 < 1 \quad (6.87)$$

ensures that all zeros of the mth-order feed-forward PARCOR prediction error filter will fall inside the unit circle

$$|z_{m,j}|^2 < 1 \ ; \qquad 1 \le j \le m \quad (6.88)$$

and therefore corresponds to a *stable* feed-back PARCOR ladder form with poles located *inside* the unit circle. Since we have assumed the stability of the order m-1 filter, it is clear that condition (6.87) holds for all m from m=1 to m=p. The proof is by induction. Strictly speaking, the feed-back PARCOR ladder form is stable *if and only if*

$$|K_m|^2 < 1 \quad \text{for} \quad 1 \le m \le p \ . \quad (6.89)$$

6.8 Burg's Harmonic Mean PARCOR Ladder Algorithm

A physical process $x(t)$ of finite length is *never* completely stationary. It might represent "ongoing" or transient parts of a signal that correspond to estimated ladder reflection coefficients of magnitude *greater* than unity. However we have seen that the feed-back PARCOR ladder form synthesis filter operates stably if and only if all K's are of magnitude *less* than unity. For this reason, in many applications where the K's are later used in a synthesis filter, one is interested in obtaining *stable* coefficient sets with all K's of magnitude smaller than unity, independent of the characteristics of the observed process.

One such procedure that guarantees at least the *wide-sense stability* of the synthesis filter, independent of the estimated signal, is *Burg's harmonic mean PARCOR ladder algorithm* [6.10, 11]. The basic idea in Burg's technique is the following. Assume that we compute the coefficient K_m as the *harmonic mean* of the forward and backward ladder reflection coefficients as

$$K_m = \frac{2 K_m^f K_m^b}{K_m^f + K_m^b} \quad . \tag{6.90}$$

Then, by recalling that

$$K_m^f = - \frac{e_{m-1}^T(t) \, r_{m-1}(t-1)}{r_{m-1}^T(t-1) \, r_{m-1}(t-1)} \qquad \text{and} \tag{6.91a}$$

$$K_m^b = - \frac{e_{m-1}^T(t) \, r_{m-1}(t-1)}{e_{m-1}^T(t) \, e_{m-1}(t)} \quad , \tag{6.91b}$$

one sees that a substitution of (6.91a,b) into (6.90) gives

$$K_m = - \frac{2 \, e_{m-1}^T(t) \, r_{m-1}(t-1)}{r_{m-1}^T(t-1) \, r_{m-1}(t-1) + e_{m-1}^T(t) \, e_{m-1}(t)} \quad . \tag{6.92}$$

It is mentioned, for completeness, that the same result is obtained when K_m is determined such that the *sum* of the squared forward/backward prediction errors is minimized:

$$\frac{\partial}{\partial(K_m)} \left(r_{m-1}^T(t-1) \, r_{m-1}(t-1) \right.$$

$$\left. + e_{m-1}^T(t) \, e_{m-1}(t) \right) = 0 \quad \longrightarrow \quad K_m \; ; \qquad 1 \le m \le p \quad . \tag{6.93}$$

It remains to demonstrate that with this choice (6.92), $|K_m| < 1$ is always satisfied.

Proof. Recall the fundamental geometrical property

$$e^T r = |e| \, |r| \, \cos \varphi \quad . \tag{6.94}$$

We may check that Burg's formula (6.92) is of the type

$$K_m = - \frac{2 |e| \, |r| \, \cos \varphi}{|e|^2 + |r|^2} \quad . \tag{6.95}$$

Since

$$|\cos \varphi| \leq 1 , \tag{6.96}$$

it remains to show that

$$\frac{2 |e| \, |r|}{|e|^2 + |r|^2} \leq 1 . \tag{6.97}$$

Clearly, condition (6.97) is only satisfied when the denominator expression is *greater than or equal* to the numerator expression of (6.97). Hence, as a proof of (6.97), it is sufficient to demonstrate that the *difference* between the denominator and the numerator terms of (6.97) is *greater than or equal to zero.*

$$|e|^2 + |r|^2 - 2 |e| \, |r| = \left(|r| - |e| \right)^2 \geq 0 . \tag{6.98}$$

From these considerations it follows that

$$|K_m| = 1 \qquad for \quad |r| = |e| \quad and \quad \cos \varphi = 1$$

$$which \ implies \quad r = e \tag{6.99a}$$

$$|K_m| < 1 \qquad otherwise. \tag{6.99b}$$

End of proof.

This result of our stability considerations requires a short discussion. Equation (6.99a) reveals the identity of forward and backward residual vectors $r = e$ is only possible for a *constant* process vector, a case that is of no interest in practice. Hence, for all practical cases, condition (6.99b) holds and therefore, Burg's formula always yields stable synthesis filters. These results were obtained disregarding the effect of round-off error in practical computation. Note that $|K_m| < 1$ might be violated due to round-off error. Table 6.7 provides a summary of Burg's harmonic mean PARCOR ladder algorithm with complete computation of residual vectors.

The algorithm of Table 6.7 is somewhat inefficient because of the nested loop structure in the computation of (6.100e,f). The total operations count is therefore $O(p^2 L)$ operations, where p denotes the model order and L is the length of the data vector. We shall now investigate a technique where delayed residual vectors are simply replaced by *circularly shifted* actual residual vectors. The error induced through this approximation is mainly concentrated at the boundaries of the residual vectors and can therefore be eliminated by employing an appropriate *semi inner product*, which cuts off the first and last m

Table 6.7. Burg type harmonic mean PARCOR ladder algorithm with explicit computation of residual vectors (block estimation). This algorithm involves division. Whenever the divisor is small, set $1/x = x$

Input: $x(t), x(t-1), \ldots, x(t-p)$

FOR $i = 0, 1, 2, \ldots, p$ *Initialize:*

$\quad e_0(t-i) = r_0(t-i) = x(t-i)$

FOR $m = 1, 2, \ldots, p$

$$E_{m-1} = e_{m-1}^T(t) \, e_{m-1}(t) \qquad\qquad (6.100a)$$

$$R_{m-1} = r_{m-1}^T(t-1) \, r_{m-1}(t-1) \qquad\qquad (6.100b)$$

$$C_{m-1} = e_{m-1}^T(t) \, r_{m-1}(t-1) \qquad\qquad (6.100c)$$

$$K_m = -2 \, C_{m-1} \Big/ \Big(E_{m-1} + R_{m-1} \Big) \qquad\qquad (6.100d)$$

$\quad$ FOR $j = 0, 1, 2, \ldots, p-m$

$$e_m(t-j) = e_{m-1}(t-j) + K_m r_{m-1}(t-1-j) \qquad\qquad (6.100e)$$

$$r_m(t-j) = r_{m-1}(t-1-j) + K_m e_{m-1}(t-j) \qquad\qquad (6.100f)$$

---

components in the residual vectors when the order recursion progresses to higher stages. This technique finally results in an algorithm of only $O(pL)$ operations. A listing of this algorithm is available in Table 6.8.

Let $r'_m(t)$ be an auxiliary backward residual vector defined as

$$r'_m(t) = r_m(t+m) \; . \qquad\qquad (6.101)$$

Then

$$r'_{m-1}(t) = r_{m-1}(t+m-1) \; , \qquad\qquad (6.102a)$$

$$r'_{m-1}(t-m) = r_{m-1}(t-1) \; . \qquad\qquad (6.102b)$$

One finds that the PARCOR ladder vector order recursions (6.34a,b) can be rewritten in terms of the auxiliary backward residual vector $r'_m(t)$

$$e_m(t) = e_{m-1}(t) + K_m r'_{m-1}(t-m) \; , \qquad\qquad (6.103a)$$

$$r'_m(t) = r'_{m-1}(t) + K_m e_{m-1}(t+m) \; , \qquad\qquad (6.103b)$$

with the initial condition

$$r'_0(t) = e_0(t) = x(t) \ . \tag{6.104}$$

Let S_m be the L×L matrix that circularly shifts a vector m components *forward*, and let S_{-m} be the corresponding L×L matrix that circularly shifts a vector m components *backward*:

$$S_m = \begin{bmatrix} & & 1 & & & & & \\ & & & 1 & & & & \\ & & & & \ddots & & & \\ & & & & & & 1 & \\ 1 & & & & & & & \\ & 1 & & & & & & \\ & & \ddots & & & & & \\ & & & 1 & & & & \end{bmatrix} , \tag{6.105a}$$

$$S_{-m} = \begin{bmatrix} & & & & 1 & & & \\ & & & & & 1 & & \\ & & & & & & \ddots & \\ & & & & & & & 1 \\ 1 & & & & & & & \\ & 1 & & & & & & \\ & & \ddots & & & & & \\ & & & & & & & 1 \end{bmatrix} , \tag{6.105b}$$

where all unspecified elements are zero. These shift matrices are related to each other via the properties

$$S_m^T = S_{-m} \ , \tag{6.106a}$$

$$S_m^T S_m = S_{-m}^T S_{-m} = S_m S_{-m} = S_{-m} S_m = I \ , \tag{6.106b}$$

where I is the identity matrix of dimension L×L.

Utilizing the shift matrices S_m and S_{-m} , we may introduce the approximations

$$r'_{m-1}(t-m) \approx S_m r'_{m-1}(t) \ , \tag{6.107a}$$

$$e_{m-1}(t+m) \approx S_{-m} e_{m-1}(t) \ . \tag{6.107b}$$

These approximations are justified from the assumption of a stationary process $\mathbf{x}(t)$ characterized by a time-invariant sequence of PARCOR ladder coefficients. Using the approximations (6.107a,b), it is easy to see that the modified PARCOR ladder vector order recursions (6.103a,b) take on the form

$$\mathbf{e}_m = \mathbf{e}_{m-1} + K_m \mathbf{S}_m \mathbf{r}'_{m-1} , \tag{6.108a}$$

$$\mathbf{r}'_m = \mathbf{r}'_{m-1} + K_m \mathbf{S}_{-m} \mathbf{e}_{m-1} , \tag{6.108b}$$

where the time index can now be omitted.

The Burg estimator for K_m may be derived by evaluating the expression

$$\partial/\partial K_m \left(\mathbf{e}_m^T \mathbf{e}_m + \mathbf{r}'^T_m \mathbf{r}'_m \right) \overset{!}{=} 0 , \tag{6.109}$$

which gives

$$K_m = - \frac{2 \mathbf{e}_{m-1}^T \mathbf{S}_m \mathbf{r}'_{m-1}}{\mathbf{e}_{m-1}^T \mathbf{e}_{m-1} + \mathbf{r}'^T_{m-1} \mathbf{r}'_{m-1}} . \tag{6.110}$$

Clearly, this estimator (6.110) will give fairly poor estimation results if the observed data vector $\mathbf{x}(t)$ is not invariant with respect to the circular shift operation. However, this is practically always the case. This problem can be circumvented by introducing a *semi inner product* of two vectors $\mathbf{x}$ and $\mathbf{y}$ as follows.

Definition. (Semi inner product.)

Let $\mathbf{x}$ and $\mathbf{y}$ be two vectors with L components

$$\mathbf{x} = \left[x_0, x_1, \ldots , x_{L-1} \right]^T , \tag{6.111a}$$

$$\mathbf{y} = \left[y_0, y_1, \ldots , y_{L-1} \right]^T . \tag{6.111b}$$

We define the *semi inner product* $|\mathbf{x}, \mathbf{y}|_m$ as

$$|\mathbf{x}, \mathbf{y}|_m = x_m y_m + x_{m+1} y_{m+1} + \cdots + x_{L-m-1} y_{L-m-1} . \tag{6.112}$$

This definition of the semi inner product "rejects" m components at the boundaries of the involved vectors, thereby omitting the nonstationary boundary regions in the residual vectors generated by the vector order recursions (6.108a,b). The modified harmonic mean estimator for K_m hence becomes

$$K_m = - \frac{2 \left| e_{m-1}, S_m r'_{m-1} \right|_m}{\left| e_{m-1}, e_{m-1} \right|_m + \left| S_m r'_{m-1}, S_m r'_{m-1} \right|_m} . \qquad (6.113)$$

Table 6.8 provides a summary of this type of PARCOR ladder algorithm with explicit computation of residual vectors and reduced computational complexity based on the circular shift approximation (6.107a,b) and the corresponding modified Burg harmonic mean estimator (6.113).

Table 6.8. Burg type harmonic mean PARCOR ladder algorithm with explicit computation and circular shift approximation of residual vectors. This algorithm involves division. Whenever the divisor is small, set $1/x = x$

Input: $x = \left[x_0, x_1, \ldots, x_{L-1} \right]^T$

Initialize residual vectors:

$e_0 = r'_0 = x$; $\varepsilon_0 = \rho_0 = \left[0 \ldots 0 \right]^T$

FOR m = 1, 2, . . . , p   *compute:*

  FOR j = L-m, L-m+1, . . . , L-1   *circular shift:*

   $\rho_{m-1}(j) = r'_{m-1}(j-L+m)$ (6.114a)

   $\varepsilon_{m-1}(j-L+m) = e_{m-1}(j)$ (6.114b)

  FOR j = 0, 1, . . . , L-m-1

   $\rho_{m-1}(j) = r'_{m-1}(j+m)$ (6.114c)

   $\varepsilon_{m-1}(j+m) = e_{m-1}(j)$ (6.114d)

   $E_{m-1} = \sum_{j=m}^{L-m-1} \left(e_{m-1}(j) \right)^2$ (6.114e)

   $R_{m-1} = \sum_{j=m}^{L-m-1} \left(\rho_{m-1}(j) \right)^2$ (6.114f)

   $C_{m-1} = \sum_{j=m}^{L-m-1} e_{m-1}(j) \, \rho_{m-1}(j)$ (6.114g)

   $K_m = - 2 \, C_{m-1} \Big/ \left(E_{m-1} + R_{m-1} \right)$ (6.114h)

  FOR j = 0, 1, . . . , L-1

   $e_m(j) = e_{m-1}(j) + K_m \rho_{m-1}(j)$ (6.114i)

   $r'_m(j) = r'_{m-1}(j) + K_m \varepsilon_{m-1}(j)$ (6.114j)

6.9 Determination of Model Order

As mentioned earlier, the application of ladder-form prediction error filters allows an easy variation of the model order, since subsequent stages of the ladder form are orthogonal or *"decoupled"*. Next we discuss some useful rules for determination of the appropriate model (or prediction error filter) order; in other words, we introduce some criteria that help to decide when it is appropriate to stop the orthogonalization process. We shall discuss two criteria. They can be traced back to some early work of Akaike (see also [6.12]).

In order to determine the most suitable filter order m, one has to take into account that both the record (window) length L and the statistical properties of the analyzed time series will affect the choice of m_{opt}. It turns out that in the majority of practical measurements, the optimum value of m lies in the range from $m_{opt} = 0.05$ L to $m_{opt} = 0.2$ L.

For the proper choice of the optimal value of the filter order, denoted by m_{opt}, we may use one of several objective criteria. These criteria are based on the *sample moments* of the given time series, and they are evaluated sucessively at each value of m. When m attains the optimal value m_{opt}, the order recursive orthogonalization process may be terminated automatically. We next discuss two specific criteria for choosing the order of the prediction error filter.

1. Final Prediction Error (FPE) Criterion. This criterion is defined as the product of two terms. The first term expresses a relationship between the window length L and the actual model order m. The second term is directly the prediction error energy $E_m(t)$ obtained by using the prediction error filter of order m:

$$FPE(m,t) = \frac{L + m + 1}{L - m - 1} \; E_m(t) \; . \tag{6.115}$$

Since $E_m(t)$ *decreases* with increasing model order m, and the first expression in the product depending on L and m *increases* with m, the FPE(m,t) will have a minimum at some value m = m_{opt}. This value defines the optimal value for the prediction filter based on the final prediction error criterion.

2. The Akaike Error Criterion (AIC). This criterion, suggested by Akaike, is based on the minimization of the log-likelihood of the prediction error energy as a function of the predictor order m:

$$AIC(m,t) = \ln\left(E_m(t)\right) + 2p/L \; . \tag{6.116}$$

Here again, AIC(m,t) may be calculated successively as the orthogonalization process progresses to higher values of m to obtain m_{opt}, which minimizes AIC(m,t).

Note that AIC(m,t) and FPE(m,t) are asymptotically equivalent for increasing window length L, hence

$$\lim_{L \to \infty} \left\{ \ln \left[FPE(m,t) \right] \right\} = AIC(m,t) \ . \tag{6.117}$$

A more detailed treatment of these criteria can be found in [6.12]. Other forms of order determination schemes for the order estimation of autoregressive models were discussed by Boekee and Buss [6.13].

6.10 Chapter Summary

We have introduced the ladder form prediction error filter as an interesting alternative to the more obvious transversal (tapped delay line) filter structure. The ladder form arises as a natural consequence of the orthogonalization process underlying the "geometrical" solution of the Normal Equations of linear prediction as constituted by the recursion formula for orthogonal projections. One of the most interesting properties of the ladder form lies in the "decoupled" nature of subsequent filter stages, which allows an easy dynamic *variation* of the filter order. Other advantages are the low sensitivity of the ladder coefficients to round-off error and their convenient interpretation as "reflection" coefficients in many applications of modeling physical processes.

The operation of ladder forms as fixed filters with low round-off error sensitivity was discussed by Fettweis [6.14,15], Sedlmeyer and Fettweis [6.16], and Bruton [6.17].

A special case of interest arises from the stationarity assumption of the underlying (observed) process. In this case, the ladder form attains a special structure, the PARCOR ladder form, characterized by identical values of the reflection coefficients in the forward and backward paths of the filter. We discussed all important relationships between the PARCOR ladder form and its transversal structured counterpart. Originally, the PARCOR ladder form was introduced as an all-zero (analysis) filter. Interestingly, by applying Mason's rule we also obtained a ladder realization of the corresponding (inverse) all-pole synthesis filter. This structure was termed the "feed-back" PARCOR ladder form.

A large part of this chapter is devoted to transfer functions of PARCOR ladder forms. Another important topic is the stability analysis of the feed-back PARCOR ladder form. Interestingly, it turned out that the feed-back PARCOR ladder form is stable if and only if all coefficients are of magnitude smaller than unity. This leads to a simple evaluation of the stability of an all-pole synthesis filter.

Since an observed process is never completely stationary, estimated PARCOR ladder coefficients may violate the above-discussed stability condition of the synthesis filter. The problem can be circumvented by applying *Burg's harmonic mean method* in the estimation of PARCOR ladder reflection coefficients. Burg's technique guarantees that the estimated reflection coefficients will always determine a stable synthesis filter. Two versions of a PARCOR ladder algorithm with explicit calculation of residual vectors based on Burg's harmonic mean principle have been discussed. The first algorithm (Table 6.7) arises quite naturally from the time recursive explicit ladder algorithm of Table 6.1 by simply combining the computation of forward and backward reflection coefficients into a single formula, namely, Burg's harmonic mean formula. The transition to the block processing case turned out to be a formidable task, since here the computational complexity was found to be $O(p^2L)$ operations. As a solution to this problem, we discussed the circular shift approximation of residual vectors, which results in an algorithm of only $O(pL)$ operations. This algorithm was listed in Table 6.8. Some criteria for the automatic determination of the optimal prediction error filter order were discussed.

Burg's method found many applications in the area of parametric spectral estimation [6.11, 18]. Several modifications of Burg's method appeared in the literature, such as Fougere's method [6.19] for a joint optimization of the estimated reflection coefficients.

7. Levinson-Type Ladder Algorithms

In Sect. 6.8, we introduced the Burg algorithm as one possible method of solving the Normal Equations in the stationary case of linear prediction (2.48). Clearly, Burg's technique is closely related to the explicit ladder algorithm of Table 6.1. Like the explicit ladder algorithm, Burg's technique requires the *explicit* calculation of residual signal vectors. Such procedures are known to have poor numerical properties and a computational complexity that grows linearly with the data (record) length. On the other hand, the computation of ladder reflection coefficients requires knowledge only about the *residual energies*, rather than the *residual signals* themselves. This important fact gives rise to the central question, whether one can establish order recursions of residual energies, as a substitute for the cumbersome residual signal vector order recursions appearing in the explicit least-squares ladder algorithm (Table 6.1) and in the Burg type PARCOR ladder algorithm of Table 6.7.

Although the desired order recursions of the forward and backward residual energies $E_m(t)$ and $R_m(t)$ are quickly established [recall (6.23a,b)], the derivation of a similar order recursion for the prediction error covariance $C_m(t)$ turns out to be a formidable task and was postponed in Chap. 6, in order to be tackled in later chapters.

In the present chapter, we will show that the *Levinson recursion* [7.1] is one technique that facilitates an order recursive updating of the prediction error covariance. Recursions of the Levinson type require the computation of transversal predictor parameters of increasing order as *intermediate variables*. This leads first to the classical Levinson-Durbin algorithm [7.1, 2] for solving *Toeplitz systems* of linear equations. The computation of transversal predictor parameters as intermediate variables turns out to be one of the major drawbacks of the Levinson-Durbin algorithm. These quantities require a higher dynamic range than PARCOR ladder reflection coefficients. Moreover, Levinson's recursion sometimes appears to be still a fairly poorly conditioned procedure.

To remedy this problem, LeRoux and Gueguen [7.3] have suggested different types of intermediate variables, which are defined as *inner products* between the observed process vector and residual vectors in successive stages of the ladder form. These types of inner products can best be characterized as *generalized residual energies* (GREs). Several types of GREs are possible. They lead to new classes of ladder algorithms with outstanding numerical properties. These algorithms are the counterparts to Levinson type algorithms and will be presented in Chap. 8.

Such an increased numerical accuracy is often paid for by an increased number of computations per recursion. Therefore, another important direction of research is focused on the development of algorithms with *minimal computational complexity* in terms of a minimum number of operations for solving a given least-squares problem. One such technique that will be discussed in this chapter is the new class of *split Levinson algorithms* introduced by Delsarte and Genin [7.4] in 1986. These algorithms require only roughly half the number of multiplications for solving a Toeplitz system of linear equations as the classical Levinson-Durbin algorithm.

Towards the end of this chapter, we present the *Makhoul covariance ladder algorithm* [7.5] as a technique for fitting a PARCOR ladder form to a (general) symmetric system of Normal Equations which are *not* a priori known to be Toeplitz (covariance case). Makhoul's covariance ladder algorithm can be viewed as the mathematical equivalent to the Burg type algorithm of Table 6.7, where only the explicit residual vector order recursions are replaced by a type of recursion based on residual energies and transversal predictor parameters, as is the case in Levinson type algorithms.

From these considerations, it is already clear that there must exist a Levinson type counterpart to the explicit (time-recursive true least-squares) ladder algorithm of Table 6.1. Such a technique is presented in the last part of this chapter. Although the algorithm presented there is not really of practical interest, we may use the considerations to derive a new form of a particularly useful Levinson type RLS ladder algorithm. This, however, is postponed to Appendix A.1. One of the major objectives of this chapter is therefore to point out how a given least-squares linear prediction problem can be solved with quite different types of algorithms and how these families of algorithms are related to each other.

7.1 The Levinson-Durbin Algorithm

Consider the *symmetric* Toeplitz system of Normal Equations as associated with the *autocorrelation case* of linear prediction. The power normalized version of these equations has become known as the *Yule-Walker* equations [7.6,7] Recall (2.48). The Yule-Walker equations constitute the relationship between a sequence of m autocorrelation coefficients and the same number of predictor parameters of the mth-order transversal prediction filter. The Yule-Walker equations are rewritten from Chap. 2, for completeness:

$$
\begin{bmatrix}
1 & c_1 & \cdots\cdots & c_{m-1} \\
c_1 & 1 & c_1 & \vdots \\
\vdots & \ddots & \ddots & \vdots \\
\vdots & & \ddots & c_1 \\
c_{m-1} & \cdots & c_1 & 1
\end{bmatrix}
\begin{bmatrix}
a_1 \\
a_2 \\
\vdots \\
\vdots \\
a_m
\end{bmatrix}
=
\begin{bmatrix}
c_1 \\
c_2 \\
\vdots \\
\vdots \\
c_m
\end{bmatrix} .
\tag{7.1}
$$

In most cases, the autocorrelation coefficients of a stationary process are known and one is interested in the computation of the *transversal predictor parameters* $\{a_i , 1 \le i \le m\}$. In this problem, the system matrix appears to be a symmetric Toeplitz matrix determined by the *power-normalized autocorrelation sequence* $\{c_i, 0 \le i \le m\}$, where $c_0 = 1$. We may introduce the *normalized residual energies* C'_m and E'_m of stage m as defined by

$$
C'_m = C_m / \Phi_{0,0} , \tag{7.2a}
$$

$$
E'_m = E_m / \Phi_{0,0} , \tag{7.2b}
$$

where $\Phi_{0,0} = x^T x$ is the energy of the observed process defined by (2.44). The definition of E_m and C_m can be traced back to (6.18a,c). We know that the backward prediction residual energy is not required since the Toeplitz assumption of the Normal Equations forces $R_m = E_m$. See also (6.32).

With these conventions, one may check that the general forward/backward prediction residual energy recursions (6.23a,b) reduce to a single energy recursion for E'_m as follows:

$$
E'_m = E'_{m-1} + K_m C'_{m-1} \tag{7.3}
$$

where

$$
K_m = - C'_{m-1} / E'_{m-1} , \tag{7.4}
$$

which simply follows from a comparison of (6.19a,b) with (6.33). From a comparison of the impulse response of an mth-order transversal prediction error filter with a PARCOR ladder form [7.8] of order m, we have deduced the Levinson recursion (6.39). See also Table 6.4. Recall that this recursion relates the transversal predictor parameters $\{a_j^{(m)} , 1 \le j \le m\}$ of the mth-order transversal prediction error filter to the PARCOR ladder reflection coefficients $\{K_i, 1 \le i \le m\}$ as

$$
a_j^{(m)} = a_j^{(m-1)} + K_m a_{m-j}^{(m-1)} , \quad 1 \le j \le m-1 , \tag{7.5a}
$$

$$
a_m^{(m)} = - K_m , \tag{7.5b}
$$

$$
a_0^{(m)} = - 1 . \tag{7.5c}
$$

An explicit listing of Levinson's recursion was given in Table 6.5. Besides the Levinson recursion, we may note that the coefficients of the transversal predictors of increasing order m are defined by the following sequence of Yule-Walker equations of increasing dimension:

$$\begin{bmatrix} 1 \end{bmatrix} \begin{bmatrix} a_1^{(1)} \end{bmatrix} = \begin{bmatrix} c_1 \end{bmatrix} \quad , \tag{7.6a}$$

$$\begin{bmatrix} 1 & c_1 \\ c_1 & 1 \end{bmatrix} \begin{bmatrix} a_1^{(2)} \\ a_2^{(2)} \end{bmatrix} = \begin{bmatrix} c_1 \\ c_2 \end{bmatrix} \quad , \tag{7.6b}$$

$$\begin{bmatrix} 1 & c_1 & c_2 \\ c_1 & 1 & c_1 \\ c_2 & c_1 & 1 \end{bmatrix} \begin{bmatrix} a_1^{(3)} \\ a_2^{(3)} \\ a_3^{(3)} \end{bmatrix} = \begin{bmatrix} c_1 \\ c_2 \\ c_3 \end{bmatrix} \quad , \tag{7.6c}$$

$$\vdots$$

$$\begin{bmatrix} 1 & c_1 & \cdots\cdots & c_{m-1} \\ c_1 & 1 & c_1 & \vdots \\ \vdots & & \ddots & \\ & & & c_1 \\ c_{m-1} & \cdots\cdots & c_1 & 1 \end{bmatrix} \begin{bmatrix} a_1^{(m)} \\ a_2^{(m)} \\ \vdots \\ a_m^{(m)} \end{bmatrix} = \begin{bmatrix} c_1 \\ c_2 \\ \vdots \\ c_m \end{bmatrix} \quad . \tag{7.6d}$$

Solving this chain of linear systems of increasing order for the last element $\{a_j^{(j)}, 1 \le j \le m\}$, one obtains the following interesting recursive law:

$$a_1^{(1)} = - K_1 = c_1 = C_0'/E_0' \quad , \tag{7.7a}$$

$$a_2^{(2)} = - K_2 = \frac{- a_1^{(1)} c_1 + c_2}{1 - a_1^{(1)} c_1} = \frac{C_1'}{E_1'} \quad , \tag{7.7b}$$

$$a_3^{(3)} = - K_3 = \frac{- a_2^{(2)} c_1 - a_1^{(2)} c_2 + c_3}{1 - a_1^{(2)} c_1 - a_2^{(2)} c_2} = \frac{C_2'}{E_2'} \quad , \tag{7.7c}$$

$$\vdots$$

$$a_m^{(m)} = -K_m = \frac{C'_{m-1}}{E'_{m-1}}$$

$$= \frac{-a_{m-1}^{(m-1)} c_1 - a_{m-2}^{(m-1)} c_2 - \cdots - a_1^{(m-1)} c_{m-1} + c_m}{1 - a_1^{(m-1)} c_1 - a_2^{(m-1)} c_2 - \cdots - a_{m-1}^{(m-1)} c_{m-1}} \ . \tag{7.7d}$$

Evaluating the numerator expression of (7.7d), one sees that we can compute the error covariance of stage m-1 from the autocorrelation sequence $\{c_i, 1 \le i \le m\}$ and the transversal predictor parameters $\{a_i^{(m-1)}$, $1 \le i \le m-1\}$. Rewriting the *numerator expression* of (7.7d), we obtain the key relationship

$$C'_{m-1} = -\sum_{i=0}^{m-1} c_{m-i} a_i^{(m-1)} \ . \tag{7.8}$$

A similar recursion can be derived from the *denominator expression* of (7.7d) for the normalized prediction error energy

$$E'_{m-1} = -\sum_{i=0}^{m-1} c_i a_i^{(m-1)} \ . \tag{7.9}$$

Just with these expressions, one can establish a closed recursion for the computation of the PARCOR ladder reflection coefficients $\{K_i, 1 \le i \le m\}$. As became apparent in the derivation of (7.8), this recursion is based on *residual energies* and *transversal predictor parameters*, rather than on the direct computation of prediction errors. This algorithm has become known as the *Levinson-Durbin algorithm* [7.1,2]. It consists of first computing C'_{m-1} via (7.8), then the residual energy E'_{m-1} is updated via (7.3) and the PARCOR ladder reflection coefficient K_m is computed according to (7.4). As soon as K_m is known, we can compute the transversal predictor parameters of stage m by using Levinson's recursion (7.5a-c). These quantities are required in the next step of the recursion for the computation of the prediction error covariance (7.8).

Several different formulations of the Levinson-Durbin algorithm can be found. For example, a closer look at (7.4) reveals that we can express the prediction error energy in terms of the prediction error covariance and the PARCOR coefficient K_m as

$$E'_{m-1} = -C'_{m-1}/K_m \ . \tag{7.10}$$

Substitution of this expression into (7.3) gives

$$E'_m = -\frac{C'_{m-1}}{K_m} + K_m C'_{m-1} = C'_{m-1}\left(K_m - 1/K_m\right) \ . \tag{7.11}$$

This finally results in the modified recursion

$$E'_m = E'_{m-1}\left(1 - K_m^2\right) \ . \tag{7.12}$$

Exploiting (7.12), the Levinson-Durbin algorithm can be conveniently formulated as a *three-term recurrence relation*

$$K_m = \left(E'_{m-1}\right)^{-1} \sum_{i=0}^{m-1} c_{m-i} a_i^{(m-1)} \ , \tag{7.13a}$$

$$E'_m = E'_{m-1}\left(1 - K_m^2\right) \ , \tag{7.13b}$$

$$a_j^{(m)} = a_j^{(m-1)} + K_m a_{m-j}^{(m-1)} \ ; \quad 1 \le j \le m-1 \ . \tag{7.13c}$$

The complete Levinson-Durbin algorithm is listed in Table 7.1. This algorithm requires $m^2 + O(m)$ multiplications and $m^2 + O(m)$ additions for computing the PARCOR coefficients from a given sequence of autocorrelation coefficients.

7.2 Computing the Autocorrelation Coefficients from the PARCOR Ladder Reflection Coefficients – The "Inverse" Levinson-Durbin Algorithm

From previous considerations, we have seen that the transversal predictor parameters $\{a_j^{(m)}, 1 \le j \le m\}$, the PARCOR ladder reflection coefficients $\{K_j, 1 \le j \le m\}$ and the sequence of autocorrelation coefficients $\{c_j, 1 \le j \le m\}$ are *equivalent* representations of a stationary autoregressive process. They can be *converted* into each other. In Chap. 6, we presented the algorithms for converting a's into K's and vice versa. Clearly, the conversion of c's into K's is accomplished by the Levinson-Durbin algorithm (Table 7.1).

We are now interested in the *inverse* procedure, namely, the conversion of a given sequence of PARCOR ladder reflection coefficients back into a sequence of autocorrelation coefficients. For this purpose, we investigate the Yule-Walker equations of the mth-order model (7.6d). The last equation in this system can be written as

$$c_m = \sum_{i=1}^{m} c_{m-i} a_i^{(m)} \ . \tag{7.14}$$

Obviously, with (7.14) one can easily establish a recursion for the compu-

Table 7.1. Levinson-Durbin algorithm for converting a sequence of auto-correlation coefficients $\{c_i, 1 \le i \le p\}$ into a sequence of PARCOR ladder reflection coefficients $\{K_m, 1 \le m \le p\}$. This algorithm involves division. Whenever the divisor is small and positive, set $1/x = x$. When the divisor is negative, set $1/x = 0$

Input: $c_i, 1 \le i \le p$

Initialize:

$K_1 = -c_1$ $\qquad\qquad$ $E_1' = (1 - K_1^2)$

$a_0^{(1)} = -1$ $\qquad\qquad$ $a_1^{(1)} = -K_1$

FOR $m = 2, 3, \ldots, p$

$$K_m = \left(E_{m-1}'\right)^{-1} \sum_{i=0}^{m-1} c_{m-i} a_i^{(m-1)}$$

$$E_m' = E_{m-1}'\left(1 - K_m^2\right)$$

FOR $j = 1, 2, \ldots, m-1$

$$a_j^{(m)} = a_j^{(m-1)} + K_m a_{m-j}^{(m-1)}$$

$$a_0^{(m)} = -1 \qquad\qquad a_m^{(m)} = -K_m$$

---

tation of the autocorrelation coefficients from the PARCOR ladder reflection coefficients as follows. We first use the Levinson recursion for a computation of the transversal predictor parameters from the PARCOR ladder reflection coefficients (Table 6.5). After the set of trans-versal predictor parameters of stage m has been computed, one can evaluate (7.14) to obtain the desired autocorrelation coefficient c_m. The complete algorithm is listed in Table 7.2.

Table 7.2. "Inverse" Levinson-Durbin algorithm for computing the sequence of autocorrelation coefficients $\{c_m, 1 \le m \le p\}$ from the PARCOR ladder reflection coefficients $\{K_m, 1 \le m \le p\}$

Input: $K_m, 1 \le m \le p$

Initialize:

$a_0^{(1)} = -1$ $a_1^{(1)} = -K_1$

$c_0 = 1$ $c_1 = a_1^{(1)}$

FOR $m = 2, 3, \ldots, p$

$\quad$ FOR $j = 1, 2, \ldots, m-1$

$\qquad a_j^{(m)} = a_j^{(m-1)} + K_m a_{m-j}^{(m-1)}$

$\quad a_0^{(m)} = -1$ $a_m^{(m)} = -K_m$

$\quad c_m = \sum_{i=1}^{m} c_{m-i} a_i^{(m)}$

7.3 Some More Properties of Toeplitz Systems and the Levinson-Durbin Algorithm

An investigation of the deep relationships of Toeplitz forms and the related Levinson-Durbin algorithm can be of great interest. Consider a normalized Toeplitz matrix $T^{(m)}$ of rank $m+1$

$$T^{(m)} = \begin{bmatrix} 1 & c_1 & \cdots\cdots & c_m \\ c_1 & 1 & c_1 & \vdots \\ \vdots & \ddots & \ddots & \vdots \\ \vdots & & \ddots & c_1 \\ c_m & \cdots\cdots & c_1 & 1 \end{bmatrix} . \tag{7.15}$$

and an $(m+1)$-dimensional column vector $r^{(m)}$

$$r^{(m)} = \left[r_0^{(m)}, r_1^{(m)}, \ldots, r_m^{(m)} \right]^T . \tag{7.16a}$$

The column vector $r^{(m)}$ may be expressed as a polynomial in z:

$$r^{(m)}(z) = r_0^{(m)} z^0 + r_1^{(m)} z^1 + \ldots + r_m^{(m)} z^m$$

$$= \sum_{i=0}^{m} r_i^{(m)} z^i \quad . \tag{7.16b}$$

For a given polynomial

$$r^{(m)}(z) = \sum_{i=0}^{m} r_i^{(m)} z^i \quad , \tag{7.17a}$$

a *reciprocal polynomial* $\overset{\vee}{r}{}^{(m)}(z)$ is defined as follows:

$$\overset{\vee}{r}{}^{(m)}(z) = \sum_{i=0}^{m} r_{m-i}^{(m)} z^i \quad . \tag{7.17b}$$

Introducing the *exchange matrix* of rank m+1

$$\mathbf{J} = \begin{bmatrix} & & & 1 \\ & & \cdot^{\cdot^{\cdot^{\cdot}}} & \\ & 1 & & \\ 1 & & & \end{bmatrix} \quad , \tag{7.18}$$

one can relate the corresponding *reciprocal vector* $\overset{\vee}{\mathbf{r}}{}^{(m)}$ to $\mathbf{r}^{(m)}$ as follows:

$$\overset{\vee}{\mathbf{r}}{}^{(m)} = \mathbf{J}\,\mathbf{r}^{(m)} \quad . \tag{7.19}$$

It will be useful for forthcoming considerations to distinguish between *symmetric* and *antisymmetric* vectors.

Definition of symmetric and antisymmetric vectors

A vector $\mathbf{r}^{(m)}$, possibly given by its polynomial representation $r^{(m)}(z)$, is termed a *symmetric* vector, iff

$$\mathbf{r}^{(m)} = \overset{\vee}{\mathbf{r}}{}^{(m)} \quad \text{or, equivalently,} \quad r^{(m)}(z) = \overset{\vee}{r}{}^{(m)}(z) \quad . \tag{7.20a}$$

A vector $\mathbf{r}^{(m)}$, possibly given by its polynomial representation $r^{(m)}(z)$, is termed an *antisymmetric* vector, iff

$$\mathbf{r}^{(m)} = -\overset{\vee}{\mathbf{r}}{}^{(m)} \quad \text{or, equivalently,} \quad r^{(m)}(z) = -\overset{\vee}{r}{}^{(m)}(z) \quad . \tag{7.20b}$$

In either of the two cases, the vector $r^{(m)}$ is completely specified by

m/2 elements, if m is *even*,

(m+1)/2 elements, if m is *odd*.

Next we note that there exists the following *exchange invariance property* of a real symmetric Toeplitz matrix $T^{(m)}$:

$$J \, T^{(m)} J \;=\; T^{(m)} . \tag{7.21}$$

Additionally, we may note that the Levinson-Durbin algorithm, as consitituted by the three-term recurrence relation (7.13a-c), can be expressed using the symmetric/antisymmetric vector notation as

$$K_m = \left(E'_{m-1} \right)^{-1} \sum_{i=0}^{m-1} c_{m-i} \, a_i^{(m-1)} , \tag{7.22a}$$

$$E'_m = E'_{m-1}\left(1 - K_m^2\right) , \tag{7.22b}$$

$$a^{(m)} = \begin{bmatrix} a^{(m-1)} \\[1em] 0 \end{bmatrix} + K_m \begin{bmatrix} 0 \\[1em] \overset{\vee}{a}{}^{(m-1)} \end{bmatrix} , \tag{7.22c}$$

where

$$a^{(m)} = \left[a_0^{(m)}, a_1^{(m)}, \ldots, a_{m-1}^{(m)}, a_m^{(m)} \right]^T , \tag{7.23a}$$

$$a^{(m-1)} = \left[a_0^{(m-1)}, a_1^{(m-1)}, \ldots, a_{m-1}^{(m-1)} \right]^T , \tag{7.23b}$$

$$\overset{\vee}{a}{}^{(m-1)} = J \, a^{(m-1)} . \tag{7.23c}$$

We have discovered that the Levinson recursion (7.13c) can be interpreted as a *vector recursion* where the actual parameter vector $a^{(m)}$ of stage m is computed as a *linear combination* of its predecessor $a^{(m-1)}$ and the *reverse-ordered* predecessor $\overset{\vee}{a}{}^{(m-1)}$ of stage m-1. This important fact can shed some light on Levinson's recursion.

We wish to rewrite equation (7.22c) more explicitly :

$$
\begin{bmatrix}
a_0^{(m)} \\
a_1^{(m)} \\
a_2^{(m)} \\
\vdots \\
\vdots \\
a_{m-1}^{(m)} \\
a_m^{(m)}
\end{bmatrix}
=
\begin{bmatrix}
a_0^{(m-1)} \\
a_1^{(m-1)} \\
a_2^{(m-1)} \\
\vdots \\
\vdots \\
a_{m-1}^{(m-1)} \\
0
\end{bmatrix}
+ K_m
\begin{bmatrix}
0 \\
a_{m-1}^{(m-1)} \\
\vdots \\
\vdots \\
a_2^{(m-1)} \\
a_1^{(m-1)} \\
a_0^{(m-1)}
\end{bmatrix}
,
\tag{7.24}
$$

where

$$
a_0^{(m)} = -1 \qquad for \qquad 1 \le m \le p .
\tag{7.25}
$$

From (7.24 and 25), it follows that

$$
a_m^{(m)} = -K_m \qquad for \qquad 1 \le m \le p .
\tag{7.26}
$$

On the other hand, one may express Levinson's recursion in terms of a *polynomial recursion* as

$$
a^{(m)}(z) = a^{(m-1)}(z) + K_m z \overset{\vee}{a}^{(m-1)}(z) .
\tag{7.27}
$$

This gives rise to the following polynomial representation of the Levinson–Durbin algorithm:

$$
K_m = \left(E'_{m-1} \right)^{-1} \sum_{i=0}^{m-1} c_{m-i} a_i^{(m-1)} ,
\tag{7.28a}
$$

$$
E'_m = E'_{m-1}\left(1 - K_m^2\right) ,
\tag{7.28b}
$$

$$
a^{(m)}(z) = a^{(m-1)}(z) + K_m z \overset{\vee}{a}^{(m-1)}(z) .
\tag{7.28c}
$$

Besides this representation of the Levinson–Durbin algorithm as a polynomial recursion in z, we may discover several more interesting properties of the Levinson–Durbin algorithm, as shown below.

From equation (7.9), in consideration of (7.15 and 23a), it follows that

$$
T^{(m)} a^{(m)} = \left[-E'_m, 0 \ldots 0 \right]^T .
\tag{7.29}
$$

Moreover, one readily finds that

$$
\mathbf{T}^{(m)} \begin{bmatrix} \mathbf{a}^{(m-1)} \\ \\ 0 \end{bmatrix} = \left[-E'_{m-1}, 0 \ldots 0, \sum_{i=0}^{m-1} c_{m-i} a_i^{(m-1)} \right]^T . \tag{7.30}
$$

Exploiting the expressions (7.8) together with (7.4), one can rewrite (7.30) in terms of the *normalized prediction error energy* and the *PARCOR ladder reflection coefficient* of stage m as

$$
\mathbf{T}^{(m)} \begin{bmatrix} \mathbf{a}^{(m-1)} \\ \\ 0 \end{bmatrix} = \left[-E'_{m-1}, 0 \ldots 0, K_m E'_{m-1} \right]^T . \tag{7.31}
$$

On the other hand,

$$
\mathbf{T}^{(m)} \begin{bmatrix} 0 \\ \\ \mathbf{a}^{(m-1)} \end{bmatrix} = \left[\sigma_{m-1}, -E'_{m-1}, 0 \ldots 0 \right]^T , \tag{7.32}
$$

where

$$
\sigma_{m-1} = \sum_{i=0}^{m-1} c_{i+1} a_i^{(m-1)} . \tag{7.33}
$$

Evaluation of polynomial (7.27) at location $z = 1$ gives

$$
a^{(m)}(1) = a^{(m-1)}(1) + K_m \overset{\vee}{a}{}^{(m-1)}(1) . \tag{7.34}
$$

From the definition of a *reciprocal* polynomial (7.17a,b), it is obvious that

$$
a^{(m-1)}(1) = \overset{\vee}{a}{}^{(m-1)}(1) , \tag{7.35}
$$

and therefore (7.34) changes to

$$
a^{(m)}(1) = \left[1 + K_m \right] a^{(m-1)}(1) , \tag{7.36}
$$

or, in a *nonrecursive* formulation,

$$
a^{(m)}(1) = \prod_{i=1}^{m} \left(1 + K_i \right) . \tag{7.37}
$$

7.4 Split Levinson Algorithms

Continuing the considerations about symmetric and antisymmetric vector recursion representations of the Levinson-Durbin algorithm, it can be shown that this algorithm is *redundant* in complexity. It can be split into *two* algorithms, only one of which needs to be processed. Such *"split Levinson" algorithms* have been introduced recently by Delsarte and Genin [7.4] and Krishna [7.9]. Split Levinson algorithms allow the inversion of a symmetric Toeplitz system with roughly *half* the number of multiplications as required in the classical Levinson-Durbin algorithm.

The development of split algorithms stems from the distinct property that *a real symmetric Toeplitz matrix transforms real symmetric vectors into real symmetric vectors* and, consequently, *transforms real antisymmetric vectors into real antisymmetric vectors* [7.10]. In this way, a special Toeplitz system, where the right–hand–side vector is, for example, symmetric, is a priori known to have a symmetric solution, and, of course, this solution can be computed much faster than the solution of a general real symmetric Toeplitz system of the same order.

Starting from this important observation, Delsarte and Genin had the key idea of embedding a general Toeplitz system of order m into a Toeplitz system of order m+1 with the right–hand–side vector being symmetric (or antisymmetric), as described above. This was only possible since it can be shown that an order m+1 symmetric Toeplitz system with a symmetric right–hand–side vector can be related to a general order m symmetric Toeplitz system so that both systems are *identical* representations of the same problem. The interesting point is, however, that the symmetric order m+1 system can be solved *much faster* than the general order m system, as mentioned above. From these introductory considerations, it is obvious that the starting point in the derivation of split algorithms is the definition of the right–hand–side vector of the special order m+1 system. Clearly, we have several possibilities. For example, Delsarte and Genin have used the convention

$$\mathbf{T}^{(m)}\mathbf{p}^{(m)} = \left[\tau^{(m)}, 0 \ldots 0, \tau^{(m)}\right]^{\mathrm{T}}, \tag{7.38a}$$

in the symmetric case and

$$\mathbf{T}^{(m)}\mathbf{p}^{*(m)} = \left[\tau^{*(m)}, 0 \ldots 0, -\tau^{*(m)}\right]^{\mathrm{T}} \tag{7.38b}$$

in the antisymmetric case. But there are several other possibilities, such as

$$\mathbf{T}^{(m)}\mathbf{q}^{(m)} = \left[r^{(m)}, r^{(m)}, \ldots, r^{(m)}\right]^{\mathrm{T}} \tag{7.39}$$

or

$$\mathbf{T}^{(m)}\mathbf{q}^{*(m)} = \left[\, +r^{(m)}, \; -r^{(m)}, \; +r^{(m)}, \, \ldots \, , \; -r^{(m)}, \; +r^{(m)} \,\right]^{T}. \quad (7.40)$$

For each such initial guess, a related split algorithm can be derived. It remains to mention that the algorithms derived from such different initial assumptions turn out to be fairly similar in their final structure.

7.4.1 Delsarte's Algorithm

In the following, we describe the split Levinson algorithm introduced by Delsarte and Genin [7.4]. Only the symmetric case (7.38a) will be discussed explicitly. The first step towards this algorithm is the insight that [recall (7.29)]

$$\mathbf{T}^{(m)}\left[\mathbf{a}^{(m)} + \overset{\vee}{\mathbf{a}}^{(m)}\right] = \left[-E_m', \, 0 \ldots 0, \, -E_m'\right]^{T}. \quad (7.41)$$

Since the real symmetric Toeplitz matrix $\mathbf{T}^{(m)}$ converts real symmetric vectors into real symmetric vectors, it is clear that the *split transversal parameter vector* $\mathbf{p}^{(m)}$ defined as

$$\mathbf{p}^{(m)} = \begin{bmatrix} \mathbf{a}^{(m-1)} \\ \\ 0 \end{bmatrix} + \begin{bmatrix} 0 \\ \\ \overset{\vee}{\mathbf{a}}^{(m-1)} \end{bmatrix} , \quad (7.42)$$

where

$$\lambda^{(m)}\mathbf{p}^{(m)} = \mathbf{a}^{(m)} + \overset{\vee}{\mathbf{a}}^{(m)} , \quad (7.43)$$

will also be symmetric, and hence the Toeplitz system (7.41) can be solved much faster than the general Yule-Walker equations (7.1). However, the solution of (7.1) is implicitly given by the solution of (7.41), thus leading to a faster solution of the Yule-Walker equations. We shall explore the latter statement by noting that [from (7.22c) and (7.42, 43)]

$$\lambda^{(m)}\mathbf{p}^{(m)} = \mathbf{p}^{(m)} + K_m\,\mathbf{p}^{(m)} , \quad (7.44)$$

or

$$\lambda^{(m)} = 1 + K_m . \quad (7.45)$$

Next, one introduces a *symmetric intermediate right-hand-side vector* determined by $\tau^{(m)}$ and

$$\mathbf{T}^{(m)}\mathbf{p}^{(m)} = \left[\tau^{(m)}, \, 0 \ldots 0, \, \tau^{(m)}\right]^{T} . \quad (7.46)$$

Combining (7.46) with (7.43) and (7.41),

$$\lambda^{(m)} T^{(m)} p^{(m)} = T^{(m)} \left[a^{(m)} + \overset{\vee}{a}{}^{(m)} \right]$$

$$= \left[-E'_m, 0 \ldots 0, -E'_m \right]^T, \qquad (7.47)$$

and again using (7.46), we have

$$\lambda^{(m)} \left[\tau^{(m)}, 0 \ldots 0, \tau^{(m)} \right]^T$$

$$= \left[-E'_m, 0 \ldots 0, -E'_m \right]^T. \qquad (7.48)$$

The vector equation (7.48) is completely determined by a single scalar equation, namely

$$\lambda^{(m)} \tau^{(m)} = -E'_m . \qquad (7.49)$$

Introducing the partitioning schemes

$$T^{(m)} \begin{bmatrix} p^{(m-1)} \\ \\ 0 \end{bmatrix} = \left[\tau^{(m-1)}, 0 \ldots 0, \tau^{(m-1)}, \gamma^{(m-1)} \right]^T, \qquad (7.50a)$$

$$T^{(m)} \begin{bmatrix} 0 \\ \\ p^{(m-1)} \end{bmatrix} = \left[\gamma^{(m-1)}, \tau^{(m-1)}, 0 \ldots 0, \tau^{(m-1)} \right]^T, \qquad (7.50b)$$

$$T^{(m)} \begin{bmatrix} 0 \\ p^{(m-2)} \\ 0 \end{bmatrix} = \left[\gamma^{(m-2)}, \tau^{(m-2)}, 0 \ldots 0, \tau^{(m-2)}, \gamma^{(m-2)} \right]^T, \qquad (7.51)$$

and the *split Levinson recursion*

$$p^{(m)} = \begin{bmatrix} p^{(m-1)} \\ \\ 0 \end{bmatrix} + \begin{bmatrix} 0 \\ \\ p^{(m-1)} \end{bmatrix} - \alpha^{(m)} \begin{bmatrix} 0 \\ p^{(m-2)} \\ 0 \end{bmatrix}, \qquad (7.52)$$

we note that a premultiplication of (7.52) with $T^{(m)}$, in consideration of (7.50a,b) and (7.51), gives

$$
\begin{bmatrix} \tau^{(m)} \\ 0 \\ \vdots \\ \vdots \\ \vdots \\ 0 \\ \tau^{(m)} \end{bmatrix} = \begin{bmatrix} \tau^{(m-1)} \\ 0 \\ \vdots \\ 0 \\ \tau^{(m-1)} \\ \gamma^{(m-1)} \end{bmatrix} + \begin{bmatrix} \gamma^{(m-1)} \\ \tau^{(m-1)} \\ 0 \\ \vdots \\ 0 \\ \tau^{(m-1)} \end{bmatrix} - \alpha^{(m)} \begin{bmatrix} \gamma^{(m-2)} \\ \tau^{(m-2)} \\ 0 \\ \vdots \\ 0 \\ \tau^{(m-2)} \\ \gamma^{(m-2)} \end{bmatrix} . \quad (7.53)
$$

The vector order recursion (7.53) is completely determined by two scalar equations

$$
\tau^{(m)} = \tau^{(m-1)} + \gamma^{(m-1)} - \alpha^{(m)} \gamma^{(m-2)} , \tag{7.54a}
$$

$$
0 = \tau^{(m-1)} - \alpha^{(m)} \tau^{(m-2)} , \tag{7.54b}
$$

where $\alpha^{(m)}$ is the *split reflection coefficient* and $\gamma^{(*)}$ is the *split residual covariance*. From (7.54b), one immediately obtains

$$
\alpha^{(m)} = \tau^{(m-1)} \Big/ \tau^{(m-2)} , \tag{7.55}
$$

and from (7.46) and (7.15), one sees

$$
\tau^{(m)} = \sum_{i=0}^{m} c_i \, p_i^{(m)} . \tag{7.56}
$$

Equation (7.43) can be solved for $\overset{\vee}{a}$ of the order (m-1) problem,

$$
\overset{\vee}{a}^{(m-1)} = \lambda^{(m-1)} p^{(m-1)} - a^{(m-1)} , \tag{7.57}
$$

and by substituting (7.57) into (7.42), we have

$$
p^{(m)} - \lambda^{(m-1)} \begin{bmatrix} 0 \\ \\ p^{(m-1)} \end{bmatrix} = \begin{bmatrix} a^{(m-1)} \\ \\ 0 \end{bmatrix} - \begin{bmatrix} 0 \\ \\ a^{(m-1)} \end{bmatrix} . \tag{7.58}
$$

Premultiplication of (7.58) by $T^{(m)}$, in consideration of (7.50b), (7.46) and (7.31 and 32) yields

$$
\begin{bmatrix}
\tau^{(m)} \\
0 \\
\vdots \\
\vdots \\
0 \\
\tau^{(m)}
\end{bmatrix}
- \lambda^{(m-1)}
\begin{bmatrix}
\gamma^{(m-1)} \\
\tau^{(m-1)} \\
0 \\
\vdots \\
0 \\
\tau^{(m-1)}
\end{bmatrix}
=
\begin{bmatrix}
-E'_{m-1} \\
0 \\
\vdots \\
\vdots \\
0 \\
K_m E'_{m-1}
\end{bmatrix}
-
\begin{bmatrix}
\sigma_{m-1} \\
-E'_{m-1} \\
0 \\
\vdots \\
0 \\
0
\end{bmatrix}
. \qquad (7.59)
$$

From (7.59), we may extract the relations

$$
- \lambda^{(m-1)} \, \tau^{(m-1)} = E'_{m-1} \quad , \qquad\qquad (7.60a)
$$

$$
\tau^{(m)} - \lambda^{(m-1)} \, \tau^{(m-1)} = K_m E'_{m-1} \quad . \qquad\qquad (7.60b)
$$

Substituting (7.60a) into (7.60b) gives

$$
\tau^{(m)} = (1 - K_m) \, \lambda^{(m-1)} \, \tau^{(m-1)} \quad . \qquad\qquad (7.61)
$$

Exploiting (7.45), we may express $\lambda^{(m-1)}$ in terms of K_{m-1} to obtain

$$
\tau^{(m)} = (1 - K_m)(1 + K_{m-1}) \, \tau^{(m-1)} \quad . \qquad\qquad (7.62)
$$

Using again (7.55), the split reflection coefficient $\alpha^{(m+1)}$ can be expressed in terms of conventional PARCOR coefficients as

$$
\alpha^{(m+1)} = (1 - K_m)(1 + K_{m-1}) \quad , \qquad\qquad (7.63)
$$

or equivalently,

$$
K_m = 1 - \frac{\alpha^{(m+1)}}{1 + K_{m-1}} \quad , \qquad\qquad (7.64)
$$

and hence the algorithm is complete. One first computes the order update of $\mathbf{p}^{(m)}$ using the split Levinson recursion (7.52). This recursion requires only $(m/2)$ multiplications due to the symmetric property of $\mathbf{p}^{(m)}$. The intermediate quantity $\tau^{(m)}$ is computed using (7.56). Here, again, only $(m/2)$ multiplications are needed because of the symmetry of the split transversal parameter vector $\mathbf{p}^{(m)}$. The split reflection coefficient $\alpha^{(m)}$ is computed via (7.55), while the computation of K_m via (7.64) concludes the order recursive algorithm. Table 7.3 provides a summary of this algorithm, which has become known as Delsarte's split Levinson algorithm.

Table 7.3. Delsarte's original split Levinson algorithm based on the *symmetric* intermediate solution vector $\mathbf{p}^{(m)}$. Equation (7.65f) is the computation of the transversal predictor parameters $\mathbf{a}^{(m-1)}$ at a desired order m. As (7.65f) is not a part of the order recursion, its computation remains *optional*. Note also that (7.65a) requires only m/2 computations due to the symmetric property of $\mathbf{p}^{(m)}$. This algorithm involves division. Whenever the divisor is small, set $1/x = x$

Input: $c_1, c_2, \ldots, c_p$

Initialize: $\mathbf{p}^{(0)} = \begin{bmatrix} -2 \end{bmatrix}$ $\mathbf{p}^{(1)} = \begin{bmatrix} -1, -1 \end{bmatrix}^T$

$\quad\quad\quad\quad \tau^{(1)} = -1 - c_1$

$\quad\quad\quad\quad \alpha^{(2)} = 1 + c_1$ $K_1 = -c_1$

FOR m = 2, 3, 4, . . .

$$
\mathbf{p}^{(m)} = \begin{bmatrix} \mathbf{p}^{(m-1)} \\ \\ 0 \end{bmatrix} + \begin{bmatrix} 0 \\ \\ \mathbf{p}^{(m-1)} \end{bmatrix} - \alpha^{(m)} \begin{bmatrix} 0 \\ \mathbf{p}^{(m-2)} \\ 0 \end{bmatrix}
\tag{7.65a}
$$

$$
\mathbf{m = odd:} \quad \tau^{(m)} = \sum_{i=0}^{(m-1)/2} \left(c_i + c_{m-i} \right) p_i^{(m)}
\tag{7.65b}
$$

$$
\mathbf{m = even:} \quad \tau^{(m)} = c_{m/2}\, p_{m/2}^{(m)} + \sum_{i=0}^{(m-2)/2} \left(c_i + c_{m-i} \right) p_i^{(m)}
\tag{7.65c}
$$

$$
\alpha^{(m+1)} = \frac{\tau^{(m)}}{\tau^{(m-1)}}
\tag{7.65d}
$$

$$
K_m = 1 - \frac{\alpha^{(m+1)}}{(1 + K_{m-1})}
\tag{7.65e}
$$

If desired (at any stage m) **compute sequentially:**

$$
\begin{bmatrix} \mathbf{a}^{(m-1)} \\ \\ 0 \end{bmatrix} = \mathbf{p}^{(m)} - (1 + K_{m-1}) \begin{bmatrix} 0 \\ \\ \mathbf{p}^{(m-1)} \end{bmatrix} + \begin{bmatrix} 0 \\ \\ \mathbf{a}^{(m-1)} \end{bmatrix}
\tag{7.65f}
$$

7.4.2 Krishna's Algorithm

Besides the more "natural" approach of Delsarte, who assumed

$$\mathbf{T}^{(m)}\mathbf{p}^{(m)} = \left[\tau^{(m)}, 0 \ldots 0, \tau^{(m)} \right]^T , \tag{7.66}$$

since

$$\mathbf{T}^{(m)} \left[\mathbf{a}^{(m)} + \overset{\vee}{\mathbf{a}}^{(m)} \right] = \left[-E_m', 0 \ldots 0, -E_m' \right]^T \tag{7.67}$$

[recall (7.38a) and (7.41)], an alternative split Levinson algorithm can be established by using the "constant vector" initial guess (7.39). This algorithm is attributed to Krishna [7.9,11]. Again we concentrate on the symmetric case. The antisymmetric case follows a very similar derivation. Consider the scalar quantity $r^{(m)}$ determining the constant right-hand-side vector of the special Toeplitz system (7.39). This quantity is adjusted such that

$$q_0^{(m)} = -1 . \tag{7.68}$$

It is emphasized that the system (7.39) has the fundamental property that the solution vector $\mathbf{q}^{(m)}$ is *symmetric* and hence it follows from (7.68) that

$$q_m^{(m)} = -1 . \tag{7.69}$$

Furthermore, we may introduce the following *asymmetric* partitioning schemes

$$\mathbf{T}^{(m)} \begin{bmatrix} \mathbf{q}^{(m-1)} \\ \\ 0 \end{bmatrix} = \left[r^{(m-1)}, r^{(m-1)}, \ldots, r^{(m-1)}, \gamma^{(m-1)} \right]^T , \tag{7.70a}$$

$$\mathbf{T}^{(m)} \begin{bmatrix} 0 \\ \\ \mathbf{q}^{(m-1)} \end{bmatrix} = \left[\gamma^{(m-1)}, r^{(m-1)}, \ldots, r^{(m-1)}, r^{(m-1)} \right]^T \tag{7.70b}$$

and the *symmetric* partitioning scheme

$$\mathbf{T}^{(m)} \begin{bmatrix} 0 \\ \mathbf{q}^{(m-2)} \\ 0 \end{bmatrix} = \left[\gamma^{(m-2)}, r^{(m-2)}, \ldots, r^{(m-2)}, \gamma^{(m-2)} \right]^T . \tag{7.71}$$

Suppose now that again a split Levinson recursion for $\mathbf{q}^{(m)}$ can be established by using $\mathbf{q}^{(m-1)}$ and $\mathbf{q}^{(m-2)}$ as follows:

$$
\mathbf{q}^{(m)} = \begin{bmatrix} \mathbf{q}^{(m-1)} \\ \\ 0 \end{bmatrix} + \begin{bmatrix} 0 \\ \\ \mathbf{q}^{(m-1)} \end{bmatrix} - \alpha^{(m)} \begin{bmatrix} 0 \\ \mathbf{q}^{(m-2)} \\ 0 \end{bmatrix} . \tag{7.72}
$$

It remains to determine the scalar quantity $\alpha^{(m)}$ such that the recursion (7.72) is satisfied. Premultiplication of both sides of (7.72) with the system matrix $\mathbf{T}^{(m)}$ gives

$$
\begin{bmatrix} r^{(m)} \\ r^{(m)} \\ \vdots \\ r^{(m)} \\ r^{(m)} \end{bmatrix} = \begin{bmatrix} r^{(m-1)} \\ r^{(m-1)} \\ \vdots \\ r^{(m-1)} \\ \gamma^{(m-1)} \end{bmatrix} + \begin{bmatrix} \gamma^{(m-1)} \\ r^{(m-1)} \\ \vdots \\ r^{(m-1)} \\ r^{(m-1)} \end{bmatrix} - \alpha^{(m)} \begin{bmatrix} \gamma^{(m-2)} \\ r^{(m-2)} \\ \vdots \\ r^{(m-2)} \\ \gamma^{(m-2)} \end{bmatrix} . \tag{7.73}
$$

A quick inspection of (7.73) reveals that this vector equation is completely specified by only *two* linear and independent *scalar* equations, namely

$$
r^{(m)} = r^{(m-1)} + \gamma^{(m-1)} - \alpha^{(m)} \gamma^{(m-2)} , \tag{7.74a}
$$

$$
r^{(m)} = 2 r^{(m-1)} - \alpha^{(m)} r^{(m-2)} . \tag{7.74b}
$$

Substituting (7.74a) into (7.74b) finally gives the desired condition on $\alpha^{(m)}$ such that recursion (7.73) is fulfilled :

$$
r^{(m-1)} + \gamma^{(m-1)} - \alpha^{(m)} \gamma^{(m-2)} = 2 r^{(m-1)} - \alpha^{(m)} r^{(m-2)} \tag{7.75}
$$

or, equivalently,

$$
\alpha^{(m)} = \frac{r^{(m-1)} - \gamma^{(m-1)}}{r^{(m-2)} - \gamma^{(m-2)}} . \tag{7.76}
$$

As soon as $\alpha^{(m)}$ has been computed via (7.76), the expression (7.74b) can serve as a recursion for the scalar quantity $r^{(m)}$ that determines the right-hand-side vector of the special Toeplitz system (7.39).

The vector order recursion (7.73) can alternatively be expressed as a polynomial recursion

$$
q^{(m)}(z) = (1 + z) q^{(m-1)}(z) - \alpha^{(m)} z q^{(m-2)}(z) . \tag{7.77}
$$

The quantities $\gamma^{(m-1)}$ and $\gamma^{(m-2)}$ can be obtained from (7.70b and 71):

$$\gamma^{(m-1)} = \sum_{i=0}^{m-1} c_{i+1} q_i^{(m-1)} \quad , \tag{7.78a}$$

$$\gamma^{(m-2)} = \sum_{i=0}^{m-2} c_{i+1} q_i^{(m-2)} \quad . \tag{7.78b}$$

After the quantities $\alpha^{(m)}$, $\gamma^{(m-1)}$ and $\gamma^{(m-2)}$ have been determined, the order recursive scheme for updating $\mathbf{q}^{(m)}$ can be summarized as follows. First compute $\gamma^{(m-1)}$ via (7.78a). $\gamma^{(m-2)}$ is obtained from an appropriate stack. Then compute $\alpha^{(m)}$ via (7.76). Finally, compute $\mathbf{q}^{(m)}$ via (7.72). The right-hand-side vector $\mathbf{r}^{(m)}$ can be updated using either (7.74a) or (7.74b).

A closer look at the order recursion (7.72) reveals that the resulting updated vector $\mathbf{q}^{(m)}$ will be *symmetric* if $\mathbf{q}^{(m-1)}$ and $\mathbf{q}^{(m-2)}$ are symmetric vectors. Hence, the symmetry of $\mathbf{q}^{(m)}$ will hold through all stages m, from $m=1$ to $m=p$. (Proof by induction.) As a consequence, the computation of (7.78a) requires only $0.5 p^2 + O(p)$ multiplications and $p^2 + O(p)$ additions.

Next, it is of interest to explore the relationships between the desired transversal predictor polynomial $a^{(m)}(z)$ and the solution of the special system (7.39) determining $q^{(m)}(z)$. A combination of this special Toeplitz system (7.39) with a scaled version of (7.70b) (scaling factor is $\lambda^{(m)}$) gives

$$\mathbf{T}^{(m)} \left(\begin{bmatrix} \mathbf{q}^{(m)} \end{bmatrix} - \lambda^{(m)} \begin{bmatrix} 0 \\ \mathbf{q}^{(m-1)} \end{bmatrix} \right) = \begin{bmatrix} r^{(m)} - \lambda^{(m)} \gamma^{(m-1)} \\ r^{(m)} - \lambda^{(m)} r^{(m-1)} \\ \vdots \\ r^{(m)} - \lambda^{(m)} r^{(m-1)} \end{bmatrix} . \tag{7.79}$$

A special case of interest appears when the scaling factor $\lambda^{(m)}$ attains the value

$$\lambda^{(m)} = \frac{r^{(m)}}{r^{(m-1)}} \quad . \tag{7.80}$$

In this case, the system (7.79) exhibits the special structure

$$\mathbf{T}^{(m)} \left(\begin{bmatrix} \mathbf{q}^{(m)} \end{bmatrix} - \lambda^{(m)} \begin{bmatrix} 0 \\ \mathbf{q}^{(m-1)} \end{bmatrix} \right) = \begin{bmatrix} r^{(m)} - \frac{r^{(m)}}{r^{(m-1)}} \gamma^{(m-1)} \\ 0 \\ \vdots \\ 0 \end{bmatrix} . \tag{7.81}$$

A comparison of this system (7.81) with (7.29) shows that

$$
\mathbf{a}^{(m)} = \begin{bmatrix} -1 \\ a_1^{(m)} \\ \vdots \\ a_{m-1}^{(m)} \\ -K_m \end{bmatrix} = \mathbf{q}^{(m)} - \lambda^{(m)} \begin{bmatrix} 0 \\ \\ \\ \mathbf{q}^{(m-1)} \end{bmatrix} \qquad (7.82a)
$$

or, in an equivalent polynomial notation,

$$
a^{(m)}(z) = q^{(m)}(z) - \lambda^{(m)} z\, q^{(m-1)}(z) , \qquad (7.82b)
$$

which is the desired relationship between the transversal predictor parameters $\mathbf{a}^{(m)}$ and the solution of the special Toeplitz system $\mathbf{q}^{(m)}$.

Furthermore, we may discover through a comparison of the right-hand-sides of (7.81) and (7.29) that

$$
- E'_m = \left(r^{(m-1)} - \gamma^{(m-1)} \right) \frac{r^{(m)}}{r^{(m-1)}} . \qquad (7.82c)
$$

Since $a_0^{(m)} = -1$, it follows from (7.82a) that $q_0^{(m)} = -1$ and consequently $q_m^{(m)} = -1$ (symmetry of $\mathbf{q}^{(m)}$). Finally, one sees that $a_m^{(m)} = -K_m$, which leads to the relations [see again (7.82a)]

$$
K_m = 1 - \lambda^{(m)} \qquad (7.83a)
$$

or, equivalently,

$$
\lambda^{(m)} = 1 - K_m . \qquad (7.83b)
$$

Substitution of (7.80) into (7.82c) results in

$$
\gamma^{(m-1)} - r^{(m-1)} = E'_m / \lambda^{(m)} . \qquad (7.84)
$$

Combination of (7.80) with (7.83b) gives

$$
\frac{r^{(m)}}{r^{(m-1)}} = 1 - K_m . \qquad (7.85)
$$

Equation (7.85) has an alternative nonrecursive formulation as follows:

$$
r^{(m)} = r^{(0)} \prod_{i=1}^{m} (1 - K_i) , \qquad (7.86)
$$

where $r^{(0)} = -c_0 = -1$.

Utilizing (7.82c), the quotient of residual energies in subsequent stages may be expressed as

$$\frac{E_m'}{E_{m-1}'} = \frac{\left(r^{(m-1)} - \gamma^{(m-1)}\right) r^{(m)} \big/ r^{(m-1)}}{\left(r^{(m-2)} - \gamma^{(m-2)}\right) r^{(m-1)} \big/ r^{(m-2)}} \quad . \tag{7.87}$$

Substituting (7.85) into (7.87) gives

$$\frac{E_m'}{E_{m-1}'} = \frac{\left(r^{(m-1)} - \gamma^{(m-1)}\right)(1 - K_m)}{\left(r^{(m-2)} - \gamma^{(m-2)}\right)(1 - K_{m-1})} \quad . \tag{7.88}$$

Recalling from (7.28b) that

$$\frac{E_m'}{E_{m-1}'} = 1 - K_m^2 \tag{7.89}$$

and taking into account (7.76), we find expression (7.88) reduces to

$$1 - K_m^2 = \alpha^{(m)} \frac{(1 - K_m)}{(1 - K_{m-1})} \tag{7.90}$$

or, more precisely,

$$(1 - K_m)(1 + K_m) = \alpha^{(m)} \frac{(1 - K_m)}{(1 - K_{m-1})} \quad ,$$

which yields the final result

$$\alpha^{(m)} = (1 + K_m)(1 - K_{m-1}) \quad . \tag{7.91}$$

Expression (7.91) can be used for the order recursive computation of K_m :

$$K_m = \frac{\alpha^{(m)}}{(1 - K_{m-1})} - 1 \quad , \tag{7.92}$$

where $\alpha^{(m)}$ is computed via (7.76).

Concluding this discussion, we note that (7.78a, 76, 74b and 72) consititute a closed recursion for the order recursive computation of the PARCOR coefficients $\{K_m, 1 \le m \le p\}$. The equations (7.78a and 72) require $m^2/4 + O(m)$ multiplications each. This gain in speed compared to the original Levinson-Durbin algorithm was achieved due to the *symmetric* nature of the intermediate solution vector $\mathbf{q}^{(m)}$. The transversal predictor coefficients $\mathbf{a}^{(m)}$ can be computed via (7.82a and 83b), if desired. These computations can be *omitted* in those cases where one is interested in the PARCOR ladder reflection coefficients only.

Exact initialization procedure

We will next develop the *exact initialization procedure* for the "split Levinson" algorithm described above. From (7.68 and 69), it follows that we can start the algorithm by setting

$$\mathbf{q}^{(0)} = \begin{bmatrix} -1 \end{bmatrix} \quad , \tag{7.93a}$$

$$\mathbf{q}^{(1)} = \begin{bmatrix} -1 \\ -1 \end{bmatrix} \quad . \tag{7.93b}$$

From this initial guess and (7.78a,b) it follows that

$$\gamma^{(0)} = -c_1 \quad , \tag{7.94a}$$

$$\gamma^{(1)} = -c_1 - c_2 \quad . \tag{7.94b}$$

Exploiting the relations (7.70b and 71), one sees that

$$r^{(0)} = -1 \quad , \tag{7.95a}$$

$$r^{(1)} = -1 - c_1 \quad . \tag{7.95b}$$

Substitution of (7.94a,b) and (7.95a,b) into (7.76) yields the initial value of $\alpha^{(2)}$:

$$\alpha^{(2)} = (c_2 - 1)/(c_1 - 1) \quad . \tag{7.96}$$

Finally, we deduce from (7.7a) that

$$K_1 = -c_1 \quad , \tag{7.97}$$

which concludes the exact initialization procedure. Table 7.4 provides a summary of this split Levinson algorithm.

7.4.3 Relationships Between Krishna's Algorithm and Delsarte's Algorithm (Symmetric Case)

As mentioned earlier, different split Levinson algorithms can be derived upon a different definition of the right-hand-side vector of the special intermediate Toeplitz system with a symmetric solution vector. Several choices have been discussed at the beginning of Sect. 7.4. For the two cases (7.38a) (symmetric pulse right-hand-side) and (7.39) (symmetric constant right-hand-side) complete algorithms have been derived. These

Table 7.4. Krishna's split Levinson algorithm based on the *symmetric* intermediate solution vector $\mathbf{q}^{(m)}$. Equation (7.98g) is the computation of the transversal predictor parameters $\mathbf{a}^{(m)}$ at a desired order m. As (7.98g) is not a part of the order recursion, its computation remains *optional*. Note also that (7.98f) requires only m/2 computations due to the symmetry property of $\mathbf{q}^{(m)}$. This algorithm involves division. Whenever the divisor is small, set $1/x = x$

Input: $c_1, c_2, \ldots, c_p$

Initialize: $\mathbf{q}^{(0)} = \begin{bmatrix} -1 \end{bmatrix}$ $\mathbf{q}^{(1)} = \begin{bmatrix} -1, & -1 \end{bmatrix}^T$

$r^{(0)} = -1$ $r^{(1)} = -1 - c_1$

$\gamma^{(0)} = -c_1$ $K_1 = -c_1$

FOR m = 2, 3, 4, . . . , p

$$\text{m = even:}\quad \gamma^{(m-1)} = \sum_{i=0}^{(m-2)/2} \left(c_{i+1} + c_{m-i} \right) q_i^{(m-1)} \tag{7.98a}$$

$$\text{m = odd:}\quad \gamma^{(m-1)} = c_{(m+1)/2}\, q_{(m-1)/2}^{(m-1)}$$

$$+ \sum_{i=0}^{(m-3)/2} \left(c_{i+1} + c_{m-i} \right) q_i^{(m-1)} \tag{7.98b}$$

$$\alpha^{(m)} = \frac{r^{(m-1)} - \gamma^{(m-1)}}{r^{(m-2)} - \gamma^{(m-2)}} \tag{7.98c}$$

$$K_m = \frac{\alpha^{(m)}}{(1 - K_{m-1})} - 1 \tag{7.98d}$$

$$r^{(m)} = 2\, r^{(m-1)} - \alpha^{(m)} r^{(m-2)} \tag{7.98e}$$

$$\mathbf{q}^{(m)} = \begin{bmatrix} \mathbf{q}^{(m-1)} \\ \\ 0 \end{bmatrix} + \begin{bmatrix} 0 \\ \\ \mathbf{q}^{(m-1)} \end{bmatrix} - \alpha^{(m)} \begin{bmatrix} 0 \\ \mathbf{q}^{(m-2)} \\ 0 \end{bmatrix} \tag{7.98f}$$

If desired (at any stage m) compute:

$$\mathbf{a}^{(m)} = \begin{bmatrix} -1 \\ a_1^{(m)} \\ \vdots \\ a_{m-1}^{(m)} \\ -K_m \end{bmatrix} = \mathbf{q}^{(m)} + (K_m - 1) \begin{bmatrix} 0 \\ \\ \mathbf{q}^{(m-1)} \end{bmatrix} \tag{7.98g}$$

---

algorithms, as proposed by Delsarte and Krishna, can be related to each other. It is therefore the purpose of this part to investigate the relationships between the symmetric versions of Krishna's algorithm (Table 7.4) and Delsarte's original split Levinson algorithm (Table 7.3). Later in this chapter, we will also derive the relationships between Krishna's *symmetric* case (Table 7.4) and Delsarte's *antisymmetric* case.

Suppose we define an alternative intermediate solution vector $\mathbf{q}^{(m)'}$ via the following relation.

$$q^{(m)}(z) = r^{(m)} q^{(m)'}(z) \quad . \tag{7.99}$$

Substitution of (7.99) into the polynomial recursion (7.82b) gives

$$a^{(m)}(z) = r^{(m)} q^{(m)'}(z) - \lambda^{(m)} r^{(m-1)} z \, q^{(m-1)'}(z) \quad . \tag{7.100}$$

Evaluation of the polynomial (7.100) at $z = 1$ yields

$$a^{(m)}(1) = r^{(m)} \sum_{i=0}^{m} q_i^{(m)'} - \lambda^{(m)} r^{(m-1)} \sum_{i=0}^{m-1} q_i^{(m-1)'} \quad . \tag{7.101}$$

Incorporation of (7.37) together with (7.83b and 86) results in an expression that relates the PARCOR coefficients to the intermediate solution vector $\mathbf{q}^{(m)'}$:

$$\prod_{i=1}^{m} (1 + K_i) = -c_0 \prod_{i=1}^{m} (1 - K_i) \sum_{i=0}^{m} q_i^{(m)'}$$
$$+ (1 - K_m) c_0 \prod_{i=1}^{m-1} (1 - K_i) \sum_{i=0}^{m-1} q_i^{(m-1)'} \quad . \tag{7.102}$$

Expression (7.102) can be rewritten as (note: $c_0 = 1$)

$$\prod_{i=1}^{m} (1 + K_i) = \prod_{i=1}^{m} (1 - K_i) \left[-\sum_{i=0}^{m} q_i^{(m)'} + \sum_{i=0}^{m-1} q_i^{(m-1)'} \right] \tag{7.103}$$

or, equivalently,

$$-\sum_{i=0}^{m} q_i^{(m)'} + \sum_{i=0}^{m-1} q_i^{(m-1)'} = \frac{\prod_{i=1}^{m} (1 + K_i)}{\prod_{i=1}^{m} (1 - K_i)} \quad . \tag{7.104}$$

Moreover, a *nonrecursive* expression of the sum over the $q_i^{(m)'}$'s exists as follows:

$$\sum_{i=0}^{m} q_i^{(m)'} = 1 - \sum_{j=1}^{m} \prod_{i=1}^{j} \frac{(1 + K_i)}{(1 - K_i)} \quad . \tag{7.105}$$

Clearly, for $m = 0$ we have

$$q^{(0)'} = \frac{1}{c_0} = 1 \quad . \tag{7.106}$$

A quick check of our initial definition of the intermediate symmetric Toeplitz system (7.39), together with the substitution (7.99) reveals that $q^{(m)'}$ is the solution to the intermediate Toeplitz system (7.39) with all elements of the right-hand-side vector set to 1 as follows:

$$\mathbf{T}^{(m)} \mathbf{q}^{(m)'} = \begin{bmatrix} 1, 1, \ldots, 1 \end{bmatrix}^{T} . \tag{7.107}$$

Moreover

$$\sum_{i=0}^{m} q_i^{(m)'} = \begin{bmatrix} 1, 1, \ldots, 1 \end{bmatrix} \mathbf{q}^{(m)'}$$

$$= \begin{bmatrix} 1, 1, \ldots, 1 \end{bmatrix} \mathbf{T}^{(m)^{-1}} \begin{bmatrix} 1, 1, \ldots, 1 \end{bmatrix}^{T} . \tag{7.108}$$

But since [recall (7.99)]

$$q^{(m)'}(z) = \left(1/r^{(m)}\right) q^{(m)}(z) \quad , \tag{7.109}$$

it follows that (evaluation at $z = 1$)

$$\sum_{i=0}^{m} q_i^{(m)} = r^{(m)} \sum_{i=0}^{m} q_i^{(m)'} \quad . \tag{7.110}$$

Now, utilizing expression (7.105) together with (7.110) results in an interesting relationship between the solution vector $\mathbf{q}^{(m)}$ and the PARCOR ladder reflection coefficients

$$\sum_{i=0}^{m} q_i^{(m)} = r^{(m)} \left[1 - \sum_{j=1}^{m} \prod_{i=1}^{j} \frac{(1 + K_i)}{(1 - K_i)} \right]$$

$$= - \prod_{i=1}^{m} (1 - K_i) \left[1 - \sum_{j=1}^{m} \prod_{i=1}^{j} \frac{(1 + K_i)}{(1 - K_i)} \right] \tag{7.111a}$$

or, equivalently,

$$\sum_{i=0}^{m} q_i^{(m)} = - \prod_{i=1}^{m} (1 - K_i) + \sum_{j=1}^{m} \left[\prod_{i=1}^{j} (1 + K_i) \prod_{l=j+1}^{m} (1 - K_l) \right] . \tag{7.111b}$$

On the other hand, an evaluation of (7.77) for z = 1 gives

$$\sum_{i=0}^{m} q_i^{(m)} = 2 \sum_{i=0}^{m-1} q_i^{(m-1)} - \alpha^{(m)} \sum_{i=0}^{m-2} q_i^{(m-2)} \quad . \tag{7.112}$$

Taking into account that $\lambda^{(m-1)} = r^{(m-1)}/r^{(m-2)}$ [see also (7.80)], and using (7.70a,b and 71), we may write

$$\mathbf{T}^{(m)} \mathbf{p}^{(m)} = \mathbf{T}^{(m)} \left(\begin{bmatrix} \mathbf{q}^{(m-1)} \\ \\ 0 \end{bmatrix} + \begin{bmatrix} 0 \\ \\ \mathbf{q}^{(m-1)} \end{bmatrix} - 2\lambda^{(m-1)} \begin{bmatrix} 0 \\ \mathbf{q}^{(m-2)} \\ 0 \end{bmatrix} \right)$$

$$= \begin{bmatrix} r^{(m-1)} \\ r^{(m-1)} \\ \vdots \\ r^{(m-1)} \\ \gamma^{(m-1)} \end{bmatrix} + \begin{bmatrix} \gamma^{(m-1)} \\ r^{(m-1)} \\ \vdots \\ r^{(m-1)} \\ r^{(m-1)} \end{bmatrix} - 2\lambda^{(m-1)} \begin{bmatrix} \gamma^{(m-2)} \\ r^{(m-2)} \\ \vdots \\ r^{(m-2)} \\ \gamma^{(m-2)} \end{bmatrix}$$

$$= \begin{bmatrix} \tau^{(m)} \\ 0 \\ \vdots \\ 0 \\ \tau^{(m)} \end{bmatrix} \quad . \tag{7.113}$$

Clearly, the vector equation (7.113) is uniquely determined by the following two *scalar equations*

$$\tau^{(m)} = r^{(m-1)} + \gamma^{(m-1)} - 2\lambda^{(m-1)}\gamma^{(m-2)} \quad , \tag{7.114a}$$

$$0 = 2 r^{(m-1)} - 2\lambda^{(m-1)} r^{(m-2)} \quad , \tag{7.114b}$$

where (7.114b) is the expression equivalent to (7.80) and (7.114a) determines the constant $\tau^{(m)}$ appearing in the special-type right-hand-side vector of (7.113). With this choice of $\tau^{(m)}$ one finds a relationship between Delsarte's intermediate solution vector $\mathbf{p}^{(m)}$ and Krishna's intermediate solution vector $\mathbf{q}^{(m)}$ as

$$
\mathbf{p}^{(m)} = \begin{bmatrix} \mathbf{q}^{(m-1)} \\ \\ 0 \end{bmatrix} + \begin{bmatrix} 0 \\ \\ \mathbf{q}^{(m-1)} \end{bmatrix} - 2\,\lambda^{(m-1)} \begin{bmatrix} 0 \\ \mathbf{q}^{(m-2)} \\ 0 \end{bmatrix} . \tag{7.115}
$$

We may rewrite (7.113) to obtain

$$
\mathbf{T}^{(m)}\mathbf{p}^{(m)} = \left[\tau^{(m)},\, 0 \ldots 0,\, \tau^{(m)} \right]^{T} , \tag{7.116}
$$

which has brought us back to the symmetric approach of Delsarte and Genin [7.4].

7.4.4 Relationships Between Krishna's Algorithm and Delsarte's Algorithm (Antisymmetric Case)

We may continue the discussion by considering the relationships between Krishna's algorithm listed in Table 7.4 and Delsarte's *antisymmetric* approach (7.38b). For this purpose, rearrange (7.114a) in the form

$$
\tau^{(m)} = \gamma^{(m-1)} - r^{(m-1)} + 2r^{(m-1)} - 2\gamma^{(m-2)} \frac{r^{(m-1)}}{r^{(m-2)}} . \tag{7.117}
$$

Noting that from (7.82c)

$$
2r^{(m-1)} - 2\gamma^{(m-2)} \frac{r^{(m-1)}}{r^{(m-2)}} = 2\frac{r^{(m-1)}}{r^{(m-2)}} \left(r^{(m-2)} - \gamma^{(m-2)} \right)
$$

$$
= -2\,E'_{m-1} , \tag{7.118}
$$

we may rewrite (7.117)

$$
\tau^{(m)} = \gamma^{(m-1)} - r^{(m-1)} - 2\,E'_{m-1} \tag{7.119}
$$

or, using (7.89),

$$
\tau^{(m)} = \gamma^{(m-1)} - r^{(m-1)} - \frac{2\,E'_{m}}{1 - K_{m}^{2}} . \tag{7.120}
$$

Combining (7.85) with (7.82c and 120) gives

$$
\tau^{(m)} = \frac{-E'_{m}}{1 + K_{m}} . \tag{7.121}
$$

Subtracting (7.70b) from (7.70a) gives

$$
T^{(m)} \left(\begin{bmatrix} q^{(m-1)} \\ \\ 0 \end{bmatrix} - \begin{bmatrix} 0 \\ \\ q^{(m-1)} \end{bmatrix} \right) = \begin{bmatrix} r^{(m-1)} \\ r^{(m-1)} \\ \vdots \\ r^{(m-1)} \\ \gamma^{(m-1)} \end{bmatrix} - \begin{bmatrix} \gamma^{(m-1)} \\ r^{(m-1)} \\ \vdots \\ r^{(m-1)} \\ r^{(m-1)} \end{bmatrix}
$$

$$
= \left(r^{(m-1)} - \gamma^{(m-1)} \right) \begin{bmatrix} 1 \\ 0 \\ \vdots \\ \\ 0 \\ -1 \end{bmatrix} . \tag{7.122}
$$

Recalling Delsarte's *antisymmetric* intermediate solution vector $p^{*(m)}$ (7.38b) and setting

$$
p^{*(m)} = \begin{bmatrix} q^{(m-1)} \\ \\ 0 \end{bmatrix} - \begin{bmatrix} 0 \\ \\ q^{(m-1)} \end{bmatrix} \tag{7.123}
$$

with the alternative polynomial expression

$$
p^{*(m)}(z) = (1 - z) q^{(m-1)}(z) \tag{7.124}
$$

in combination with

$$
\tau^{*(m)} = r^{(m-1)} - \gamma^{(m-1)} \tag{7.125}
$$

reveals that expression (7.122) represents Delsarte's antisymmetric approach [recall (7.38b)]

$$
T^{(m)} p^{*(m)} = \left[\tau^{*(m)}, 0 \ldots 0, - \tau^{*(m)} \right]^T . \tag{7.126}
$$

The expression (7.123) is the desired relationship between the intermediate solution vectors in the case of Krishna's algorithm (Table 7.4) and the split Levinson algorithm resulting from Delsarte's antisymmetric approach. All these algorithms resulting from the different symmetry assumptions have a very similar algorithm structure. Since split Levinson algorithms are relatively new techniques, their numerical accuracy compared to the traditional Levinson–Durbin algorithm is still fairly unexplored. Nevertheless, in Chap. 8, we introduce the *split Schur* algorithms as a different class of split Levinson algorithms, which is related to the traditional counterparts of Delsarte and Krishna in the same way as the LeRoux-Gueguen algorithm is related to the Levinson-Durbin algorithm.

7.5   A Levinson-Type Least-Squares Ladder Estimation Algorithm

We shall return to the problem of computing the reflection coefficients $\{K_m^f(t), K_m^b(t), 1 \le m \le p\}$ from an observed process $x(t)$ in the *nonstationary* case. In Table 6.1, we have listed an algorithm that solves this problem via a procedure that requires the *explicit* computation of prediction error vectors after each stage of the ladder form. This "explicit" ladder algorithm has two major drawbacks. First, its computational complexity grows with the length of the data vector $x(t)$, and second, the explicit computation of prediction errors is known to be an ill-conditioned procedure, especially when the prediction errors are used as inputs in successive stages of an order recursive algorithm. On the other hand, we have seen, in Chap. 6, that the computation of the ladder reflection coefficients requires knowledge only about the residual *energies*. This gave rise to the idea of developing algorithms that are based on order recursions of residual *energies* rather than on order recursions of residual *vectors*. We discovered in the early part of this chapter that Levinson's recursion is an easy way to accomplish this, at least in the stationary (Toeplitz) case where the prediction problem was determined by the Yule-Walker equations. As an algorithmical solution, we obtained the Levinson-Durbin algorithm, which is characterized by the distinct feature that it accomplishes the updating of the PARCOR ladder form reflection coefficients by a three-term recurrence relation using the transversal predictor parameters as *intermediate recursion variables* in each stage of the algorithm.

A similar algorithm will be developed next for the nonstationary case, where the system matrix of the Normal Equations is *not* Toeplitz. This case is associated with the exact *time recursive* least-squares ladder form with different reflection coefficients in the forward and backward prediction paths. This algorithm can be viewed as the counterpart of the "explicit" least-squares ladder estimation algorithm as listed in Table 6.1. Recalling the definition of the forward/backward transversal prediction error filters in the stationary case (6.38a,b), one finds that, in the more general *nonstationary case*, these relations appear as

$$e_m(t) = - \sum_{j=0}^{m} a_j^{(m)}(t) x(t-j) \quad , \tag{7.127a}$$

$$r_m(t) = - \sum_{j=0}^{m} b_j^{(m)}(t) x(t-m+j) \quad , \tag{7.127b}$$

where $a_j^{(m)}(t)$ and $b_j^{(m)}(t)$ are the *forward/backward transversal predictor parameters* and $x(t)$ is the *observed process*.

Now it is easy to check that one can express the residual energies in terms of the observed process and the forward/backward predictor parameters as follows:

$$E_m(t) = e_m^T(t) e_m(t) = \sum_{i=0}^{m} \sum_{j=0}^{m} a_i^{(m)}(t) a_j^{(m)}(t) x^T(t-i) x(t-j) \quad , \tag{7.128a}$$

$$R_m(t-1) = r_m^T(t-1) r_m(t-1)$$

$$= \sum_{i=0}^{m} \sum_{j=0}^{m} b_i^{(m)}(t-1) b_j^{(m)}(t-1) x^T(t-1-m+i) x(t-1-m+j) \quad , \tag{7.128b}$$

$$C_m(t) = e_m^T(t) r_m(t-1)$$

$$= \sum_{i=0}^{m} \sum_{j=0}^{m} a_i^{(m)}(t) b_j^{(m)}(t-1) x^T(t-i) x(t-1-m+j) \quad . \tag{7.128c}$$

Recalling the definition of the *covariance matrix* of the observed process

$$\Phi_{i,j}(t) = x^T(t-i) x(t-j) \quad , \tag{7.129}$$

one may express (7.128a-c) in terms of the elements of the covariance matrix as

$$E_m(t) = \sum_{i=0}^{m} \sum_{j=0}^{m} a_i^{(m)}(t) a_j^{(m)}(t) \Phi_{i,j}(t) \quad , \tag{7.130a}$$

$$R_m(t-1) = \sum_{i=0}^{m} \sum_{j=0}^{m} b_i^{(m)}(t-1) b_j^{(m)}(t-1) \Phi_{m+1-i, m+1-j}(t) \quad , \tag{7.130b}$$

$$C_m(t) = \sum_{i=0}^{m} \sum_{j=0}^{m} a_i^{(m)}(t) b_j^{(m)}(t-1) \Phi_{m+1-i, j}(t) \quad . \tag{7.130c}$$

These expressions, together with the definition of the ladder reflection coefficients (6.19a,b) and the Levinson type recursion of Table 6.3

establish a closed recursion for the *order recursive computation* of the ladder reflection coefficients by using residual energies and transversal predictor parameters as intermediate recursion variables. Clearly, the recursions (7.130 a,b) may be replaced by the simpler relations (6.22a,b) or (6.23a,b), which do the same job more efficiently in a smaller number of arithmetic operations. This Levinson type least-squares ladder algorithm is listed in Table 7.5.

Table 7.5. Time recursive Levinson type least-squares ladder algorithm. This algorithm involves division. Whenever the divisor is small and positive, set $1/x = x$. When the divisor is negative, set $1/x = 0$

FOR $t = 0, 1, 2, \ldots$

$\quad$ *Input:* $\quad \Phi_{i,j}(t) ; 0 \le i,j \le p$

$\quad$ Initialize: $E_0(t) = \Phi_{0,0}(t) \quad R_0(t-1) = \Phi_{1,1}(t) \quad C_0(t) = \Phi_{0,1}(t)$

$\qquad\qquad K_1^f(t) = - C_0(t)/R_0(t-1) \qquad K_1^b(t) = - C_0(t)/E_0(t)$

$\qquad\qquad a_0^{(0)}(t) = - 1 \qquad\qquad b_0^{(0)}(t) = - 1$

$\quad$ FOR $m = 1, 2, 3, \ldots, p - 1$

$\qquad$ FOR $j = 1, 2, 3, \ldots, m - 1$

$$a_j^{(m)}(t) = a_j^{(m-1)}(t) + K_m^f(t) b_{j-1}^{(m-1)}(t-1) \tag{7.131a}$$

$$b_j^{(m)}(t) = b_{j-1}^{(m-1)}(t-1) + K_m^b(t) a_j^{(m-1)}(t) \tag{7.131b}$$

$$a_0^{(m)}(t) = - 1 \qquad a_m^{(m)}(t) = - K_m^f(t) \tag{7.131c}$$

$$b_0^{(m)}(t) = - K_m^b(t) \qquad b_m^{(m)}(t) = - 1 \tag{7.131d}$$

$$E_m(t) = E_{m-1}(t) + K_m^f(t) C_{m-1}(t) \tag{7.131e}$$

$$R_m(t) = R_{m-1}(t-1) + K_m^b(t) C_{m-1}(t) \tag{7.131f}$$

$$C_m(t) = \sum_{i=0}^{m} \sum_{j=0}^{m} a_i^{(m)}(t) b_j^{(m)}(t-1) \Phi_{m+1-i, j}(t) \tag{7.131g}$$

$$K_{m+1}^f(t) = - C_m(t)/R_m(t-1) \tag{7.131h}$$

$$K_{m+1}^b(t) = - C_m(t)/E_m(t) \tag{7.131i}$$

The Levinson type least-squares ladder algorithm of Table 7.5 may be viewed as the counterpart of the "explicit" ladder algorithm of Table 6.1. The number of arithmetic operations is now *independent* of the length of the process vector $\mathbf{x}(t)$. The recursion is based on *residual energies* and *transversal predictor parameters*, which serve as *intermediate recursion variables*. A significant drawback of this algorithm is that it requires $O(p^3)$ computations per recursion due to (7.131g). It is intuitively clear, however, that we may succeed in *speeding-up* equation (7.131g) significantly. This was achieved shortly after the manuscript of this book was completed. See Appendix A.1, where a new algorithm named ARRAYLAD 1 is presented, which is just this speeded-up counterpart of the algorithm listed in Table 7.5.

7.6 The Makhoul Covariance Ladder Algorithm

After these considerations, it is now an easy task to derive *Makhoul's covariance ladder algorithm* as published in his 1977 paper [7.5]. While the Levinson type algorithm of Table 7.5 was interpreted as the counterpart of the "explicit" ladder algorithm of Table 6.1 using only residual energies and transversal predictor parameters as intermediate variables, the Makhoul covariance ladder algorithm is the counterpart to the Burg type algorithm listed in Table 6.7. Clearly, just like this Burg type algorithm, Makhoul's method attempts to fit a PARCOR ladder form on a process that was *not* assumed to be stationary a priori. But in contrast to this Burg type algorithm, which requires the explicit computation of prediction error vectors, the Makhoul covariance ladder algorithm is based entirely on *transversal predictor parameters* and *residual energy recursions*.

Again we define the forward and backward transversal prediction error filters. In the case of the PARCOR ladder form, we know that the forward and backward prediction error filters have identical and time-invariant coefficient sets and hence

$$\mathbf{e}_m(t) = - \sum_{j=0}^{m} a_j^{(m)} \mathbf{x}(t-j) \quad , \tag{7.132a}$$

$$\mathbf{r}_m(t) = - \sum_{j=0}^{m} a_j^{(m)} \mathbf{x}(t-m+j) \quad . \tag{7.132b}$$

Just as in the more general nonstationary case, the residual energies may be expressed in terms of transversal predictor parameters and elements of the covariance matrix, where the time index has now been omitted and the index pair {i,j} denotes linearly shifted segments of a given data record

$$E_m = e_m^T e_m = \sum_{i=0}^{m} \sum_{j=0}^{m} a_i^{(m)} a_j^{(m)} \, x^T(-i) \, x(-j) \quad , \tag{7.133a}$$

$$R_m = r_m^T r_m = \sum_{i=0}^{m} \sum_{j=0}^{m} a_i^{(m)} a_j^{(m)} \, x^T(-1-m+i) \, x(-1-m+j) \quad , \tag{7.133b}$$

$$C_m = e_m^T r_m = \sum_{i=0}^{m} \sum_{j=0}^{m} a_i^{(m)} a_j^{(m)} \, x^T(-i) \, x(-1-m+j) \quad , \tag{7.133c}$$

or, equivalently,

$$E_m = \sum_{i=0}^{m} \sum_{j=0}^{m} a_i^{(m)} a_j^{(m)} \, \Phi_{i,j} \quad , \tag{7.134a}$$

$$R_m = \sum_{i=0}^{m} \sum_{j=0}^{m} a_i^{(m)} a_j^{(m)} \, \Phi_{m+1-i, \, m+1-j} \quad , \tag{7.134b}$$

$$C_m = \sum_{i=0}^{m} \sum_{j=0}^{m} a_i^{(m)} a_j^{(m)} \, \Phi_{m+1-i, \, j} \quad . \tag{7.134c}$$

The PARCOR ladder reflection coefficient is finally computed via Burg's formula (6.92), which may be rewritten in terms of the residual energies

$$K_{m+1} = - \frac{2 \, C_m}{E_m + R_m} \quad . \tag{7.135}$$

Table 7.6 provides a summary of this technique, which has become known as Makhoul's covariance ladder algorithm. Similarly to the exact procedure listed in Table 7.5, we may use the easier recursion (7.131e) for a computation of the forward prediction residual energy.

7.7 Chapter Summary

This chapter presented the class of ladder algorithms where the recursions are based on *residual energies* and *transversal predictor parameters* [7.12]. The transversal predictor parameters serve as important *intermediate variables* in this class of ladder algorithms. The Levinson recursion appeared as a fundamental tool, in that it relates the transversal predictor parameters to the ladder reflection coefficients in each stage of the order recursive algorithms. A large part of this chapter was devoted to the *pure stationary case* characterized by a *Toeplitz* system matrix of the Normal Equations. Besides the classical Levinson-Durbin algorithm, we discussed some new forms, termed *split Levinson algorithms*, which allow the inversion of a Toeplitz system with only *half*

Table 7.6. Makhoul covariance ladder algorithm (block processing) for computing the PARCOR coefficients $K_1, K_2 \ldots K_p$ from a given *covariance matrix* Φ. This algorithm involves division. Whenever the divisor is small and positive, set $1/x = x$. When the divisor is negative, set $1/x = 0$

Input: $\Phi_{i,j}\,;\; 0 \le i,j \le p$

Initialize: $E_0 = \Phi_{0,0}$; $R_0 = \Phi_{1,1}$; $C_0 = \Phi_{0,1}$

$\qquad K_1 \;\; = -\,2\,C_0\big/(E_0 + R_0)$

$\qquad a_0^{(1)} = -\,1 \qquad\qquad a_1^{(1)} = -\,K_1$

FOR $m = 1, 2, 3, \ldots, p - 1$

$\qquad$ FOR $j = 1, 2, 3, \ldots, m - 1$

$$a_j^{(m)} = a_j^{(m-1)} + K_m\, a_{m-j}^{(m-1)} \tag{7.136a}$$

$$a_0^{(m)} = -\,1 \qquad\qquad a_m^{(m)} = -\,K_m \tag{7.136b}$$

$$E_m = E_{m-1} + K_m\, C_{m-1} \tag{7.136c}$$

$$R_m = \sum_{i=0}^{m}\sum_{j=0}^{m} a_i^{(m)}\, a_j^{(m)}\, \Phi_{m+1-i,\,m+1-j} \tag{7.136d}$$

$$C_m = \sum_{i=0}^{m}\sum_{j=0}^{m} a_i^{(m)}\, a_j^{(m)}\, \Phi_{m+1-i,\,j} \tag{7.136e}$$

$$K_{m+1} = -\,\frac{2\,C_m}{E_m + R_m} \tag{7.136f}$$

---

the number of multiplications as in the case of the classical Levinson-Durbin algorithm.

We have also demonstrated how a time recursive least-squares ladder algorithm can be developed using Levinson's recursion. In the block processing case involving the PARCOR ladder form, a simplification of this algorithm is available in terms of the *Makhoul covariance ladder algorithm*, which is simply the Levinson type representation of the Burg type PARCOR ladder algorithm of Table 6.7.

Some more practical comments seem to be in order. If infinite precision arithmetic is assumed, it is clear that Levinson type algorithms perform identically to their counterparts based on an explicit computation of prediction error vectors. The situation changes as soon as finite precision arithmetic is used. In this case, it may happen that the Levinson type algorithms produce *negative* forward or backward residual energies $E_m(t)$ and $R_m(t)$. This effect is due to a possible *propagation of round-off errors* in the order recursions of residual energies. The residual energies must *always be greater than or equal to zero* by definition, and hence we must *set them to zero* in the case that they attain negative values due to numerical errors. It is clear that this effect is not possible in the algorithms of the "explicit" type since the explicit computation of residual energies *prevents* negative results, independent of the numerical accuracy of processing.

A major drawback of the algorithms listed in Tables 7.5 and 7.6 is that these routines require $O(p^3)$ operations per recursion. Moreover, the computation of Levinson type recursions and the transversal predictor parameters may require a considerably higher dynamic range than the ladder reflection coefficients themselves. These difficulties lead us to think about alternative recursion variables. Such refined techniques will be presented in the next chapter. See also Appendix A.1 for a more refined recently developed version of a computationally more efficient time recursive least-squares ladder algorithm of the Levinson type.

8. Covariance Ladder Algorithms

The ladder algorithms presented in Chap. 7 were based on the recursive computation of transversal predictor parameters, which serve as *intermediate recursion variables*. This type of ladder algorithm employed the Levinson recursion for a conversion between the transversal predictor parameters and the ladder reflection coefficients. Although the ladder reflection coefficients are known to have a lower variance than the transversal predictor parameters, this advantage is compensated by the Levinson recursion, which requires a high dynamic range for the computation of the intermediate transversal predictor parameters. In fact, the intermediate transversal predictor parameters are *unbounded* quantities, i.e., they may attain large values, depending on the data. This makes the fixed-point implementation of Levinson type algorithms quite a difficult task.

Therefore, great effort has gone into the development of alternative computation schemes based on intermediate recursion variables that are *bounded* in their values. This chapter is devoted to alternative covariance ladder algorithms where the Levinson recursion is replaced by recursions in appropriate inner product spaces, called *"generalized residual energies"* (GREs), thereby *avoiding* the explicit computation of unbounded transversal predictor parameters. Interestingly, it will be shown that two types of such inner product spaces exist, each resulting in bounded ladder recursions facilitating an easy fixed-point implementation.

The first class of recursions is based on inner products of residual vectors and the observed process vector. The first algorithm of this type was introduced by LeRoux and Gueguen [8.1] in 1977. Subsequently, Cumani's 1982 paper "On a covariance ladder algorithm for linear prediction" [8.2] introduced a second class of inner product recursions, where the inner products are formed exclusively between residual vectors. The observed process vector was not used to form inner products in Cumani's approach.

At first glance, it seemed that the approaches of LeRoux-Gueguen and Cumani tackled two completely different problems, as the LeRoux-Gueguen algorithm is just an alternative formulation of the Levinson-Durbin algorithm whereas the Cumani covariance ladder algorithm is an alternative formulation of the Makhoul covariance ladder algorithm [8.3]. It will be shown in this chapter, however, that these classical algorithms are just special cases of two much larger classes of covariance ladder algorithms, which also include the solution of the true recursive least-squares (RLS) covariance ladder estimation problem. All these algorithms, as discussed in this chapter, have a *common algorithm structure*. They are initialized from the *covariance information* of the observed process at each time step. The computation of the ladder reflection coefficients is then accomplished by a *pure order recursive* procedure. Therefore, algorithms of this class are sometimes named *pure order recursive ladder algorithms* (PORLAs). The time recursion is restricted to the updating of the covariance information. *Higher-order recursive windows*, as discussed in Chap. 2, can easily be incorporated in the covariance matrix time recursion. In this way, the steady-state and tracking behavior of the algorithms can be tuned quite accurately according to the requirements of a specific application. Since the ladder form is constructed completely anew in each time step, round-off error propagation in time is impossible in this class of algorithm, and hence the estimated parameters will not attain a bias from accumulated round-off noise.

PORLA computation schemes in the true recursive least-squares case were first seriously studied by Strobach [8.4]. He has demonstrated that Cumani's approach is just a special case of a much broader class of covariance ladder algorithms. Strobach has derived the first recursive least-squares PORLA algorithm [8.5], exploiting several interesting properties of inner product recursions in the recursive case. Most recently, Sokat and Strobach [8.6] and Sokat [8.7] have demonstrated that the same ideas of generalization hold in the case of the other class of inner products that originally appeared in the LeRoux–Gueguen algorithm. Following Strobach's ideas by just using the GREs of LeRoux and Gueguen [8.1], Sokat obtained a completely new class of PORLA computation schemes including the LeRoux–Gueguen algorithm as a special case of approximation. All PORLA computation schemes have a complexity of $O(p^2)$ computations, thus leading to a *triangular array implementation* quite similar to what we discussed in the case of the recursive QR decomposition in Chap. 4. The implementation of PORLA computation schemes has been discussed in [8.8]. Additionally, new forms of PORLA algorithms are presented in the Appendix of this book. We have also developed the particularly interesting triangular array structures of these new PORLA algorithms. See the Appendix for details.

8.1 The LeRoux-Gueguen Algorithm

We shall start our discussion of PORLA computation schemes with the classical LeRoux-Gueguen algorithm [8.1] as the first representative of this class of covariance ladder algorithms. Recall the Levinson-Durbin algorithm [8.9,10], as stated by the three-term recurrence relation (7.13a-c), where the normalized forward residual energy E'_{m-1} was defined as

$$E'_{m-1} = \frac{e^T_{m-1}(t) e_{m-1}(t)}{x^T(t) x(t)} \,. \tag{8.1}$$

From (7.132a) one knows that

$$e_{m-1}(t) = - \sum_{j=0}^{m-1} a_j^{(m-1)} x(t-j) \,. \tag{8.2}$$

Additionally, one sees that [recall (2.24)]

$$e^T_{m-1}(t) e_{m-1}(t) = x^T(t) e_{m-1}(t) \tag{8.3}$$

and therefore

$$E'_{m-1} = \frac{x^T(t) e_{m-1}(t)}{x^T(t) x(t)} = - \sum_{j=0}^{m-1} a_j^{(m-1)} \frac{x^T(t) x(t-j)}{x^T(t) x(t)}$$

$$= - \sum_{j=0}^{m-1} a_j^{(m-1)} c_j \,. \tag{8.4}$$

The fundamental step of LeRoux and Gueguen was the introduction of a *"generalized"* residual energy $E'_{m-1,i}$ as follows:

$$E'_{m-1,i} = \frac{x^T(t-i) e_{m-1}(t)}{x^T(t) x(t)} \tag{8.5}$$

or, using (8.4),

$$E'_{m-1,i} = - \sum_{j=0}^{m-1} a_j^{(m-1)} \frac{x^T(t-i) x(t-j)}{x^T(t) x(t)} \,. \tag{8.6}$$

Note that in the *stationary* (Toeplitz) case,

$$\frac{\Phi_{i,j}(t)}{x^T(t) x(t)} = \frac{x^T(t-i) x(t-j)}{x^T(t) x(t)} = \frac{\Phi_{i-i,\,j-i}(t)}{x^T(t) x(t)} = c_{j-i} \,, \tag{8.7}$$

where Φ is the covariance matrix of the observed process (2.26). Expression (8.7) can be used to rewrite (8.6) as

$$E'_{m-1,i} = -\sum_{j=0}^{m-1} a_j^{(m-1)} c_{j-i} \quad . \tag{8.8}$$

Note also that from (7.8)

$$C'_{m-1} = -\sum_{j=0}^{m-1} a_j^{(m-1)} c_{m-j} \quad , \tag{8.9}$$

and by taking into account

$$c_j = \frac{x^T(t)x(t-j)}{x^T(t)x(t)} = \frac{x^T(t+j)x(t)}{x^T(t+j)x(t+j)} = \frac{x^T(t+j)x(t)}{x^T(t)x(t)} = c_{-j} \tag{8.10}$$

one finds

$$C'_{m-1} = E'_{m-1,\,m} \quad . \tag{8.11}$$

According to (8.11), the covariance C'_{m-1} appears as the generalized residual energy $E'_{m-1,\,m}$. Substituting the Levinson recursion (7.13c) into the definition of the generalized residual energy $E'_{m,i}$ (8.8), one obtains

$$E'_{m,i} = -\sum_{j=0}^{m} a_j^{(m)} c_{j-i} = -\sum_{j=0}^{m} \left[a_j^{(m-1)} + K_m a_{m-j}^{(m-1)} \right] c_{j-i}$$

$$= -\sum_{j=0}^{m} a_j^{(m-1)} c_{j-i} - K_m \sum_{j=0}^{m} a_{m-j}^{(m-1)} c_{j-i} \quad . \tag{8.12}$$

Since $a_m^{(m-1)} = 0$, it is clear that

$$-\sum_{j=0}^{m} a_j^{(m-1)} c_{j-i} = E'_{m-1,i} \quad , \tag{8.13a}$$

$$-\sum_{j=0}^{m} a_{m-j}^{(m-1)} c_{j-i} = -\sum_{j=0}^{m} a_j^{(m-1)} c_{m-j-i} = E'_{m-1,\,m-i} \quad . \tag{8.13b}$$

This allows (8.12) to be rewritten as an *order recursion* of generalized residual energies (GREs) as follows:

$$E'_{m,i} = E'_{m-1,i} + K_m E'_{m-1,\,m-i} \quad . \tag{8.14}$$

Clearly,

$$E'_{m,0} = E'_m \ . \tag{8.15}$$

This leads to the following *three-term recurrence relation* as an alternative scheme for computing the Levinson-Durbin algorithm with a recursion *solely based on GREs:*

$$K_m = - \left(E'_{m-1,0}\right)^{-1} E'_{m-1,m} \ , \tag{8.16a}$$

$$E'_{m,0} = E'_{m-1,0} \ (1 - K_m^2) \ , \tag{8.16b}$$

$$E'_{m,i} = E'_{m-1,i} + K_m E'_{m-1,m-i} \ . \tag{8.16c}$$

The algorithm, as constituted by (8.16a-c) has become known as the LeRoux-Gueguen algorithm [8.1].

Obviously, the first two equations (8.16a,b) are identical with the original Levinson-Durbin algorithm (7.13a-c). But the Levinson recursion (7.13c) has now been replaced by a recursion of GREs (8.16c). It will be shown later in this chapter that the GREs are *bounded* quantities, whereas the transversal predictor parameters, as appearing in the Levinson recursion, may grow unbounded with increasing order. Therefore, (8.16c) can be implemented with fixed-point arithmetic without much danger of overflows.

Defining a second GRE by

$$R'_{m-1,i} = E'_{m-1,m-i} \ , \tag{8.17}$$

one can rewrite (8.16c) as

$$E'_{m,i} = E'_{m-1,i} + K_m \ R'_{m-1,i} \ . \tag{8.18}$$

But since

$$R'_{m,i} = E'_{m,m+1-i} \ , \tag{8.19}$$

one notes also that

$$R'_{m,i} = E'_{m,m+1-i} = E'_{m-1,m+1-i} + K_m \ E'_{m-1,m-(m+1-i)}$$

$$= E'_{m-1,m+1-i} + K_m \ E'_{m-1,i-1}$$

$$= R'_{m-1,i-1} + K_m \ E'_{m-1,i-1} \ . \tag{8.20}$$

With (8.17, 18 and 20), another recurrence relation can be stated:

$$K_m = - R'_{m-1,0} \Big/ E'_{m-1,0} \quad , \tag{8.21a}$$

$$E'_{m,i} = E'_{m-1,i} + K_m R'_{m-1,i} \quad , \tag{8.21b}$$

$$R'_{m,i} = R'_{m-1,i-1} + K_m E'_{m-1,i-1} \quad , \tag{8.21c}$$

which is simply another representation of the LeRoux– Gueguen algorithm. We note that at stage $m = 0$, the algorithm is initialized from the *auto-correlation sequence* $c_1, c_2, \ldots, c_p$ of the observed process as

$$E'_{0,i} = c_i \quad , \tag{8.22a}$$

$$R'_{0,i} = E'_{0,1-i} = c_{1-i} = c_{i-1} \quad . \tag{8.22b}$$

Table 8.1 provides a summary of the LeRoux–Gueguen algorithm.

Table 8.1. Algorithm of LeRoux and Gueguen: a "fixed-point computation of partial correlation coefficients". This algorithm involves division. Whenever the divisor is small and positive, set $1/x = x$. When the divisor is negative, set $1/x = 0$

Input: $c_0 = 1, c_1, c_2, \ldots, c_p$

FOR $i = 0, 1, 2, \ldots, p$ *initialize:*

$\quad E'_{0,i} = c_i$

$\quad R'_{0,i} = c_{i-1}$ *note:* $c_{-1} = c_1$

FOR $m = 1, 2, \ldots, p$

$\quad K_m = - R'_{m-1,0} \Big/ E'_{m-1,0}$

$\quad E'_{m,0} = E'_{m-1,0} + K_m R'_{m-1,0}$

$\quad$ FOR $i = m+1, m+2, \ldots, p$

$\quad\quad E'_{m,i} = E'_{m-1,i} + K_m R'_{m-1,i}$

$\quad\quad R'_{m,i} = R'_{m-1,i-1} + K_m E'_{m-1,i-1}$

$\quad R'_{m,0} = E'_{m,m+1}$

8.1.1 Bounds on GREs

Next it is of interest to investigate the bounds of GREs. Note that it is sufficient to investigate just $E'_{m,i}$, since $R'_{m,i}$ can be represented by the E''s according to (8.19). Recall the definitions

$$E'_{m,0} = \frac{x^T(t)e_m(t)}{x^T(t)x(t)} \quad , \tag{8.23a}$$

$$E'_{m,i} = \frac{x^T(t-i)\,e_m(t)}{x^T(t)x(t)} \quad . \tag{8.23b}$$

According to the Cauchy-Schwarz inequality [8.11], we may write

$$\left(\frac{x^T(t-i)\,e_m(t)}{x^T(t)x(t)} \right)^2 \leq \frac{x^T(t-i)x(t-i)}{x^T(t)x(t)} \; \frac{e_m^T(t)e_m(t)}{x^T(t)x(t)} \quad . \tag{8.24}$$

Note:

$$x^T(t-i)x(t-i) = x^T(t)x(t) \quad , \tag{8.25a}$$

$$e_m^T(t)e_m(t) = x^T(t)e_m(t) \quad . \tag{8.25b}$$

It follows that

$$\left(E'_{m,i} \right)^2 \leq c_0 \, E'_{m,0} \quad . \tag{8.26}$$

Taking into account that [recall (8.16b)]

$$E'_{m,0} = E'_{m-1,0}\,(1 - K_m^2) \quad , \tag{8.27}$$

where the value of K_m is bounded on

$$-1 \leq K_m \leq 1 \quad , \tag{8.28}$$

it is easy to check that

$$E'_{0,0} = c_0 \quad , \tag{8.29a}$$

$$E'_{m,0} \leq E'_{m-1,0} \qquad for \quad 1 \leq m \leq p \quad , \tag{8.29b}$$

hence $c_0 E'_{m,0} \leq c_0^2$ and therefore

$$E'_{m,i} \leq 1 \qquad \forall \quad i, m \quad . \tag{8.30}$$

As an interesting result, it turns out that all GREs appearing in the LeRoux-Gueguen algorithm will be bounded on a value less than or equal to unity, hence facilitating an easy implementation of this algorithm on a fixed-point processor. Moreover, inequality (8.29b) reveals that the GREs in successive stages of the order recursive algorithm will have *decreasing* or at least *constant* values. Hence, it should also be possible to apply a *rescaling* in higher stages of the algorithm in order to obtain a better precision of the GREs at higher stages of the algorithm.

8.2 The Cumani Covariance Ladder Algorithm

We have discussed the LeRoux-Gueguen algorithm where the Levinson recursion on (unbounded) transversal predictor parameters was conveniently replaced by a recursion on (bounded) GREs yielding a robust computation scheme for a fixed-point implementation of the Levinson-Durbin algorithm. In 1983, Cumani [8.2] presented a similar idea to replace the Levinson recursion in the Makhoul covariance ladder algorithm [8.3]. Interestingly, Cumani introduces three new GREs as follows:

$$E_{m,i,j}(t) = e_m^T(t-i)e_m(t-j) \quad , \tag{8.31a}$$

$$R_{m,i,j}(t) = r_m^T(t-1-i)r_m(t-1-j) \quad , \tag{8.31b}$$

$$C_{m,i,j}(t) = e_m^T(t-i)\, r_m(t-1-j) \quad , \tag{8.31c}$$

for $0 \le m \le p, \quad 0 \le i,j \le p-1$.

In contrast to the approach of LeRoux and Gueguen, these new GREs are defined as *inner products of residual vectors* rather than inner products between the observed process vector and the residual vectors. Clearly, these new GREs as defined by (8.31a-c) are *matrix-valued* quantities. They can be interpreted as *covariance matrices of residual vectors* in successive stages of the ladder form.

Note that the residual vectors appearing in the new GREs can be expressed as the convolution of the input process and the impulse response of the associated transversal predictor. According to (7.132a,b), we have

$$e_m(t-i) = - \sum_{k=0}^{m} a_k^{(m)} x(t-k-i) \quad , \tag{8.32a}$$

$$r_m(t-j) = - \sum_{l=0}^{m} a_l^{(m)} x(t-m+1-j) \quad . \tag{8.32b}$$

Substitution of the expressions (8.32a,b) into the GREs (8.31a-c) gives

$$E_{m,i,j}(t) = \sum_{k=0}^{m} \sum_{l=0}^{m} a_k^{(m)} a_l^{(m)} \mathbf{x}^T(t-k-i)\mathbf{x}(t-l-j) \quad , \tag{8.33a}$$

$$R_{m,i,j}(t) = \sum_{k=0}^{m} \sum_{l=0}^{m} a_k^{(m)} a_l^{(m)} \mathbf{x}^T(t-i-1-m+k)\mathbf{x}(t-j-1-m+l) \quad , \tag{8.33b}$$

$$C_{m,i,j}(t) = \sum_{k=0}^{m} \sum_{l=0}^{m} a_k^{(m)} a_l^{(m)} \mathbf{x}^T(t-k-i) \mathbf{x}(t-j-1-m+l) \quad . \tag{8.33c}$$

Note that the inner products of delayed process vectors as appearing in the expressions (8.33a-c) can be interpreted as elements of the covariance matrix of the observed process:

$$\Phi_{k+1, 1+j}(t) = \mathbf{x}^T(t-k-i)\mathbf{x}(t-l-j) \quad , \tag{8.34a}$$

$$\Phi_{m+1-k+i, m+1-1+j}(t) = \mathbf{x}^T(t-i-1-m+k)\mathbf{x}(t-j-1-m+l) \quad , \tag{8.34b}$$

$$\Phi_{k+i, m+1-1+j}(t) = \mathbf{x}^T(t-k-i) \mathbf{x}(t-j-1-m+l) \quad . \tag{8.34c}$$

Substituting the Levinson recursion (7.13c) into the expressions (8.33a-c) yields a set of recursions of GREs as follows:

$$E_{m,i,j}(t) = E_{m-1,i,j}(t) + K_m\big(C_{m-1,i,j}(t) + C_{m-1,j,i}(t)\big)$$
$$+ K_m^2 R_{m-1,i,j}(t) \quad , \tag{8.35a}$$

$$R_{m,i,j}(t) = R_{m-1,i+1,j+1}(t) + K_m\big(C_{m-1,i+1,j+1}(t) + C_{m-1,j+1,i+1}(t)\big)$$
$$+ K_m^2 E_{m-1,i+1,j+1}(t) \quad , \tag{8.35b}$$

$$C_{m,i,j}(t) = C_{m-1,i,j+1}(t) + K_m\big(E_{m-1,i,j+1}(t) + R_{m-1,i,j+1}(t)\big)$$
$$+ K_m^2 C_{m-1,j+1,i}(t) \quad . \tag{8.35c}$$

These well-conditioned matrix order recursions can be used to replace the unbounded Levinson recursion in the Makhoul covariance ladder algorithm. Note that the "conventional" (scalar) residual energies appear as just the upper-left elements of the respective GREs:

$$E_m(t) = E_{m,0,0}(t) \quad , \tag{8.36a}$$

$$R_m(t-1) = R_{m,0,0}(t) \quad , \tag{8.36b}$$

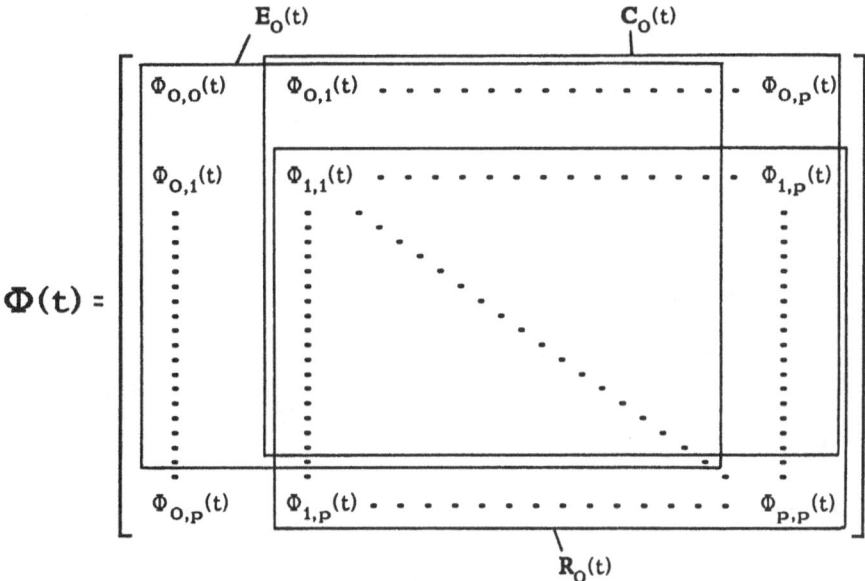

Fig. 8.1. Initialization of GREs from the input data covariance matrix $\Phi(t)$ at stage zero of the Cumani covariance ladder algorithm

$$C_m(t) = C_{m,0,0}(t) \quad . \tag{8.36c}$$

Moreover, the GRE matrices in stage zero of the new algorithm appear as certain submatrices of the input data covariance matrix as follows:

$$E_{0,i,j}(t) = \Phi_{i,j}(t) \quad , \tag{8.37a}$$

$$R_{0,i,j}(t) = \Phi_{i+1,j+1}(t) \quad , \tag{8.37b}$$

$$C_{0,i,j}(t) = \Phi_{i,j+1}(t) \quad . \tag{8.37c}$$

This scheme of initialization is illustrated in Fig. 8.1.

Table 8.2 provides a summary of this algorithm, which has become known as the Cumani covariance ladder algorithm [8.2]. This algorithm can be viewed as a fixed-point implementation of the well-known Makhoul covariance ladder algorithm [8.3].

Note that in the algorithm of Table 8.2, Burg's formula (6.92) [8.12] has been expressed in terms of the upper-left elements of the GREs which are simply the "conventional" scalar residual energies.

Table 8.2. Cumani covariance ladder algorithm (block processing) for computing the PARCOR coefficients K_1, K_2, . . . , K_p from a given covariance matrix. This algorithm involves division. Whenever the divisor is small and positive, set $1/x = x$. When the divisor is negative, set $1/x = 0$

Input: $\Phi_{i,j}$; $0 \leq i,j \leq p$

FOR i = 0, 1, 2, . . . , p - 1   *initialize:*

FOR j = 0, 1, 2, . . . , p - 1   *initialize:*

$$E_{0,i,j} = \Phi_{i,j} \tag{8.38a}$$

$$R_{0,i,j} = \Phi_{i+1,j+1} \tag{8.38b}$$

$$C_{0,i,j} = \Phi_{i,j+1} \tag{8.38c}$$

FOR m = 1, 2, . . . , p

$$K_m = - \frac{2\,C_{m-1,0,0}}{E_{m-1,0,0} + R_{m-1,0,0}} \tag{8.39}$$

FOR i = 0, 1, 2, . . . , p - m - 1

FOR j = 0, 1, 2, . . . , p - m - 1

$$E_{m,i,j} = E_{m-1,i,j} + K_m\left(C_{m-1,i,j} + C_{m-1,j,i}\right)$$
$$+ K_m^2 R_{m-1,i,j} \tag{8.40a}$$

$$R_{m,i,j} = R_{m-1,i+1,j+1} + K_m\left(C_{m-1,i+1,j+1} + C_{m-1,j+1,i+1}\right)$$
$$+ K_m^2 E_{m-1,i+1,j+1} \tag{8.40b}$$

$$C_{m,i,j} = C_{m-1,i,j+1} + K_m\left(E_{m-1,i,j+1} + R_{m-1,i,j+1}\right)$$
$$+ K_m^2 C_{m-1,j+1,i} \tag{8.40c}$$

8.3 Recursive Covariance Ladder Algorithms

The covariance ladder algorithms discussed so far are basically *block estimation* or "block processing" techniques. In the block processing case, the autocorrelation vector (windowed case) or the covariance matrix (unwindowed case) is determined from a set of measurements available in the observation (data) vector. In the windowed case, the LeRoux-Gueguen algorithm may be used to determine the desired PARCOR ladder reflection coefficients from the autocorrelation vector, whereas in the unwindowed case, the Cumani covariance ladder algorithm can be used to determine the PARCOR ladder reflection coefficients [8.13] from a given covariance matrix. It is now of interest to investigate how these classical block processing schemes can be extended to the *recursive* case, i.e., to the case where the covariance information is *updated* each time a new sample of the observed process is captured. Inspecting Table 8.2, we note that Cumani's algorithm is a computationally intensive technique requiring $O(p^3)$ operations for the determination of a set of PARCOR coefficients.

In the following part of this chapter, we investigate several interesting properties of GREs in the time recursive case. These properties were first seriously studied by Strobach [8.4]. They lead to a class of recursive covariance ladder algorithms with outstanding numerical properties and a reduced complexity of $O(p^2)$ operations.

8.3.1 Recursive Least-Squares Using Generalized Residual Energies

So far, the GREs have been used for the derivation of *block processing* covariance ladder algorithms assuming a PARCOR ladder form. In an early paper [8.5] Strobach showed how GREs can be used to derive exact recursive least-squares (RLS) covariance ladder algorithms. For this purpose, we may rewrite the vector order recursions (6.15a,b) determining the least-squares ladder form as

$$e_0(t) = r_0(t) = x(t) \ , \tag{8.41a}$$

$$e_m(t) = e_{m-1}(t) + K_m^f(t)r_{m-1}(t-1) \ , \tag{8.41b}$$

$$r_m(t) = r_{m-1}(t-1) + K_m^b(t)e_{m-1}(t) \ , \tag{8.41c}$$

for $1 \leq m \leq p$,

where

$$K_m^f(t) = - C_{m-1}(t)\big/R_{m-1}(t-1) \ , \tag{8.42a}$$

$$K_m^b(t) = - C_{m-1}(t)\big/E_{m-1}(t) \ . \tag{8.42b}$$

In order to derive true least-squares ladder recursions based on GREs, we may consider the following "expansion" of the vector order recursions (8.41a-c):

$$\mathbf{e}_0(t-i) = \mathbf{x}(t-i) \ ; \qquad \mathbf{r}_0(t-j) = \mathbf{x}(t-j) \ , \tag{8.43a}$$

$$\mathbf{e}_m(t-i) = \mathbf{e}_{m-1}(t-i) + K_m^f(t-i) \, \mathbf{r}_{m-1}(t-1-i) \ , \tag{8.43b}$$

$$\mathbf{r}_m(t-j) = \mathbf{r}_{m-1}(t-1-j) + K_m^b(t-j) \, \mathbf{e}_{m-1}(t-j) \ . \tag{8.43c}$$

By stating all possible inner products of *"shifted"* residual vectors

$$
\begin{aligned}
\mathbf{e}_m^T(t-i)\mathbf{e}_m(t-j) \ = \ & \mathbf{e}_{m-1}^T(t-i) \, \mathbf{e}_{m-1}(t-j) \\
& + K_m^f(t-i) \, \mathbf{r}_{m-1}^T(t-1-i) \, \mathbf{e}_{m-1}(t-j) \\
& + K_m^f(t-j) \, \mathbf{e}_{m-1}^T(t-i) \, \mathbf{r}_{m-1}(t-1-j) \\
& + K_m^f(t-i) \, K_m^f(t-j) \, \mathbf{r}_{m-1}^T(t-1-i) \, \mathbf{r}_{m-1}(t-1-j) \ , \tag{8.44a}
\end{aligned}
$$

$$
\begin{aligned}
\mathbf{r}_m^T(t-1-i)\mathbf{r}_m(t-1-j) = \ & \mathbf{r}_{m-1}^T(t-2-i) \, \mathbf{r}_{m-1}(t-2-j) \\
& + K_m^b(t-1-i) \, \mathbf{e}_{m-1}^T(t-1-i) \, \mathbf{r}_{m-1}(t-2-j) \\
& + K_m^b(t-1-j) \, \mathbf{r}_{m-1}^T(t-2-i) \, \mathbf{e}_{m-1}(t-1-j) \\
& + K_m^b(t-1-i) K_m^b(t-1-j) \, \mathbf{e}_{m-1}^T(t-1-i) \mathbf{e}_{m-1}(t-1-j) , \tag{8.44b}
\end{aligned}
$$

$$
\begin{aligned}
\mathbf{e}_m^T(t-i)\mathbf{r}_m(t-1-j) \ = \ & \mathbf{e}_{m-1}^T(t-i) \, \mathbf{r}_{m-1}(t-2-j) \\
& + K_m^f(t-i) \, \mathbf{r}_{m-1}^T(t-1-i) \, \mathbf{r}_{m-1}(t-2-j) \\
& + K_m^b(t-1-j) \, \mathbf{e}_{m-1}^T(t-i) \, \mathbf{e}_{m-1}(t-1-j) \\
& + K_m^f(t-i) \, K_m^b(t-1-j) \, \mathbf{r}_{m-1}^T(t-1-i) \, \mathbf{e}_{m-1}(t-1-j) , \tag{8.44c}
\end{aligned}
$$

one sees that the inner products associated with stage m and stage m−1 can be expressed in terms of GREs. Thus, substituting the GREs (8.31a-c) into (8.44a-c) yields an exact least-squares recursion of GREs :

$$
\begin{aligned}
E_{m,i,j}(t) = \ & E_{m-1,i,j}(t) + K_m^f(t-i) C_{m-1,j,i}(t) + K_m^f(t-j) C_{m-1,i,j}(t) \\
& + K_m^f(t-i) \, K_m^f(t-j) \, R_{m-1,i,j}(t) \ , \tag{8.45a}
\end{aligned}
$$

$$R_{m,i,j}(t) = R_{m-1,i+1,j+1}(t) + K_m^b(t-1-i)\,C_{m-1,i+1,j+1}(t)$$

$$+ K_m^b(t-1-j)\,C_{m-1,j+1,i+1}(t)$$

$$+ K_m^b(t-1-i)\,K_m^b(t-1-j)\,E_{m-1,i+1,j+1}(t) \quad , \tag{8.45b}$$

$$C_{m,i,j}(t) = C_{m-1,i,j+1}(t) + K_m^f(t-i)\,R_{m-1,i,j+1}(t)$$

$$+ K_m^b(t-1-j)\,E_{m-1,i,j+1}(t)$$

$$+ K_m^f(t-i)\,K_m^b(t-1-j)\,C_{m-1,j+1,i}(t) \quad . \tag{8.45c}$$

Obviously, the reflection coefficients can be computed from the upper-left elements of the GREs as

$$K_m^f(t) = -\,C_{m-1,0,0}(t)\big/R_{m-1,0,0}(t) \quad , \tag{8.46a}$$

$$K_m^b(t) = -\,C_{m-1,0,0}(t)\big/E_{m-1,0,0}(t) \quad . \tag{8.46b}$$

Just the recursions (8.45a–c) together with (8.46a,b) and the initialization scheme of GREs from the covariance matrix of the observed process (8.37a–c) already establish a closed recursion for the true least-squares computation of the reflection coefficients. Interestingly, the recursions of GREs (8.45a–c) thus obtained contain the reflection coefficients of the previous stage m–1 at successively delayed time steps.

8.3.2 Strobach's Algorithm

A closer look at the recursions (8.45a–c) reveals that this "direct form" of the exact recursive least-squares covariance ladder algorithm based on GREs has the disadvantage of a computational complexity of $O(p^3)$. This problem can be circumvented by a better utilization of the useful and interesting properties of GREs.

Since the covariance matrix of the observed process is always symmetric,

$$\Phi_{i,j}(t) = \Phi_{j,i}(t) \quad , \tag{8.47}$$

the GREs $E_0(t)$ and $R_0(t)$ at stage m=0 will also be symmetric as can be seen from the initialization scheme (Fig. 8.1). This gives rise to a first theorem:

Theorem. *(Symmetric property of GREs).* The GREs $E_m(t)$ and $R_m(t)$ are symmetric matrices satisfying the relations

$$E_{m,i,j}(t) = E_{m,j,i}(t) \; , \tag{8.48a}$$

$$R_{m,i,j}(t) = R_{m,j,i}(t) \; , \tag{8.48b}$$

for $0 \le m \le p; \quad 0 \le i,j \le p \; .$

Proof. From (8.47 and 38a,b), we know

$$E_{0,i,j}(t) = E_{0,j,i}(t) \; , \tag{8.49a}$$

$$R_{0,i,j}(t) = R_{0,j,i}(t) \; . \tag{8.49b}$$

Permuting indices i and j in (8.45a,b), we obtain

$$E_{m,j,i}(t) = E_{m-1,j,i}(t) + K_m^f(t-i)C_{m-1,j,i}(t)$$
$$+ K_m^f(t-j) C_{m-1,i,j}(t)$$
$$+ K_m^f(t-i) K_m^f(t-j) R_{m-1,j,i}(t) \; , \tag{8.50a}$$

$$R_{m,j,i}(t) = R_{m-1,j+1,i+1}(t) + K_m^b(t-1-i) C_{m-1,i+1,j+1}(t)$$
$$+ K_m^b(t-1-j) C_{m-1,j+1,i+1}(t)$$
$$+ K_m^b(t-1-i)K_m^b(t-1-j) E_{m-1,j+1,i+1}(t) \; . \tag{8.50b}$$

Suppose now that the GREs of stage m-1 are symmetric (proof by induction),

$$E_{m-1,i,j}(t) = E_{m-1,j,i}(t) \; , \tag{8.51a}$$

$$R_{m-1,i,j}(t) = R_{m-1,j,i}(t) \; , \tag{8.51b}$$

and substitute (8.51a,b) into (8.50a,b), yielding

$$E_{m,j,i}(t) = E_{m-1,i,j}(t) + K_m^f(t-i)C_{m-1,j,i}(t) + K_m^f(t-j) C_{m-1,i,j}(t)$$
$$+ K_m^f(t-i) K_m^f(t-j) R_{m-1,i,j}(t) \tag{8.52a}$$

$$R_{m,j,i}(t) = R_{m-1,i+1,j+1}(t) + K_m^b(t-1-i) C_{m-1,i+1,j+1}(t)$$
$$+ K_m^b(t-1-j) C_{m-1,j+1,i+1}(t)$$
$$+ K_m^b(t-1-i)K_m^b(t-1-j) E_{m-1,i+1,j+1}(t) \; . \tag{8.52b}$$

The right-hand sides of (8.52a,b) are identical to the right-hand sides of (8.45a,b) and therefore (8.48a,b) holds for all stages m, from m = 0 to m = p.

End of proof.

The GRE $C_m(t)$ is not symmetric. Until now, we have only considered the symmetry of GREs in *consecutive stages* of the algorithm at a *current time step*. A second interesting property of GREs can be discovered by investigating the interdependence of GREs of a fixed stage m at subsequent time steps. This gives rise to a second theorem.

Theorem. *(Shift invariance property of GREs)*. The GREs $E_m(t)$, $R_m(t)$ and $C_m(t)$ are related to their predecessors $E_m(t-1)$, $R_m(t-1)$ and $C_m(t-1)$ according to the following relations:

$$E_{m,i+1,j+1}(t) = E_{m,i,j}(t-1) \quad , \tag{8.53a}$$

$$R_{m,i+1,j+1}(t) = R_{m,i,j}(t-1) \quad , \tag{8.53b}$$

$$C_{m,i+1,j+1}(t) = C_{m,i,j}(t-1) \quad , \tag{8.53c}$$

$$for \quad 0 \le m \le p ; \quad 0 \le i,j \le p \quad .$$

Proof. Consider the GREs at time step t-1,

$$E_{m,i,j}(t-1) = e_m^T(t-1-i)e_m(t-1-j) \quad , \tag{8.54a}$$

$$R_{m,i,j}(t-1) = r_m^T(t-2-i)r_m(t-2-j) \quad , \tag{8.54b}$$

$$C_{m,i,j}(t-1) = e_m^T(t-1-i)r_m(t-2-j) \quad . \tag{8.54c}$$

We note that the right-hand-side terms in (8.54a-c) can also be expressed in terms of GREs at the current time step t,

$$E_{m,i+1,j+1}(t) = e_m^T(t-1-i)e_m(t-1-j) \quad , \tag{8.55a}$$

$$R_{m,i+1,j+1}(t) = r_m^T(t-2-i)r_m(t-2-j) \quad , \tag{8.55b}$$

$$C_{m,i+1,j+1}(t) = e_m^T(t-1-i)r_m(t-2-j) \quad . \tag{8.55c}$$

The right-hand sides of (8.54a-c and 55a-c) are identical, and therefore (8.53a-c) holds for all stages m, from m = 0 to m = p.

End of proof.

The shift-invariance property of GREs (8.53a-c) represents the very useful result that a GRE at time step t arises from the diagonal down-shifted elements of the same GRE at the *previous time step*. Figure 8.2 illustrates this very useful shift invariance property of GREs. Only the first row and first column of each GRE must be updated at each time step. Utilizing the symmetric property of GREs (8.48a,b), we conclude that only the updating of four recursion vectors of length $(p - m -1)$ is required in this recursive least-squares ladder algorithm, in contrast to the updating of three recursion matrices of dimension $(p - m -1) \times (p - m -1)$ if the symmetric and shift invariance properties of GREs are not exploited. With this step, the computational complexity of the recursive covariance ladder algorithm based on GREs reduces to $O(p^2)$ computations per recursion. Only the following vector order recursions must be computed:

$$E_{m,0,j}(t) = E_{m-1,0,j}(t) + K_m^f(t)C_{m-1,j,0}(t) + K_m^f(t-j)\,C_{m-1,0,j}(t)$$
$$+ K_m^f(t)\,K_m^f(t-j)\,R_{m-1,0,j}(t) \quad , \tag{8.56a}$$

$$R_{m,0,j}(t) = R_{m-1,1,j+1}(t) + K_m^b(t-1)C_{m-1,1,j+1}(t)$$
$$+ K_m^b(t-1-j)\,C_{m-1,j+1,1}(t)$$
$$+ K_m^b(t-1)K_m^b(t-1-j)\,E_{m-1,1,j+1}(t) \quad , \tag{8.56b}$$

$$C_{m,0,j}(t) = C_{m-1,0,j+1}(t) + K_m^f(t)\,R_{m-1,0,j+1}(t)$$
$$+ K_m^b(t-1-j)\,E_{m-1,0,j+1}(t)$$
$$+ K_m^f(t-i)\,K_m^b(t-1-j)\,C_{m-1,j+1,0}(t) \quad , \tag{8.56c}$$

$$C_{m,j,0}(t) = C_{m-1,j,1}(t) + K_m^f(t-j)\,R_{m-1,1,j}(t)$$
$$+ K_m^b(t-1)\,E_{m-1,1,j}(t)$$
$$+ K_m^f(t-j)\,K_m^b(t-1)\,C_{m-1,1,j}(t) \quad . \tag{8.56d}$$

The algorithm based on the recursions (8.56a-d) has been named the *pure order recursive ladder algorithm* (PORLA). This terminology emerges from the fact that PORLA constructs the reflection coefficients of the true least-squares ladder form by a pure order recursive procedure employing the GREs as *intermediate variables*. The GREs are initialized from the covariance matrix of the observed process. See again Fig. 8.1.

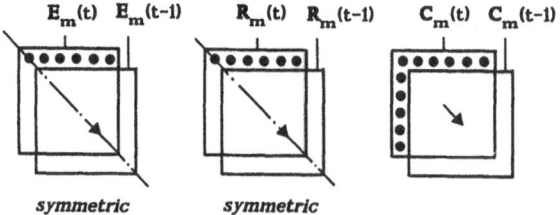

Fig. 8.2. Efficient computation of GREs utilizing the symmetric property and the shift invariance property of GREs

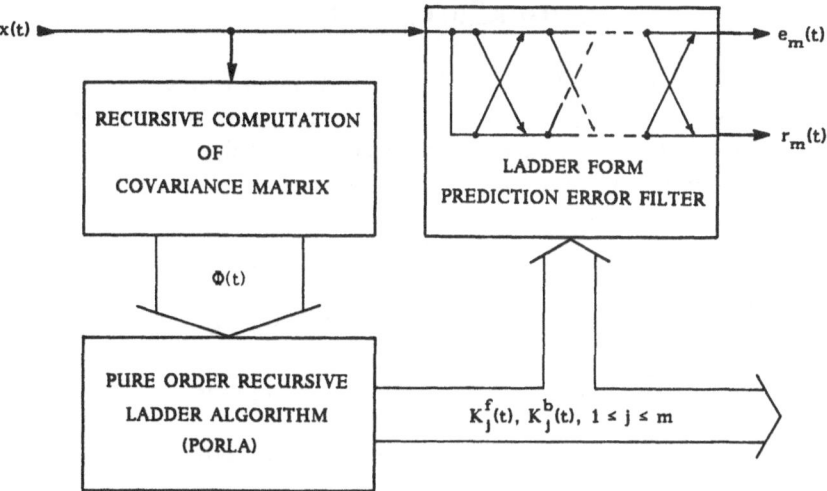

Fig. 8.3. Principle of structure of PORLA RLS algorithms: Tracking of a time-varying process $\{x(t)\}$ by recursively updated second-order (covariance) information $\Phi(t)$, computation of reflection coefficients $\{K_m^f(t), K_m^b(t); 1 \le m \le p\}$ by a pure order recursive procedure (PORLA) from the updated covariance information, and ladder-form prediction error filter (optional)

The time recursion is restricted to the updating of the time-recursively computed covariance matrix. See also Fig. 8.3 for an illustration of the principal structure of the PORLA RLS algorithm. At this point, we can take advantage of the recursive windowing algorithms presented in Chap. 2. With a variation of these windowing techniques, the algorithm can be tuned quite accurately to obtain any desired steady-state or tracking behavior. Moreover, the PORLA RLS algorithm can serve as a useful tool in those estimation problems where only the covariance (second-order) information, rather than the process itself, is available.

As can be seen from (8.56b,d), this early version of PORLA requires the stacking of four GREs. Rewriting the recursions (8.56b,d) in a slightly different form:

$$R_{m,0,j}(t+1) = R_{m-1,0,j}(t) + K_m^b(t)C_{m-1,0,j}(t) + K_m^b(t-j)C_{m-1,j,0}(t)$$

$$+ K_m^b(t)K_m^b(t-j)E_{m-1,0,j}(t) \quad , \tag{8.57a}$$

$$C_{m,j+1,0}(t+1) = C_{m-1,j,0}(t) + K_m^f(t-j)R_{m-1,0,j}(t) + K_m^b(t)E_{m-1,0,j}(t)$$

$$+ K_m^f(t-j)K_m^b(t)C_{m-1,0,j}(t) \quad , \tag{8.57b}$$

only the variables $R_{m,0,j}(t+1)$ and $C_{m,j+1,0}(t+1)$ must be stacked. Table 8.3 summarizes this version of PORLA. As demonstrated in [8.5], this algorithm has outstanding numerical properties. In Table 8.3, we use the following simplified notations for convenience:

$$E(j) = E_{m,0,j}(t) \quad , \qquad\qquad RR(m-1,j) = R_{m,0,j}(t+1) \quad ,$$

$$R(j) = R_{m,0,j}(t) \quad , \qquad\qquad CC(m-1,j) = C_{m,j,0}(t+1) \quad ,$$

$$C(j) = C_{m,0,j}(t) \quad , \qquad\qquad KF(m,j) = K_m^f(t-j) \quad ,$$

$$CT(j) = C_{m,j,0}(t) \quad , \qquad\qquad KB(m,j) = K_m^b(t-j) \quad ,$$

8.3.3 Approximate PORLA Computation Schemes

In many applications of adaptive signal processing, the observed signal and the underlying model parameters change slowly and smoothly compared to the data sample rate. In those cases, *approximate* ladder algorithms can reach almost the same performance as a true least-squares solution. The PORLA recursions offer the possibility of incorporation of a wide variety of different stationarity assumptions. In this way, five approximate covariance ladder algorithms have been derived. Interestingly, it turns out that one of these algorithms is identical with the Cumani covariance ladder algorithm discussed earlier in this chapter. In the presented recursive formulation, however, the algorithm requires far fewer computations than Cumani's traditional approach. This was achieved through a utilization of shift invariance and symmetric properties of GREs. Finally, at the highest degree of approximation, we obtain a PORLA computation scheme for the Levinson-Durbin algorithm. This algorithm can be viewed as a counterpart of the LeRoux-Gueguen algorithm, as it offers an alternative method for a fixed-point implementation of the Levinson-Durbin algorithm.

Table 8.3. Strobach's recursive least-squares PORLA computation technique. The algorithm is initialized from the time-recursively updated covariance matrix of the observed process and computes the ladder reflection coefficients at each time step by a purely order recursive procedure. The algorithm involves division. Whenever the divisor is small and positive, set $1/x = x$. When the divisor is negative, set $1/x = 0$

FOR   t = 0, 1, 2, . . .

   Input: $\Phi_{i,j}(t)$; $0 \le i,j \le p$

   FOR   j = 0, 1, 2, . . . , p - 1   *initialize:*

      $EM(j) = \Phi_{0,j}(t)$ $R(j) = \Phi_{1,j+1}(t)$

      $CM(j) = \Phi_{0,j+1}(t)$ $CT(j) = \Phi_{1,j}(t)$

   $KF(1,0) = - CM(0)\big/R(0)$ $KB(1,0) = - CM(0)\big/EM(0)$

   FOR   m = 1, 2, . . . , p - 1
     FOR   j = 0, 1, . . . , p - m - 1

       $E(j)$ $=$ $EM(j) + KF(m,0)CT(j) + KF(m,j)CM(j)$

              $+ KF(m,0)KF(m,j)R(j)$

       $RR(m-1,j) = R(j) + KB(m,0)CM(j) + KB(m,j)CT(j)$

               $+ KB(m,0)KB(m,j)EM(j)$

       $C(j)$ $=$ $CM(j+1) + KF(m,0)R(j+1) + KB(m,j+1)EM(j+1)$

               $+ KF(m,0)KB(m,j+1)CT(j+1)$

     $CC(m,0) = C(0)$

     FOR   j = 0, 1, . . . , p - m - 1

       $CC(m-1,j+1) = CT(j) + KF(m,j)R(j) + KB(m,0)EM(j)$

                 $+ KF(m,j)KB(m,0)CM(j)$

       $R(j) = RR(m,j)$
       $CT(j) = CC(m,j)$
       $EM(j) = E(j)$
       $CM(j) = C(j)$

     $KF(m+1,0) = - C(0)\big/R(0)$ $KB(m+1,0) = - C(0)\big/E(0)$

   FOR   m = p -1, p-2, . . . , 1   *increment stack pointers:*
     FOR   j = p-1, p-2, . . . , 0

       $RR(m,j) = RR(m-1,j)$
       $CC(m,j) = CC(m-1,j)$
       $KF(m,j+1) = KF(m,j)$
       $KB(m,j+1) = KB(m,j)$

There are four basic approximations that arise as a natural consequence of stationarity. Several of these approximations can be combined in a meaningful way in order to obtain an algorithm with a *desired behavior*. Five examples of approximate PORLA algorithms are given.

Approximation 1: Piecewise time-invariant reflection coefficients

In the case of stationarity of the observed process, the reflection coefficients can be assumed to be *piecewise time invariant* as follows:

$$K_m^f(t-i) = K_m^f(t) \quad , \tag{8.58a}$$

$$K_m^b(t-j) = K_m^b(t) \quad . \tag{8.58b}$$

The application of this approximation allows us to omit the stack of reflection coefficients which was required in the true least-squares PORLA computation scheme as listed in Table 8.3.

Approximation 2: Toeplitz structure of GREs

In the case of stationarity, the structure of the input data covariance matrix is Toeplitz [8.14].

$$\Phi_{i+1,j+1}(t) = \Phi_{i,j}(t) \quad . \tag{8.59}$$

The GREs at stage zero will share this property, as can be seen from the initialization scheme in Fig. 8.1. Now it would be of interest to see whether the GREs remain Toeplitz during order recursions. Considering the GRE $E_m(t)$ for the moment and taking into account that the reflection coefficients are set time-invariant according to approximation 1, we obtain

$$E_{m,i,j}(t) = E_{m-1,i,j}(t) + K_m^f(t)\big(C_{m-1,j,i}(t) + C_{m-1,i,j}(t)\big)$$
$$+ K_m^{f\,2}(t)R_{m-1,i,j}(t) \quad , \tag{8.60a}$$

$$E_{m,i+1,j+1}(t) = E_{m-1,i+1,j+1}(t) + K_m^f(t)\big(C_{m-1,j+1,i+1}(t) + C_{m-1,i+1,j+1}(t)\big)$$
$$+ K_m^{f\,2}(t)R_{m-1,i+1,j+1}(t) \quad . \tag{8.60b}$$

Suppose now that the GREs of stage m-1 are Toeplitz:

$$E_{m-1,i+1,j+1}(t) = E_{m-1,i,j}(t) \quad , \tag{8.61a}$$

$$R_{m-1,i+1,j+1}(t) = R_{m-1,i,j}(t) \quad , \tag{8.61b}$$

$$C_{m-1,i+1,j+1}(t) = C_{m-1,i,j}(t) \quad , \tag{8.61c}$$

and substitute (8.61a-c) into (8.60b), yielding

$$E_{m,i+1,j+1}(t) = E_{m-1,i,j}(t) + K_m^f(t)\Big(C_{m-1,j,i}(t) + C_{m-1,i,j}(t)\Big)$$
$$+ K_m^{f\,2}(t)R_{m-1,i,j}(t) \quad . \tag{8.62}$$

The right-hand sides of (8.62) and (8.60a) are identical. Therefore, we conclude that

$$E_{m,i+1,j+1}(t) = E_{m,i,j}(t) \quad , \tag{8.63a}$$

and similarly

$$R_{m,i+1,j+1}(t) = R_{m,i,j}(t) \quad , \tag{8.63b}$$

$$C_{m,i+1,j+1}(t) = C_{m,i,j}(t) \quad , \tag{8.63c}$$

which justifies the Toeplitz assumption for all stages m, from $m=0$ to $m=p$.

The Toeplitz property of GREs represents a useful substitute for the true LS shift invariance property of GREs (8.53a-c). This approximation allows us to omit the stack of GREs. Therefore, the Toeplitz approach requires an amount of storage space that depends only *linearly* on p, whereas the storage space of the exact PORLA technique grows with the *square* of p.

Approximation 3: Burg's method

In the unwindowed (covariance) case of linear prediction, where one is interested in a PARCOR model of the observed process, Burg's method [8.12] is frequently applied. See Chap. 6 for a more detailed discussion of Burg's approach. Clearly, Burg's formula (6.92) can be expressed in terms of GREs as

$$K_m(t) = \frac{-2\,C_{m-1,0,0}(t)}{E_{m-1,0,0}(t) + R_{m-1,0,0}(t)} \quad . \tag{8.64}$$

Approximation 4: Identical forward/backward residual energies

An even more striking approximation assumes that the forward and backward residual energies are identical in each stage of the algorithm. This is satisfied in the case of a stationary process and a PARCOR ladder form with identical reflection coefficients in the forward and backward predictors. (See Sect. 6.3 for a proof that a stationary process corresponds to identical energies and coefficients in a PARCOR ladder form.) Clearly, this is also the underlying assumption in the Levinson-Durbin algorithm and, in terms of GREs, we may express this approximation as

$$E_{m,i,j}(t) = R_{m,i,j}(t) \ . \tag{8.65}$$

In the following, we scan some useful combinations of the discussed approximations and present the summaries of the resulting approximate PORLA computation techniques. Subsequently, the relationships between the obtained simplified schemes and some classical approaches are explained. Five approximate PORLA schemes, named PORLA 1 to PORLA 5 have been obtained. Table 8.4 illustrates the relationships between the underlying approximations and the algorithms. Tables 8.5 - 9 summarize the algorithms.

Table 8.4. Approximate PORLA computation techniques and the underlying approximations. PORLA 2 is the recursive case of Cumani's covariance ladder algorithm, which can be viewed as a fixed-point computation of the Makhoul covariance ladder algorithm. PORLA 5 is a fixed-point computation of the Levinson-Durbin algorithm

Approximation	Algorithm				
	PORLA 1	PORLA 2	PORLA 3	PORLA 4	PORLA 5
1. Time-invariant K's	*	*	*	*	*
2. Toeplitz GRE's			*	*	*
3. Burg's method		*		*	
4. $E_m(t) = R_m(t)$					*

Table 8.5. Approximate algorithm PORLA 1. This algorithm involves division. Whenever the divisor is small and positive, set $1/x = x$. When the divisor is negative, set $1/x = 0$

FOR $t = 0, 1, 2, \ldots$

> *Input:* $\Phi_{i,j}(t)$; $0 \le i,j \le p$
>
> FOR $j = 0, 1, 2, \ldots, p-1$ *initialize:*
>
> > $EM(j) = \Phi_{0,j}(t)$ $R(j) = \Phi_{1,j+1}(t)$
> >
> > $CM(j) = \Phi_{0,j+1}(t)$ $CT(j) = \Phi_{1,j}(t)$
>
> $KF(1) = -CM(0)\big/R(0)$ $KB(1) = -CM(0)\big/EM(0)$
>
> FOR $m = 1, 2, \ldots, p-1$
> > FOR $j = 0, 1, \ldots, p-m-1$
> >
> > > $E(j) \quad = \quad EM(j) + KF(m)\big(CT(j) + CM(j)\big) + KF(m)KF(m)R(j)$
> > >
> > > $RR(m-1,j) = R(j) + KB(m)\big(CT(j) + CM(j)\big) + KB(m)KB(m)EM(j)$
> > >
> > > $C(j) \quad = \quad CM(j+1) + KF(m)R(j+1) + KB(m)EM(j+1)$
> > > $\qquad\qquad\qquad + KF(m)\,KB(m)\,CT(j+1)$
> >
> > $CC(m,0) = C(0)$
> >
> > FOR $j = 0, 1, \ldots, p-m-1$
> >
> > > $CC(m-1,j+1) = CT(j) + KF(m)R(j) + KB(m)\,EM(j)$
> > > $\qquad\qquad\qquad + KF(m)\,KB(m)\,CM(j)$
> > > $R(j) = RR(m,j)$
> > > $CT(j) = CC(m,j)$
> > > $EM(j) = E(j)$
> > > $CM(j) = C(j)$
> >
> > $KF(m+1) = -C(0)\big/R(0)$ $KB(m+1) = -C(0)\big/E(0)$
>
> FOR $m = p-1, p-2, \ldots, 1$ *increment stack pointers:*
> > FOR $j = p-1, p-2, \ldots, 0$
> >
> > > $RR(m,j) = RR(m-1,j)$
> > > $CC(m,j) = CC(m-1,j)$

Table 8.6. Summary of PORLA 2. One may take advantage of identical expressions. This algorithm involves division. Whenever the divisor is small and positive, set $1/x = x$. When the divisor is negative, set $1/x = 0$

FOR $t = 0, 1, 2, \ldots$

> *Input:* $\Phi_{i,j}(t)$; $0 \le i,j \le p$
>
> FOR $j = 0, 1, 2, \ldots, p-1$ *initialize:*
>
> > $EM(j) = \Phi_{0,j}(t)$ $R(j) = \Phi_{1,j+1}(t)$
> >
> > $CM(j) = \Phi_{0,j+1}(t)$ $CT(j) = \Phi_{1,j}(t)$
>
> $K(1)$ $= -2\,CM(0)\Big/\Big(EM(0) + R(0)\Big)$
>
> FOR $m = 1, 2, \ldots, p-1$
>
> > FOR $j = 0, 1, \ldots, p-m-1$
> >
> > > $E(j) = EM(j) + K(m)\Big(CT(j) + CM(j)\Big) + K(m)K(m)R(j)$
> > > $RR(m-1,j) = R(j) + K(m)\Big(CT(j) + CM(j)\Big) + K(m)K(m)EM(j)$
> > > $C(j) = CM(j+1) + K(m)\Big(R(j+1) + EM(j+1)\Big) + K(m)K(m)CT(j+1)$
> >
> > $CC(m,0) = C(0)$
> >
> > FOR $j = 0, 1, \ldots, p-m-1$
> >
> > > $CC(m-1,j+1) = CT(j) + K(m)\Big(R(j) + EM(j)\Big) + K(m)K(m)CM(j)$
> > >
> > > $R(j) = RR(m,j)$
> > > $CT(j) = CC(m,j)$
> > > $EM(j) = E(j)$
> > > $CM(j) = C(j)$
> >
> > $K(m+1) = -2\,C(0)\Big/\Big(E(0) + R(0)\Big)$
>
> FOR $m = p-1, p-2, \ldots, 1$ *increment stack pointers:*
>
> > FOR $j = p-1, p-2, \ldots, 0$
> >
> > > $RR(m,j) = RR(m-1,j)$
> > > $CC(m,j) = CC(m-1,j)$

Table 8.7. Summary of PORLA 3. This algorithm involves division. Whenever the divisor is small and positive, set $1/x = x$. When the divisor is negative, set $1/x = 0$

FOR   t = 0, 1, 2, . . .

$\quad$ *Input:* $\Phi_{i,j}(t)$; $0 \le i,j \le p$

$\quad$ FOR   j = 0, 1, 2, . . . , p − 1 $\qquad\qquad$ *initialize:*

$\quad\quad$ EM(j) = $\Phi_{0,j}(t)$ $\qquad$ RM(j) = $\Phi_{1,j+1}(t)$

$\quad\quad$ CM(j) = $\Phi_{0,j+1}(t)$ $\quad$ CTM(j) = $\Phi_{1,j}(t)$

$\quad$ FOR   j = 0, 1, 2, . . . , p − 1 $\qquad$ *Alternative autocorrelation*
$\qquad\qquad\qquad\qquad\qquad\qquad\qquad\qquad$ *initialization scheme:*

$\quad\quad$ EM(j) = $\Phi_{0,j}(t)$ $\qquad$ RM(j)= $\Phi_{0,j}(t)$

$\quad\quad$ CM(j) = $\Phi_{0,j+1}(t)$ $\quad$ CTM(j+1) = $\Phi_{0,j}(t)$

$\quad$ CTM(0) = $\Phi_{0,1}(t)$

$\quad$ KF(1) = − CM(0)$\big/$RM(0) $\qquad\qquad$ KB(1) = − CM(0)$\big/$EM(0)

$\quad$ FOR   m = 1, 2, . . . , p − 1
$\quad\quad$ FOR   j = 0, 1, . . . , p − m − 1

$\quad\quad\quad$ E(j) = EM(j) + KF(m)$\big($CM(j) + CTM(j)$\big)$ + KF(m)KF(m)RM(j)
$\quad\quad\quad$ R(j) = RM(j) + KB(m)$\big($CM(j) + CTM(j)$\big)$ + KB(m)KB(m)EM(j)
$\quad\quad\quad$ C(j) = CM(j+1) + KF(m)RM(j+1) + KB(m)EM(j+1)
$\quad\quad\quad\quad\quad$ + KF(m) KB(m) CTM(j+1)

$\quad\quad$ CT(0) = C(0)

$\quad\quad$ FOR   j = 0, 1, . . . , p − m − 2

$\quad\quad\quad$ CT(j+1) = CTM(j) + KF(m)R(j) + KB(m)EM(j)
$\quad\quad\quad\quad\quad$ + KF(m) KB(m)CM(j)

$\quad\quad$ FOR   j = 0, 1, . . . , p − m − 1

$\quad\quad\quad$ EM (j) = E(j)
$\quad\quad\quad$ RM(j) = R(j)
$\quad\quad\quad$ CM(j) = C(j)
$\quad\quad\quad$ CTM(j) = CT(j)

$\quad\quad$ KF(m+1) = − C(0)$\big/$R(0) $\qquad\qquad$ KB(m+1) = − C(0)$\big/$E(0)

Table 8.8. Summary of PORLA 4. One may take advantage of identical expressions. This algorithm involves division. Whenever the divisor is small and positive, set $1/x = x$. When the divisor is negative, set $1/x = 0$

FOR   t = 0, 1, 2, . . .

> *Input:* $\Phi_{i,j}(t)$; $0 \le i,j \le p$
>
> > FOR   j = 0, 1, 2, . . . , p − 1   *initialize:*
> >
> > > $EM(j) = \Phi_{0,j}(t)$ $RM(j) = \Phi_{1,j+1}(t)$
> > >
> > > $CM(j) = \Phi_{0,j+1}(t)$ $CTM(j) = \Phi_{1,j}(t)$
> >
> > FOR   j = 0, 1, 2, . . . , p − 1   *Alternative autocorrelation*
> > *initialization scheme:*
> > > $EM(j) = \Phi_{0,j}(t)$ $RM(j) = \Phi_{0,j}(t)$
> > >
> > > $CM(j) = \Phi_{0,j+1}(t)$ $CTM(j+1) = \Phi_{0,j}(t)$
> >
> > $CTM(0) = \Phi_{0,1}(t)$
> >
> > $K(1)$ $= - 2\, CM(0) \Big/ \Big(EM(0) + RM(0) \Big)$
> >
> > FOR   m = 1, 2, . . . , p − 1
> > > FOR   j = 0, 1, . . . , p − m − 1
> > >
> > > > $E(j) = EM(j) + K(m)\Big(CTM(j) + CM(j)\Big) + K(m)K(m)\,RM(j)$
> > > > $R(j) = RM(j) + K(m)\Big(CTM(j) + CM(j)\Big) + K(m)K(m)EM(j)$
> > > > $C(j) = CM(j+1) + K(m)\Big(RM(j+1) + EM(j+1)\Big) + K(m)K(m)\,CTM(j+1)$
> > >
> > > $CT(0) = C(0)$
> > >
> > > FOR   j = 0, 1, . . . , p − m − 2
> > >
> > > > $CT(j+1) = CTM(j) + K(m)\Big(RM(j) + EM(j)\Big) + K(m)K(m)\, CM(j)$
> > >
> > > FOR   j = 0, 1, . . . , p − m − 1
> > > > $EM(j) = E(j)$
> > > > $RM(j) = R(j)$
> > > > $CM(j) = C(j)$
> > > > $CTM(j) = CT(j)$
> > >
> > > $K(m+1) = - 2\, C(0) \Big/ \Big(E(0) + R(0) \Big)$

Table 8.9. Summary of PORLA 5. In the nonrecursive case, this algorithm is an alternative computation of the Levinson-Durbin algorithm. One may take advantage of identical expressions. This algorithm involves division. Whenever the divisor is small and positive, set $1/x = x$. When the divisor is negative, set $1/x = 0$

Input: $\Phi_{i,j}$; $0 \le i,j \le p$

FOR $j = 0, 1, 2, \ldots, p - 1$ *Autocorrelation initialization scheme:*

$$\left[\quad EM(j) = \Phi_{0,j} \qquad CM(j) = \Phi_{0,j+1} \qquad CTM(j+1) = \Phi_{0,j} \right.$$

$$CTM(0) = \Phi_{0,1}$$

$$K(1) \quad = - \; CM(0) \Big/ EM(0)$$

FOR $m = 1, 2, \ldots, p - 1$

$\left[\quad \right.$ FOR $j = 0, 1, \ldots, p - m - 1$

$$\left[\quad E(j) = EM(j) + K(m)\big(CTM(j) + CM(j)\big) + K(m)K(m) \, EM(j) \right.$$

$$\left. \quad C(j) = CM(j+1) + 2 \, K(m) \, EM(j+1) + K(m)K(m) \, CTM(j+1) \right.$$

$$CT(0) = C(0)$$

FOR $j = 0, 1, \ldots, p - m - 2$

$$\left[\quad CT(j+1) = CTM(j) + 2 \, K(m) \, EM(j) + K(m)K(m) \, CM(j) \right.$$

FOR $j = 0, 1, \ldots, p - m - 1$

$$\left[\quad \begin{array}{l} EM(j) = E(j) \\ CM(j) = C(j) \\ CTM(j) = CT(j) \end{array} \right.$$

$$K(m+1) = - \; C(0) \Big/ E(0)$$

8.3.4 Sokat's Algorithm – An Extension of the LeRoux-Gueguen Algorithm

In Sect. 8.3.2, we saw that a recursive least-squares covariance ladder algorithm can be obtained from a generalization of the residual energies appearing in the Cumani covariance ladder algorithm. In a similar procedure, one can extend the LeRoux-Gueguen algorithm to the time recursive least-squares case [8.6]. Based on the GREs of LeRoux and Gueguen, a completely new PORLA computation scheme has been developed. This approach is attributed to Sokat [8.7]. In fact, it can be shown that the algorithm of LeRoux and Gueguen appears as a special case of Sokat's PORLA computation technique. This section presents the basic ideas underlying this new PORLA algorithm. Quite similar to the approach discussed in Sect. 8.3.2, several approximations can be incorporated in this new PORLA method leading to schemes quite similar to those seen in the case of Strobach's algorithms. As a basic difference, Sokat's method does not require the computation of quadratic terms, and hence the new PORLA scheme is computationally more efficient. But in principle, the new algorithm has the same nice structure as associated with the PORLA technique [recall Fig. 8.3] and therefore its behavior is expected to be quite similar to the original PORLA method discussed in Sect. 8.3.2.

Consider again the vector order recursions of the true least-squares ladder form as stated in (8.41a-c) and the observed process vector defined as

$$\mathbf{x}(t) = \Big[x(t),\ x(t-1),\ x(t-2),\ \dots,\ x(t-L-1) \Big]^{T} , \tag{8.66}$$

and the growing-order subspace of past observations spanned by $\mathbf{X}_m(t)$

$$\mathbf{X}_m(t) = \Big[\mathbf{x}(t-1),\ \mathbf{x}(t-2),\ \dots,\ \mathbf{x}(t-m) \Big] . \tag{8.67}$$

Note that the forward residual vector $\mathbf{e}_m(t)$ is the orthogonal complement of the orthogonal projection of $\mathbf{x}(t)$ onto $\mathbf{X}_m(t)$ and, consequently, the backward residual vector $\mathbf{r}_m(t)$ is the orthogonal complement of the orthogonal projection of $\mathbf{x}(t-m)$ onto $\mathbf{X}_m(t+1)$. Hence (by employing the notation of orthogonal projections as introduced in Sect. 6.1), we may write

$$\mathbf{e}_m(t) = \mathbf{x}(t)\langle \mathbf{X}_m(t)\rangle , \tag{8.68a}$$

$$\mathbf{r}_m(t) = \mathbf{x}(t-m)\langle \mathbf{X}_m(t+1)\rangle . \tag{8.68b}$$

These relations ultimately emerge from the recursion formula for orthogonal projections, as given by (6.4) and (6.5 - 7). The forward and backward predicted process vectors $\hat{\mathbf{x}}_m^f(t)$ and $\hat{\mathbf{x}}_m^b(t)$ are *orthogonal projections* on the subspaces $\mathbf{X}_m(t)$ and $\mathbf{X}_m(t+1)$, respectively. It is clear that

$$\mathbf{x}(t) = \hat{\mathbf{x}}_m^f(t) + \mathbf{e}_m(t) , \tag{8.69a}$$

$$\mathbf{x}(t-m) = \hat{\mathbf{x}}_m^b(t) + \mathbf{r}_m(t) , \tag{8.69b}$$

and consequently

$$\hat{\mathbf{x}}_m^f(t)\langle \mathbf{X}_m(t)\rangle = [0\dots0]^{T} , \tag{8.70a}$$

$$\hat{\mathbf{x}}_m^b(t) \langle \mathbf{X}_m(t+1) \rangle = [0 \ldots 0]^T \quad . \tag{8.70b}$$

Obviously, the inner product between an orthogonal projection and its orthogonal complement is zero, hence

$$\mathbf{e}_m^T(t) \hat{\mathbf{x}}_m^f(t) = 0 \quad , \tag{8.71a}$$

$$\mathbf{r}_m^T(t) \hat{\mathbf{x}}_m^b(t) = 0 \quad . \tag{8.71b}$$

Combination of (8.69a,b) with (8.71a,b) yields

$$\mathbf{x}^T(t) \mathbf{e}_m(t) = \mathbf{e}_m^T(t) \mathbf{e}_m(t) \quad . \tag{8.72a}$$

Furthermore, a simple postmultiplication of (8.72a) with the operator $\mathbf{e}_m^T(t)[\mathbf{e}_m(t)\mathbf{e}_m^T(t)]^{-1} \mathbf{r}_m(t-1)$ reveals that

$$\mathbf{x}^T(t) \mathbf{r}_m(t-1) = \mathbf{e}_m^T(t) \mathbf{r}_m(t-1) \quad . \tag{8.72b}$$

An analogous expression is stated in terms of backward prediction errors where

$$\mathbf{x}^T(t-m) \mathbf{r}_m(t) = \mathbf{r}_m^T(t) \mathbf{r}_m(t) \quad . \tag{8.73a}$$

Postmultiplication of (8.73a) with the operator $\mathbf{r}_m^T(t)[\mathbf{r}_m(t)\mathbf{r}_m^T(t)]^{-1} \mathbf{e}_m(t+1)$ gives

$$\mathbf{x}^T(t-m-1) \mathbf{e}_m(t) = \mathbf{e}_m^T(t) \mathbf{r}_m(t-1) \quad . \tag{8.73b}$$

Introducing two new GREs

$$E_{m,i,j}(t) = \mathbf{x}^T(t-i) \mathbf{e}_m(t-j) \quad , \tag{8.74a}$$

$$R_{m,i,j}(t) = \mathbf{x}^T(t-i) \mathbf{r}_m(t-1-j) \quad , \tag{8.74b}$$

and premultiplying both sides of the shifted ladder vector order recursions (8.41a-c) by the shifted process vector $\mathbf{x}(t-i)$, we obtain

$$\mathbf{x}^T(t-i) \mathbf{e}_m(t-j) = \mathbf{x}^T(t-i) \mathbf{e}_{m-1}(t-j)$$
$$+ K_m^f(t-j) \mathbf{x}^T(t-i) \mathbf{r}_{m-1}(t-1-j) \quad , \tag{8.75a}$$

$$\mathbf{x}^T(t-i) \mathbf{r}_m(t-1-j) = \mathbf{x}^T(t-i) \mathbf{r}_{m-1}(t-2-j)$$
$$+ K_m^b(t-1-j) \mathbf{x}^T(t-i) \mathbf{e}_{m-1}(t-1-j) \quad . \tag{8.75b}$$

The inner product recursions thus obtained (8.75a,b) can be expressed in terms of the GREs (8.74a,b) as follows:

$$E_{m,i,j}(t) = E_{m-1,i,j}(t) + K_m^f(t-j)R_{m-1,i,j}(t) \quad , \tag{8.76a}$$

$$R_{m,i,j}(t) = R_{m-1,i,j+1}(t) + K_m^b(t-1-j)E_{m-1,i,j+1}(t) \quad . \tag{8.76b}$$

Clearly, the definition of the reflection coefficients (6.19a,b) of the least-squares ladder form can be rewritten in terms of GREs

$$K_m^f(t) = - \frac{R_{m-1,0,0}(t)}{R_{m-1,m,0}(t)} = - \frac{E_{m-1,m,0}(t)}{R_{m-1,m,0}(t)} \quad , \tag{8.77a}$$

$$K_m^b(t) = - \frac{R_{m-1,0,0}(t)}{E_{m-1,0,0}(t)} = - \frac{E_{m-1,m,0}(t)}{E_{m-1,0,0}(t)} \quad . \tag{8.77b}$$

These equations, together with the GRE recurrence relations (8.76a,b) already establish the closed recursion of Sokat's PORLA computation scheme. At stage zero of the algorithm, the new GREs will be initialized from the time-recursively updated covariance matrix as follows:

$$E_{0,i,j}(t) = \Phi_{i,j}(t) \quad , \tag{8.78a}$$

$$R_{0,i,j}(t) = \Phi_{i,j+1}(t) \quad . \tag{8.78b}$$

Obviously, the new GRE's obey a *shift invariance law* similar to (8.53a-c) in that

$$E_{m,i+1,j+1}(t) = E_{m,i,j}(t-1) \quad , \tag{8.79a}$$

$$R_{m,i+1,j+1}(t) = R_{m,i,j}(t-1) \quad . \tag{8.79b}$$

Note, however, that the GREs $\mathbf{E}_m(t)$ and $\mathbf{R}_m(t)$ in this case are *not symmetric*. This leads to an exact PORLA computation scheme requiring $2(p^2 - p)$ multiplications per recursion. The algorithm is listed in Table 8.10. We use the following simplified notations for convenience.

$$LXE(j) = E_{m,0,j}(t) \quad , \qquad KF(m,j) = K_{m+1}^f(t-j) \quad ,$$

$$LXR(j) = R_{m,0,j}(t) \quad , \qquad KB(m,j) = K_{m+1}^b(t-j) \quad ,$$

$$CXE(j) = E_{m,j,0}(t) \quad , \qquad OLDCXR(m,j) = R_{m,j-1,0}(t-1) \quad ,$$

$$CXR(j) = R_{m,j,0}(t) \quad , \qquad OLDCXE(m,j) = E_{m,j-1,0}(t-1) \quad ,$$

Table 8.10. Sokat's recursive least-squares PORLA computation technique. This algorithm involves division. Whenever the divisor is small and positive, set $1/x = x$. When the divisor is negative, set $1/x = 0$

FOR $t = 0, 1, 2, \ldots$

> *Input:* $\Phi_{i,j}(t)$; $0 \le i,j \le p$
>
> FOR $j = 0, 1, \ldots, p - 1$ *initialize:*
>
> > $LXE(j) = \Phi_{0,j}(t)$
> > $LXR(j) = \Phi_{0,j+1}(t)$
> > $CXE(j) = \Phi_{j,0}(t)$
> > $CXR(j) = \Phi_{j,1}(t)$
>
> $KF = - LXR(0)\big/CXR(1)$ $KB = - CXE(1)\big/LXE(0)$
>
> FOR $m = 1, 2, \ldots, p - 2$
>
> > FOR $j = p - m - 1, p - m - 2, \ldots, 0$ *increment stack pointers:*
> >
> > > $KF(m,j+1) = KF(m,j)$
> > > $KB(m,j+1) = KB(m,j)$
> >
> > $KF(m,0) = KF$ $KB(m,0) = KB$
> >
> > FOR $j = 0, 1, \ldots, p - m - 2$
> >
> > > $LXE(j) = LXE(j) + KF(m,j) LXR(j)$
> > > $LXR(j) = LXR(j+1) + KB(m,j+1) LXE(j+1)$
> >
> > FOR $j = p, p - 1, \ldots, m + 2$
> >
> > > $CXE(j) = CXE(j) + KF(m,0) CXR(j)$
> > > $CXR(j) = OLDCXR(m,j) + KB(m,1) OLDCXE(m,j)$
> > > $OLDCXR(m,j) = CXR(j-1)$
> > > $OLDCXE(m,j) = CXE(j-1)$
> >
> > $KF = - LXR(0)\big/CXR(m+2)$ $KB = - CXE(m+2)\big/LXE(0)$

Several approximations can be incorporated in this new PORLA scheme. These approximate algorithms may be obtained in a way very similar to that examined in the case of Strobach's algorithms.

Another interesting point that should be mentioned is that the LeRoux–Gueguen algorithm appears as a simplified version of Sokat's PORLA computation technique. To see this, we first assume that the reflection coefficients are identical and time invariant (PARCOR case):

$$K_m^f(t) = K_m^b(t) = K_m(t) . \tag{8.80}$$

Furthermore, the exact shift-invariance property (8.79a,b) is replaced by the Toeplitz assumption

$$E_{m,i+1,j+1}(t) = E_{m,i,j}(t) , \tag{8.81a}$$

$$R_{m,i+1,j+1}(t) = R_{m,i,j}(t) . \tag{8.81b}$$

By substituting these approximations into the exact recursions (8.76a,b), it is easy to see that these recursions reduce to

$$E_{m,i,j} = E_{m-1,i,j} + K_m R_{m-1,i,j} , \tag{8.82a}$$

$$R_{m,i,j} = R_{m-1,i,j+1} + K_m E_{m-1,i,j+1} . \tag{8.82b}$$

A quick inspection of (8.77b) gives

$$K_m = - \frac{R_{m-1,0,0}}{E_{m-1,0,0}} , \tag{8.83}$$

and hence the recursion is complete. We may note that the recursion over the first row or, alternatively, the first column vector of each GRE is sufficient because of the underlying Toeplitz structure.

Evaluating (8.82a,b) as a *row vector recursion*, we have

$$E_{m,0,j} = E_{m-1,0,j} + K_m R_{m-1,0,j} , \tag{8.84a}$$

$$R_{m,0,j} = R_{m-1,0,j+1} + K_m E_{m-1,0,j+1} , \tag{8.84b}$$

which corresponds to the GREs

$$E_{m,0,j} = \mathbf{x}^T(t)\mathbf{e}_m(t-j) , \tag{8.85a}$$

$$R_{m,0,j} = \mathbf{x}^T(t)\mathbf{r}_m(t-1-j) . \tag{8.85b}$$

But note that in the case of the LeRoux-Gueguen algorithm, the GRE [recall (8.5)] was defined as

$$E'_{m-1,i} = \frac{\mathbf{x}^T(t-i)\mathbf{e}_{m-1}(t)}{\mathbf{x}^T(t)\mathbf{x}(t)} . \tag{8.86}$$

The fact that this "original" GRE was normalized to the process power does not affect the principal arguments. In fact, inspection of (8.86) leads to the insight that we should also investigate the *alternative* recursion over the column GRE vectors of (8.82a,b), namely, the recursion over the i's where the j's are held fixed at $j = 0$. Hence, setting $j = 0$ in the approximate matrix order recursions (8.82a,b) and considering again the Toeplitz assumption (8.81a,b), one finds

$$E_{m,i,0} = E_{m-1,i,0} + K_m R_{m-1,i,0} \quad , \tag{8.87a}$$

$$R_{m,i,0} = R_{m-1,i-1,0} + K_m E_{m-1,i-1,0} \quad , \tag{8.87b}$$

which are indeed exactly the recursions of LeRoux and Gueguen as stated in the modified (two-GRE) form (8.21b,c). This has brought us back to the starting point of Sect. 8.1.

8.3.5 Additional Notes on Sokat's PORLA Method

Recent investigations have shown that Sokat's algorithm, as listed in Table 8.10, is redundant in complexity. To see this, we may rewrite the four GRE vector order recursions constituting the algorithm of Table 8.10 as follows:

$$E_{m,0,j}(t) = E_{m-1,0,j}(t) + K_m^f(t-j)R_{m-1,0,j}(t) \quad , \tag{8.88a}$$

$$R_{m,0,j}(t) = R_{m-1,0,j+1}(t) + K_m^b(t-1-j)E_{m-1,0,j+1}(t) \quad , \tag{8.88b}$$

$$E_{m,i,0}(t) = E_{m-1,i,0}(t) + K_m^f(t)R_{m-1,i,0}(t) \quad , \tag{8.89a}$$

$$R_{m,i,0}(t) = R_{m-1,i,1}(t) + K_m^b(t-1)E_{m-1,i,1}(t) \quad . \tag{8.89b}$$

Exploiting the shift-invariance properties of GREs (8.79a,b), we may formulate the recursion (8.89b) as a recursion on *previous* (time-delayed) GREs as follows:

$$R_{m,i,0}(t) = R_{m-1,i-1,0}(t-1) + K_m^b(t-1)E_{m-1,i-1,0}(t-1) \quad . \tag{8.90}$$

Now, by using (8.90), we may note that an order recursive computation of $R_{m,m+1,0}(t)$ exists in that

$$R_{m,m+1,0}(t) = R_{m-1,m,0}(t-1) + K_m^b(t-1)E_{m-1,m,0}(t-1) \quad . \tag{8.91}$$

According to (8.74a), the quantity $E_{m-1,m,0}(t-1)$ can be expressed as

$$E_{m-1,m,0}(t) = \mathbf{x}^T(t-m)\mathbf{e}_{m-1}(t) \quad , \tag{8.92}$$

but since [recall (8.73b)]

$$\mathbf{x}^T(t-m)\mathbf{e}_{m-1}(t) = \mathbf{e}_{m-1}^T(t)\mathbf{r}_{m-1}(t-1) \quad , \tag{8.93}$$

it is also clear that [according to (8.72b)]

$$\mathbf{e}_{m-1}^T(t)\mathbf{r}_{m-1}(t-1) = \mathbf{x}^T(t)\mathbf{r}_{m-1}(t-1) \quad . \tag{8.94}$$

On the other hand, (8.94) can be compared with (8.74b) to show that

$$\mathbf{x}^T(t)\mathbf{r}_{m-1}(t-1) = R_{m-1,0,0}(t) \quad . \tag{8.95}$$

Now, comparing (8.95) with (8.92 - 94), it is obvious that

$$E_{m-1,m,0}(t-1) = R_{m-1,0,0}(t-1) \quad , \tag{8.96}$$

and therefore recursion (8.91) can be written as

$$R_{m,m+1,0}(t) = R_{m-1,m,0}(t-1) + K_m^b(t-1)R_{m-1,0,0}(t-1) \quad . \tag{8.97}$$

Note that the appearance of the quantity $R_{m-1,m,0}(t)$ in the computation of $K_{m+1}^f(t)$ [recall (8.77a)] was the only reason for computing the recurrence relations (8.89a,b) in Sokat's original PORLA computation technique of Table 8.10. But now, it is clear that (8.97) provides an *alternative* way of computing this quantity that lies *outside* the recurrence loops. In particular, the recursion (8.97) permits the order recursive updating of the quantity $R_{m,m+1,0}(t)$ *without* the necessity of computing the recurrence relations (8.89a,b). Thus, a true RLS ladder algorithm of the PORLA type can be established that requires only $p^2 + O(p)$ multiplications and roughly the same number of additions per recursion. This algorithm is listed in Table 8.11. But before presenting the algorithm, we shall describe several additional relationships.

Note that, [according to (8.74b)], we may write

$$R_{m,m+1,0}(t) = \mathbf{x}^T(t-m-1)\mathbf{r}_m(t-1) \quad , \tag{8.98}$$

but [from (8.73a)] one finds

$$\mathbf{x}^T(t-m-1)\mathbf{r}_m(t-1) = \mathbf{r}_m^T(t-1)\mathbf{r}_m(t-1) \quad . \tag{8.99}$$

This consideration reveals that the quantity $R_{m,m+1,0}(t)$ is nothing but the conventional backward residual energy $R_m(t-1)$. Moreover, it can be readily verified that

$$E_{m,m+1,0}(t) = \mathbf{x}^T(t-m-1)\mathbf{e}_m(t) = \mathbf{e}_m^T(t)\mathbf{r}_m(t-1) = C_m(t) \quad , \tag{8.100}$$

Table 8.11. Sokat's recursive least-squares PORLA computation scheme with reduced computational complexity. This algorithm involves division. Whenever the divisor is small and positive, set $1/x = x$. When the divisor is negative, set $1/x = 0$

FOR t = 0, 1, 2, . . .

$\quad$ *Input:* $\Phi_{i,j}(t)$; $0 \le i,j \le p$

$\quad$ FOR j = 0, 1, . . . , p − 1   *initialize:*

$\qquad$ $LXE(j) = \Phi_{0,j}(t)$
$\qquad$ $LXR(j) = \Phi_{0,j+1}(t)$

$\quad$ $R_0(t) = \Phi_{1,1}(t)$

$\quad$ $KF = -\,LXR(0)\big/R_0(t)$ $\qquad\qquad$ $KB = -\,LXR(0)\big/LXE(0)$

$\quad$ FOR m = 1, 2, . . . , p − 2

$\qquad$ FOR j = p − m − 1, p − m − 2, . . . , 0   *increment stack pointers:*

$\qquad\quad$ $KF(m,j+1) = KF(m,j)$
$\qquad\quad$ $KB(m,j+1) = KB(m,j)$

$\qquad$ $KF(m,0) = KF$ $\qquad\qquad$ $KB(m,0) = KB$

$\qquad$ FOR j = 0, 1, . . . , p − m − 2

$\qquad\quad$ $LXE(j) = LXE(j) + KF(m,j)\,LXR(j)$
$\qquad\quad$ $LXR(j) = LXR(j+1) + KB(m,j+1)\,LXE(j+1)$

$\qquad$ $R_{m+1}(t) = R_m(t-1) + KB(m,0)LXR(0)$

$\qquad$ $KF = -\,LXR(0)\big/R_{m+1}(t-1)$ $\qquad$ $KB = -\,LXR(0)\big/LXE(0)$

---

which constitutes that the conventional covariance $C_m(t)$ appears as the quantity $E_{m,m+1,0}(t)$. Hence, recursion (8.91) can be interpreted as an *equivalent representation* of the conventional recursion

$$R_m(t-1) = R_{m-1}(t-2) + K_m^b(t-1)C_{m-1}(t-1) \;, \qquad (8.101)$$

which was stated much earlier (6.23b). Note that similar arguments hold for the counterpart recursion

$$E_m(t) = E_{m-1}(t) + K_m^f(t)C_{m-1}(t) \quad , \tag{8.102}$$

as previously stated in (6.23a), which is ultimately disclosed as an alternative recursion to

$$E_{m,0,0}(t) = E_{m-1,0,0}(t) + K_m^f(t)R_{m-1,0,0}(t) \quad . \tag{8.103}$$

In conclusion, we may state the interesting relationships between elements of the GRE vectors and conventional residual energies as follows:

$$E_m(t) = E_{m,0,0}(t) \quad , \tag{8.104a}$$

$$R_m(t) = R_{m,m+1,0}(t+1) \quad , \tag{8.104b}$$

$$C_m(t) = E_{m,m+1,0}(t) = R_{m,0,0}(t) \quad . \tag{8.104c}$$

We should make the point that - based on these considerations - *two* new PORLA computation schemes with reduced computational complexity can be derived. The structures of these algorithms turn out to be very similar. We provide a complete listing of one of these algorithms in Table 8.11.

8.4 Split Schur Algorithms

Just as the LeRoux-Gueguen algorithm can be viewed as the "fixed-point computation" of the classical Levinson-Durbin algorithm, the algorithms presented in this section will be the "fixed-point counterparts" of the split Levinson algorithms of Delsarte and Krishna. In particular, it will be demonstrated how the recursion variables in the Levinson-Durbin algorithm are related to the recursion variables in the split algorithms. Based on these considerations, the split algorithms can be *reformulated* using the concept of generalized residual energies (GREs) in a very similar way to that examined in the case of the LeRoux-Gueguen algorithm in the first part of this chapter. The resulting split Schur algorithms [8.15, 16] exhibit a more "direct" computation scheme with *fewer recursion variables* and a *better numerical accuracy*, while the total operations count remains roughly the same when compared to the original algorithms of Delsarte and Krishna. The terminology split "Schur" is commonly used in the signal processing lierature in honour of Schur's early work on bounded analytic functions [8.17].

We shall start our deliberations with a brief comparison of the Levinson-Durbin algorithm with the split algorithms. While the Levinson-Durbin algorithm is based on the *Levinson recurrence*

$$
\mathbf{a}^{(m)} = \begin{bmatrix} \mathbf{a}^{(m-1)} \\ 0 \end{bmatrix} + K_m \begin{bmatrix} 0 \\ \overset{v}{\mathbf{a}}{}^{(m-1)} \end{bmatrix} , \tag{8.105}
$$

the split Levinson algorithms are generally based on the *split Levinson recurrence*

$$
\mathbf{q}^{(m)} = \begin{bmatrix} \mathbf{q}^{(m-1)} \\ 0 \end{bmatrix} + \begin{bmatrix} 0 \\ \mathbf{q}^{(m-1)} \end{bmatrix} - \alpha^{(m)} \begin{bmatrix} 0 \\ \mathbf{q}^{(m-2)} \\ 0 \end{bmatrix} . \tag{8.106}
$$

On comparing (8.105) and (8.106), the similarities of these recursions become very obvious. While the set of intermediate variables $\mathbf{a}^{(m)}$ was interpreted as the coefficient set of the mth-order *transversal* prediction filter in the "conventional" case, the set $\mathbf{q}^{(m)}$ may be interpreted as the coefficients of a "fictitious" *split transversal* prediction filter. Consequently, $\alpha^{(m)}$ may be interpreted as the *split reflection coefficient* which is the counterpart to the "conventional" reflection coefficient K_m.

Following this line of thought, recurrence (8.106) can be viewed as another type of Levinson recursion because it relates reflection coefficients to transversal predictor parameter sets. Returning to the first part of this chapter, we note that the main goal in the LeRoux-Gueguen algorithm was to replace the Levinson recursion by a different form of recurrence relation, where generalized residual energies (GREs) are related to reflection coefficients. This substitution yields algorithms with a better numerical accuracy and – in the case of the LeRoux-Gueguen algorithm – also an algorithm where all recursion variables are less than 1 in magnitude.

In a similar way, the split Levinson recurrence (8.106) can be replaced by an appropriate recursion on GREs. This important step will be examined in both the Delsarte and the Krishna split Levinson algorithms.

8.4.1 A Split Schur Formulation of Krishna's Algorithm

Consider (7.70a). It is seen that

$$
c_m q_0^{(m-1)} + c_{m-1} q_1^{(m-1)} + \ldots + c_1 q_{m-1}^{(m-1)} = \gamma^{(m-1)} , \tag{8.107a}
$$

and from (7.70b), it follows that

$$
c_1 q_0^{(m-1)} + c_2 q_1^{(m-1)} + \ldots + c_m q_{m-1}^{(m-1)} = \gamma^{(m-1)} . \tag{8.107b}
$$

Clearly, the relationship between the two equations (8.107a) and (8.107b) stems from the *symmetric property* of the "split transversal predictor" parameter set $\mathbf{q}^{(m-1)}$. This reveals that in the case of split algorithms, the *energy* becomes equivalent to the *covariance*. For an illustration of the meaning of this statement, the reader should compare (8.107a) with (7.8) and (8.107b) with (7.9). More concisely, we can write

$$\gamma^{(m-1)} = \sum_{i=0}^{m-1} c_{i+1} q_i^{(m-1)} = \sum_{i=0}^{m-1} c_{m-i} q_i^{(m-1)} \quad . \tag{8.108}$$

From (8.106), we may state the *component form* of the split Levinson recursion as

$$q_i^{(m-1)} = q_i^{(m-2)} + q_{i-1}^{(m-2)} - \alpha^{(m-1)} q_{i-1}^{(m-3)} \quad . \tag{8.109}$$

Rewriting (8.108) in terms of the observed process vector $\mathbf{x}(t)$,

$$\gamma^{(m-1)} = \sum_{i=0}^{m-1} c_{i+1} q_i^{(m-1)} = \sum_{i=0}^{m-1} \frac{\mathbf{x}^T(t)\mathbf{x}(t-1-i)}{\mathbf{x}^T(t)\mathbf{x}(t)} q_i^{(m-1)} \quad , \tag{8.110}$$

we may introduce a *generalized residual energy* (GRE) of the form

$$\gamma_j^{(m-1)} = \sum_{i=0}^{m-1} \frac{\mathbf{x}^T(t-j)\,\mathbf{x}(t-1-i)}{\mathbf{x}^T(t)\mathbf{x}(t)} q_i^{(m-1)} \quad . \tag{8.111}$$

Note,

$$\frac{\mathbf{x}^T(t-j)\,\mathbf{x}(t-1-i)}{\mathbf{x}^T(t)\mathbf{x}(t)} = \frac{\Phi_{j,i+1}(t)}{\mathbf{x}^T(t)\mathbf{x}(t)} = \frac{\Phi_{j-j,i-j+1}(t)}{\mathbf{x}^T(t)\mathbf{x}(t)}$$

$$= \frac{\Phi_{0,i-j+1}(t)}{\mathbf{x}^T(t)\mathbf{x}(t)} = c_{i-j+1} \quad , \tag{8.112}$$

and therefore the GRE $\gamma_j^{(m-1)}$ may be expressed in terms of the auto-correlation sequence of the observed process as

$$\gamma_j^{(m-1)} = \sum_{i=0}^{m-1} c_{i-j+1} q_i^{(m-1)} = \sum_{i=0}^{m-1} c_{j-i-1} q_i^{(m-1)} \quad . \tag{8.113}$$

Substituting the split Levinson recursion (8.109) into (8.113) gives

$$\gamma_j^{(m-1)} = \sum_{i=0}^{m-1} c_{j-i-1} \left[q_i^{(m-2)} + q_{i-1}^{(m-2)} - \alpha^{(m-1)} q_{i-1}^{(m-3)} \right] \quad . \tag{8.114}$$

Noting that

$$\sum_{i=0}^{m-1} c_{j-i-1} q_i^{(m-2)} = \gamma_j^{(m-2)} \quad , \tag{8.115a}$$

and

$$\sum_{i=0}^{m-1} c_{j-i-1} q_{i-1}^{(m-2)} = \gamma_{j-1}^{(m-2)} , \qquad (8.115b)$$

we may express (8.114) as a recursion of GREs,

$$\gamma_j^{(m-1)} = \gamma_j^{(m-2)} + \gamma_{j-1}^{(m-2)} - \alpha^{(m-1)} \gamma_{j-1}^{(m-3)} , \qquad (8.116)$$

which is simply the GRE-based counterpart to the original formulation (8.109) which was based on the "split transversal predictor parameter set" $q^{(m)}$. The conventional "split residual energy" $\gamma^{(m-1)}$ is just the *top component* of the GRE vector $\gamma^{(m-1)}$, hence

$$\gamma^{(m-1)} = \gamma_0^{(m-1)} . \qquad (8.117)$$

Note at this point that we do not yet have a complete recursion since the computation of $\gamma_0^{(m-1)}$ requires knowledge of the quantities $\gamma_{-1}^{(m-2)}$ and $\gamma_{-1}^{(m-3)}$. Clearly, to find a way out of this difficulty, an alternative scheme must be established for computing the top component $\gamma_0^{(m-1)}$. For this purpose, note that the intermediate solution vector $q^{(m-1)}$ is symmetric and hence

$$q_i^{(m-1)} = q_{m-1-i}^{(m-1)} . \qquad (8.118)$$

This, in turn, leads to an alternative definition of the GRE $\gamma^{(m-1)}$ as follows:

$$\gamma_j^{(m-1)} = \sum_{i=0}^{m-1} c_{m-i-1} q_{i-1-i}^{(m-1)} = \sum_{i=0}^{m-1} c_{m-i-j} q_i^{(m-1)} . \qquad (8.119)$$

Using this definition, it is easy to check that one may compute $\gamma_0^{(m-1)}$ via

$$\gamma_0^{(m-1)} = \sum_{i=0}^{m-1} c_{m-i} q_i^{(m-1)} . \qquad (8.120)$$

On the other hand, an interpretation of (8.120) in terms of the original definition of GREs (8.113) shows that

$$\sum_{i=0}^{m-1} c_{m-i} q_i^{(m-1)} = \gamma_{m+1}^{(m-1)} , \qquad (8.121)$$

and therefore

$$\gamma_0^{(m-1)} = \gamma_{m+1}^{(m-1)} , \qquad (8.122)$$

which is the desired relationship for computing the top component of the GRE vector $\gamma^{(m-1)}$. Table 8.12 provides a complete summary of this split algorithm, which was termed SCHURSPLIT 1.

Table 8.12. Summary of the split algorithm SCHURSPLIT 1. As (8.123f) is not a part of the order recursion, its computation remains *optional*. This algorithm involves division. Whenever the divisor is small, set $1/x = x$

Input: $c_1, c_2, \ldots, c_p$

Initialize:

$r^{(0)} = -1;$ $r^{(1)} = -1 - c_1;$ $K_1 = -c_1$

FOR $j = 2, 3, \ldots, p+1$ *Initialize:*

$\left[\quad \gamma_j^{(0)} = -c_{j-1} \right.$

$\gamma_0^{(0)} = -c_1$

FOR $m = 2, 3, 4, \ldots, p$

$\quad$ FOR $j = p+1, p, p-1, \ldots, m+1$

$\qquad$ IF $(m\text{ .EQ. } 2)$ THEN

$$\gamma_j^{(m-1)} = \gamma_j^{(m-2)} + \gamma_{j-1}^{(m-2)} \tag{8.123a}$$

$\qquad$ ELSE

$$\gamma_j^{(m-1)} = \gamma_j^{(m-2)} + \gamma_{j-1}^{(m-2)} - \alpha^{(m-1)}\gamma_{j-1}^{(m-3)} \tag{8.123b}$$

$\qquad$ ENDIF

$$\gamma_0^{(m-1)} = \gamma_{m+1}^{(m-1)} \tag{8.123c}$$

$$\alpha^{(m)} = \frac{r^{(m-1)} - \gamma_0^{(m-1)}}{r^{(m-2)} - \gamma_0^{(m-2)}} \tag{8.123d}$$

$$r^{(m)} = 2\, r^{(m-1)} - \alpha^{(m)} r^{(m-2)} \tag{8.123e}$$

$$K_m = \frac{\alpha^{(m)}}{1 - K_{m-1}} - 1 \qquad\qquad \textit{(optional)} \tag{8.123f}$$

8.4.2 Bounds on Recursion Variables

Next it is of interest to investigate the dynamic range requirements of SCHURSPLIT 1, listed in Table 8.12. Solving (8.123f) for the split reflection coefficient $\alpha^{(m)}$, we obtain

$$\alpha^{(m)} = K_m - K_m K_{m-1} + 1 \quad . \tag{8.124}$$

Note that for a stationary AR process, the reflection coefficients are bounded on

$$-1 < K_m < +1 \quad . \tag{8.125}$$

This allows the determination of the bounds of the split reflection coefficient $\alpha^{(m)}$. For an evaluation of (8.124), it is useful to plot the linear function $\alpha^{(m)}(K_m)$ as a function of the parameter K_{m-1}. For several distinct values of the parameter K_{m-1}, one obtains the following linear relationships:

$$for \quad K_{m-1} = 1 \quad : \quad \alpha^{(m)} = 1 \quad , \tag{8.126a}$$

$$K_{m-1} = 0 \quad : \quad \alpha^{(m)} = K_m + 1 \quad , \tag{8.126b}$$

$$K_{m-1} = -1 \quad : \quad \alpha^{(m)} = 2 K_m + 1 \quad . \tag{8.126c}$$

These relationships are illustrated in Fig. 8.4.

From Fig. 8.3, it becomes apparent that $K_{m-1} = -1$ is the *worst case*, in that the split reflection coefficient attains its maximum dynamic range determined by

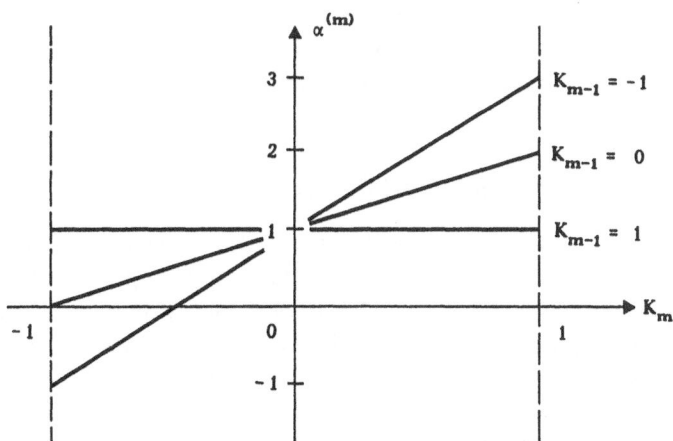

Fig. 8.4. Possible values of the split reflection coefficient $\alpha^{(m)}$ as a function of the parameters K_m and K_{m-1} in the case of SCHURSPLIT 1

$$-1 < \alpha^{(m)} < 3 \quad . \tag{8.127}$$

Another interesting point is the investigation of the *bounds* on the inter-mediate recursion variable $r^{(m)}$. From (7.85), one sees that

$$r^{(m)} = r^{(m-1)} (1 - K_m) \quad . \tag{8.128}$$

Since K_m was bounded to be less than 1 in magnitude, one finds that in the worst case (all K's tend to the value -1) the value of $r^{(m)}$ *grows linearly* according to the recursive law

$$r^{(m)} = 2 r^{(m-1)} \quad . \tag{8.129}$$

Exploiting (7.86), expression (8.129) can be brought into a nonrecursive formulation:

$$r^{(m)} = r^{(0)} \prod_{i=1}^{m} \left(1 - K_i \right) \quad ; \qquad r^{(0)} = -1 \quad . \tag{8.130}$$

Again, we see that a sequence of K's that are all arbitrarily close to -1 will determine the *worst-case* situation with the *highest possible* dynamic range requirement for the intermediate variable $r^{(m)}$, which grows *linearly* with the algorithm order. An evaluation of (8.130) yields the bound

$$-2^m < r^{(m)} < 0 \quad . \tag{8.131}$$

Although a sequence of K's with all K's arbitrarily close to -1 is very unlikely, inequality (8.131) indicates that - at least in this worst case - one has to *increase* the wordlength by one bit in each additional stage of the algorithm or to apply a normalization by a factor of one-half. This is a disadvantage when compared to the LeRoux–Gueguen algorithm where all recursion variables are strictly bounded to a value of less than 1 in magnitude.

8.4.3 A Split Schur Formulation of Delsarte's Algorithm

In a procedure similar to that examined in the case of Krishna's algorithm, we may develop a GRE-based formulation of Delsarte's original split Levinson algorithm as listed in Table 7.3. This split Schur formulation of Delsarte's algorithm was presented in [8.15]. We note that the inter-mediate recursion variable $\tau^{(m)}$ in the case of Delsarte's split Levinson algorithm can be rewritten in terms of the observed process vector $\mathbf{x}(t)$ as

$$\tau^{(m)} = \sum_{i=0}^{m} c_i \, p_i^{(m)} = \sum_{i=0}^{m} \frac{\mathbf{x}^T(t)\mathbf{x}(t-i)}{\mathbf{x}^T(t)\mathbf{x}(t)} \, p_i^{(m)} \quad . \tag{8.132}$$

This expression allows the GREs $\tau_j^{(m)}$ to be introduced as follows:

$$\tau_j^{(m)} = \sum_{i=0}^{m} \frac{\mathbf{x}^T(t-j)\mathbf{x}(t-i)}{\mathbf{x}^T(t)\mathbf{x}(t)} p_i^{(m)} \ . \tag{8.133}$$

Taking into account that

$$\frac{\mathbf{x}^T(t-j)\,\mathbf{x}(t-i)}{\mathbf{x}^T(t)\mathbf{x}(t)} = \frac{\Phi_{j,i}(t)}{\mathbf{x}^T(t)\mathbf{x}(t)} = \frac{\Phi_{j-j,i-j}(t)}{\mathbf{x}^T(t)\mathbf{x}(t)}$$

$$= \frac{\Phi_{0,i-j}(t)}{\mathbf{x}^T(t)\mathbf{x}(t)} = c_{i-j} = c_{j-i} \ , \tag{8.134}$$

expression (8.133) can be rewritten in terms of the autocorrelation sequence of input data

$$\tau_j^{(m)} = \sum_{i=0}^{m} c_{i-j}\, p_i^{(m)} = \sum_{i=0}^{m} c_{j-i}\, p_i^{(m)} \ . \tag{8.135}$$

Combination of (8.135) with the split Levinson recurrence (7.65a) gives

$$\tau_j^{(m)} = \sum_{i=0}^{m} c_{i-j} \left[p_i^{(m-1)} + p_{i-1}^{(m-1)} - \alpha^{(m)} p_{i-1}^{(m-2)} \right] \ . \tag{8.136}$$

This leads to a recursion of GREs

$$\tau_j^{(m)} = \tau_j^{(m-1)} + \tau_{j-1}^{(m-1)} - \alpha^{(m)} \tau_{j-1}^{(m-2)} \ , \tag{8.137}$$

where

$$\tau^{(m)} = \tau_0^{(m)} \ . \tag{8.138}$$

Clearly, the computation of $\tau_0^{(m)}$ requires a knowledge of $\tau_{-1}^{(m-1)}$ and $\tau_{-1}^{(m-2)}$. To solve this difficulty, we proceed in the same way as in the case of the development of SCHURSPLIT 1. Since $\mathbf{p}^{(m)}$ is a *symmetric* intermediate solution vector, one may set up an alternative definition of GREs:

$$\tau_j^{(m)} = \sum_{i=0}^{m} c_{i-j}\, p_i^{(m)} = \sum_{i=0}^{m} c_{i-j}\, p_{m-i}^{(m)} \tag{8.139}$$

or, equivalently,

$$\tau_j^{(m)} = \sum_{i=0}^{m} c_{m-i-j}\, p_i^{(m)} \ , \tag{8.140}$$

which leads to

$$\tau_0^{(m)} = \tau_m^{(m)} \, . \tag{8.141}$$

Defining the *partial* GRE vector

$$\tau_p^{(m)} = \left[\tau_0^{(m)}, \tau_1^{(m)}, \ldots, \tau_m^{(m)} \right]^T \, , \tag{8.142}$$

we are interested to see that

$$\tau_p^{(m)} = T^{(m)} p^{(m)} \, , \tag{8.143}$$

which indicates that $\tau_p^{(m)}$ is also *symmetric*, since the Toeplitz matrix $T^{(m)}$ transforms symmetric vectors into symmetric vectors. Unfortunately, we cannot take advantage of this property, since the recursion settles on the variables $\{ \tau_j^{(m)}, \, m \le j \le p \}$.

A complete listing of this split algorithm is available in Table 8.13. The algorithm was named SCHURSPLIT 2. An important property is that SCHURSPLIT 2 relies on only *two* recursion variables, namely, the vector-valued GRE $\tau^{(m)}$ and the split reflection coefficient $\alpha^{(m)}$. For this reason, SCHURSPLIT 2 is possibly the most efficient and regular currently available algorithm for solving the Yule–Walker equations.

We are finally interested in the bounds on the recursion variables of SCHURSPLIT 2. Clearly, from (8.144c), we may deduce the dependence of $\alpha^{(m+1)}$ on K_m and K_{m-1} as

$$\alpha^{(m+1)} = 1 - K_m + K_{m-1} - K_m K_{m-1} \, . \tag{8.145}$$

This is again a linear parametric function, which can be evaluated for several values of the parameter K_{m-1} as follows:

$$\text{for} \quad K_{m-1} = -1 \quad : \quad \alpha^{(m+1)} = 0 \, , \tag{8.146a}$$

$$K_{m-1} = 0 \quad : \quad \alpha^{(m+1)} = 1 - K_m \, , \tag{8.146b}$$

$$K_{m-1} = 1 \quad : \quad \alpha^{(m+1)} = 2 - 2K_m \, . \tag{8.146c}$$

These functions are plotted in Fig. 8.5.

Figure 8.4 indicates that $\alpha^{(m)}$ in the case of SCHURSPLIT 2 is bounded on

$$0 < \alpha^{(m)} < 4 \, . \tag{8.147}$$

From this result, a simple inspection of (8.144b) shows that the GRE $\tau_m^{(m)}$

Table 8.13. Summary of split algorithm SCHURSPLIT 2. As (8.144c) is not a part of the order recursion, its computation remains *optional*. This algorithm involves division. Whenever the divisor is small, set $1/x = x$

Input: $c_1, c_2, \ldots, c_p$

Initialize: $\alpha^{(2)} = 1 + c_1$; $K_1 = -c_1$

FOR $j = 2, 3, \ldots, p+1$ *Initialize:*

$$\begin{bmatrix} \tau_j^{(0)} = -2\,c_j \\ \tau_j^{(1)} = -c_j - c_{j-1} \end{bmatrix}$$

FOR $m = 2, 3, 4, \ldots, p$

$$\left[\begin{array}{l} \text{FOR } j = p, p-1, \ldots, m \\[2mm] \left[\tau_j^{(m)} = \tau_j^{(m-1)} + \tau_{j-1}^{(m-1)} - \alpha^{(m)}\,\tau_{j-1}^{(m-2)} \right. \hspace{2cm} (8.144a) \\[4mm] \alpha^{(m+1)} = \dfrac{\tau_m^{(m)}}{\tau_{m-1}^{(m-1)}} \hspace{4cm} (8.144b) \\[4mm] K_m = 1 - \dfrac{\alpha^{(m+1)}}{1 + K_{m-1}} \hspace{1cm} \textit{(optional)} \hspace{1cm} (8.144c) \end{array} \right.$$

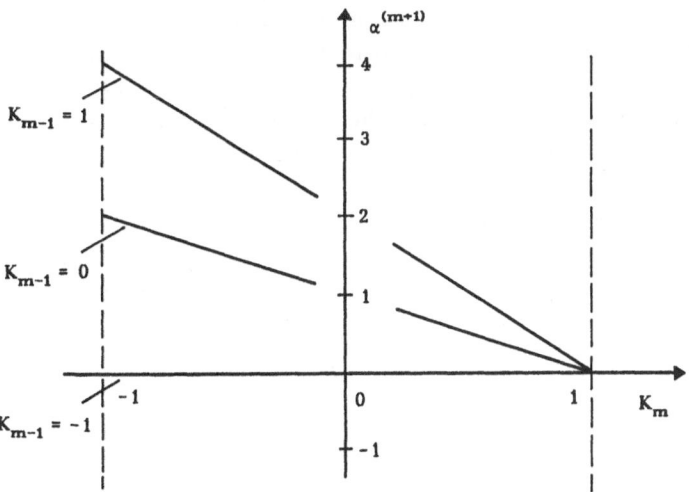

Fig. 8.5. Possible values of the split reflection coefficient $\alpha^{(m+1)}$ as a function of the variables K_m and K_{m-1}

will grow by a factor of *four* in each stage of the algorithm in the worst-case situation that all K's tend to - 1. A fixed-point implementation requires an appropriate normalization. This concludes our considerations of the class of split Schur algorithms based on the method of generalized residual energies.

8.5 Chapter Summary

This chapter has shed some light on the new class of *pure order recursive ladder algorithms* (PORLA). This class of algorithms constitutes a new and interesting way of solving the true recursive least-squares parameter estimation problem. PORLA computation schemes offer the possibility of *arbitrary recursive windowing* of the observed process combined with *outstanding numerical properties* of the ladder recursions, which are entirely based on inner products (residual energies) and reflection coefficients in contrast to conventional approaches, which rely on recursions of numerically uncertain residual signals. It has been shown that several classical covariance ladder algorithms fit in the more general class of PORLA computation schemes. Just like the recursive QR decomposition, the PORLA techniques can be implemented on triangular systolic arrays. See also the Appendix for a more detailed inspection of triangular systolic arrays in conjunction with PORLA algorithms. The PORLA method has been extended to FIR system identification [8.18] and ARMA system identification [8.19].

Besides a treatment of the recursive least-squares (RLS) problem using PORLA computation schemes, we have also shown how the PORLA method can be applied to derive the class of split Schur algorithms. Just as the LeRoux-Gueguen algorithm is a reformulation of the Levinson-Durbin algorithm in terms of GRE recursions, the newly developed algorithms SCHURSPLIT 1 and SCHURSPLIT 2 are the GRE-based counterparts to the split Levinson algorithms introduced by Krishna and Delsarte. The split Schur algorithms have better numerical properties, fewer recursion variables and a more regular algorithm structure which is well suited to a VLSI implementation. Pipelined architectures for solving Toeplitz systems of linear equations have been discussed by Kung and Hu [8.20] and You, Hu and Feng [8.21].

9. Fast Recursive Least-Squares Ladder Algorithms

Chapters 7 and 8 have illuminated the classes of ladder algorithms that are based on a pure order recursive construction of the ladder form. In this type of algorithm, the central problem appeared to be the *order recursive* updating of the covariance $C_m(t)$ according to

$$C_m(t) = C_{m-1}(t) + \ldots \quad . \tag{9.1}$$

Two solutions to the problem (9.1) have been presented. The first approach, discussed in Chap. 7, was based on the incorporation of transversal predictor parameters as intermediate recursion variables in Levinson-type algorithms. Alternatively, Chap. 8 introduced the algorithms of the PORLA type where the objective was to replace the Levinson-type recursion by inner product recursions, where inner products, also termed "generalized residual energies", were used as intermediate recursion variables, hence avoiding the explicit computation of transversal predictor parameters.

Another important class of recursive least-squares (RLS) ladder algorithms can be derived via the approach of updating the covariance $C_m(t)$ *in time* according to

$$C_m(t) = C_m(t-1) + \Delta C_m(t) \quad , \tag{9.2}$$

where $\Delta C_m(t)$ represents a *time differential* of $C_m(t)$ reflecting the change in the statistics of the input data each time a new data sample is captured. The problem of computing $\Delta C_m(t)$ in the true least-squares sense was first seriously studied by Lee and Morf. Lee [9.1] presented a unified treatment of RLS ladder algorithms of the type (9.2), employing both *geometrical* and *operator-based* approaches. Lee's geometrical considerations first explained the development of the projection operator $P_m(t)$ in a nonstationary environment when time progresses. The basic result of Lee's work is the *exact time-update theorem of projection operators*. The RLS ladder algorithms of this chapter, which can all be derived from this fundamental theorem, are *"fast"* in terms of a *linear* dependence of the computational complexity on the model order.

9.1 The Exact Time-Update Theorem of Projection Operators

In Sect. 2.2, we discussed the geometrical nature of the Normal Equations. In particular, it turned out that the projection operator $P_m(t)$ projects vectors *orthogonally* onto the subspace of past observations spanned by the *oblique* basis $X_m(t)$. When time progresses, the observation matrix will be updated and the new subspace of past observations $X_m(t+1)$ thus obtained will be a somewhat *rotated* version of its predecessor in the nonstationary case. The question is now whether it is possible to provide an *incremental update* of the projection operator $P_m(t)$ such that the *change* in the angle of subsequent subspaces is compensated so that $P_m(t+1)$, obtained by an incremental update from the "old" projection operator $P_m(t)$, again provides an exact orthogonal projection with respect to $X_m(t+1)$.

This problem is solved by the exact least-squares time update theorem of projection operators, which will be derived below. Again, we consider the simplest pre-windowed growing-order least-squares problem with an exponential weighting factor of $\lambda = 1$ and the observations

$$\mathbf{x}(t) = \left[x(t), x(t-1), \ldots, x(1), x(0), 0 \ldots 0 \right]^T , \qquad (9.3a)$$

$$\mathbf{x}(t-m) = \left[x(t-m), x(t-m-1), \ldots, x(1), x(0), 0 \ldots 0 \right]^T ; \qquad (9.3b)$$

$$1 \le m \le p ,$$

where the subspace of past observations was spanned by

$$\mathbf{X}_m(t) = \left[\mathbf{x}(t-1), \mathbf{x}(t-2), \ldots, \mathbf{x}(t-m) \right] . \qquad (9.3c)$$

Let

$$\mathbf{a}^{(m)}(t) = \left[a_1^{(m)}(t), a_2^{(m)}(t), \ldots, a_m^{(m)}(t) \right]^T \qquad (9.4)$$

be the true least-squares solution of the order m Normal Equations

$$\mathbf{a}^{(m)}(t) = \left[\mathbf{X}_m^T(t) \mathbf{X}_m(t) \right]^{-1} \mathbf{X}_m^T(t) \mathbf{x}(t) . \qquad (9.5)$$

The prediction error vector $\mathbf{e}_m(t)$ of the mth-order problem is then given by

$$\mathbf{e}_m(t) = \mathbf{x}(t) - \mathbf{X}_m(t) \mathbf{a}^{(m)}(t) . \qquad (9.6)$$

Substitution of the optimal parameter vector $\mathbf{a}^{(m)}(t)$ into the prediction error filter equation (9.6) shows that the prediction error filter determined by the optimal parameter set provides an *orthogonal* projection of $x(t)$ onto the subspace spanned by $X_m(t)$, hence

$$P_m(t) = X_m(t)\left[X_m^T(t)X_m(t)\right]^{-1} X_m^T(t) \tag{9.7}$$

and

$$e_m(t) = x(t) - P_m(t)x(t) \quad . \tag{9.8}$$

Defining the orthogonal complement $P_m^\perp(t)$

$$P_m^\perp(t) = I - P_m(t) \quad , \tag{9.9}$$

where I is the identity matrix, we see that

$$e_m(t) = P_m^\perp(t)\,x(t) \quad , \tag{9.10a}$$

$$\hat{x}_m^f(t) = P_m(t)\,x(t) \quad , \tag{9.10b}$$

where

$$x(t) = e_m(t) + \hat{x}_m^f(t) \quad . \tag{9.11}$$

Figure 9.1 illustrates this decomposition of the actual observation vector $x(t)$ into a component $\hat{x}_m^f(t)$, which lies in the subspace $X_m(t)$, and a component $e_m(t)$, which is orthogonal with respect to the subspace $X_m(t)$.

Recall the successive orthogonalization procedures (6.12a-e) and (6.13a-e) that were introduced in Chap. 6 in order to derive the least-squares ladder form. Rewriting (6.12e) and (6.13e) in a more compact subspace notation

$$e_m(t) = x(t) \langle X_m(t)\rangle \quad , \tag{9.12a}$$

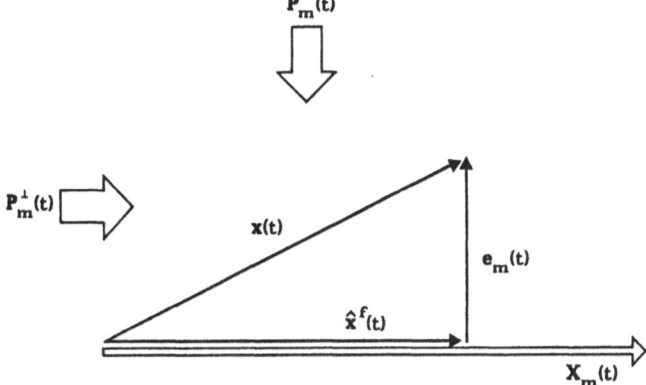

Fig. 9.1. Decomposition of the observation $x(t)$ into an "in-space" component $\hat{x}_m^f(t)$ and an orthogonal component $e_m(t)$

$$r_m(t-1) = x(t-m-1) \langle X_m(t) \rangle \quad , \tag{9.12b}$$

we see that the orthogonality relations expressed by (9.12a,b) can be stated in terms of projection operators as follows:

$$e_m(t) = P_m^\perp(t) x(t) \quad , \tag{9.13a}$$

$$r_m(t-1) = P_m^\perp(t) x(t-m-1) \quad . \tag{9.13b}$$

Moreover, the order recursion of forward/backward residual energies (6.23a,b) allows the development of order recursions of projection operators. To see this, the order recursions of residual energies (6.22a,b) are rewritten in terms of inner products

$$e_{m+1}^T(t) e_{m+1}(t) = e_m^T(t) e_m(t) - \frac{e_m^T(t) r_m(t-1) r_m^T(t-1) e_m(t)}{r_m^T(t-1) r_m(t-1)} \quad , \tag{9.14a}$$

$$r_{m+1}^T(t) r_{m+1}(t) = r_m^T(t-1) r_m(t-1) - \frac{r_m^T(t-1) e_m(t) e_m^T(t) r_m(t-1)}{e_m^T(t) e_m(t)} \quad . \tag{9.14b}$$

Exploiting the expressions (9.13a,b), together with the symmetry and idempotence property of projection operators [recall (2.19 and 20)], we may express (9.14a,b) in terms of projection operators as

$$x^T(t) P_{m+1}^\perp(t) x(t) = x^T(t) P_m^\perp(t) x(t)$$

$$- \frac{x^T(t) P_m^\perp(t) r_m(t-1) r_m^T(t-1) P_m^\perp(t) x(t)}{r_m^T(t-1) r_m(t-1)} \quad , \tag{9.15a}$$

$$x^T(t-m-1) P_{m+1}^\perp(t+1) x(t-m-1) = x^T(t-m-1) P_m^\perp(t) x(t-m-1)$$

$$- \frac{x^T(t-m-1) P_m^\perp(t) e_m(t) e_m^T(t) P_m^\perp(t) x(t-m-1)}{e_m^T(t) e_m(t)} \quad . \tag{9.15b}$$

But, according to the idempotence property of projection operators,

$$P_m^\perp(t) e_m(t) = P_m^\perp(t) P_m^\perp(t) x(t) = P_m^\perp(t) x(t) = e_m(t) \tag{9.16a}$$

and, similarly,

$$P_m^\perp(t) r_m(t-1) = P_m^\perp(t) P_m^\perp(t) x(t-m-1)$$

$$= P_m^\perp(t) x(t-m-1) = r_m(t-1) \quad . \tag{9.16b}$$

Clearly, the relations (9.16a,b) can be used to rewrite (9.15a,b) in a more condensed form as

$$x^T(t)\,P_{m+1}^{\perp}(t)\,x(t) = x^T(t)\,P_m^{\perp}(t)\,x(t)$$

$$- x^T(t)\,\frac{r_m(t-1)\,r_m^T(t-1)}{r_m^T(t-1)\,r_m(t-1)}\,x(t) \quad , \tag{9.17a}$$

$$x^T(t-m-1)\,P_{m+1}^{\perp}(t+1)\,x(t-m-1) = x^T(t-m-1)\,P_m^{\perp}(t)\,x(t-m-1)$$

$$- x^T(t-m-1)\,\frac{e_m(t)\,e_m^T(t)}{e_m^T(t)\,e_m(t)}\,x(t-m-1) \quad , \tag{9.17b}$$

which finally gives the desired order recursions of projection operators

$$P_{m+1}^{\perp}(t) = P_m^{\perp}(t) - \frac{r_m(t-1)\,r_m^T(t-1)}{R_m(t-1)} \quad , \tag{9.18a}$$

$$P_{m+1}(t) = P_m(t) + \frac{r_m(t-1)\,r_m^T(t-1)}{R_m(t-1)} \quad , \tag{9.18b}$$

$$P_{m+1}^{\perp}(t+1) = P_m^{\perp}(t) - \frac{e_m(t)\,e_m^T(t)}{E_m(t)} \quad , \tag{9.19a}$$

$$P_{m+1}(t+1) = P_m(t) + \frac{e_m(t)\,e_m^T(t)}{E_m(t)} \quad . \tag{9.19b}$$

Further, the covariance $C_m(t)$ can be expressed in terms of projection operators and observation vectors. Using (9.13a,b) and considering the symmetric and idempotence properties of projection operators, we obtain

$$C_m(t) = e_m^T(t)\,r_m(t-1) = x^T(t)\,P_m^{\perp}(t)\,x(t-m-1) \quad . \tag{9.20}$$

According to (9.20), the one time step delayed covariance $C_m(t-1)$ can be expressed by

$$C_m(t-1) = x^T(t-1)\,P_m^{\perp}(t-1)\,x(t-m-2) \quad . \tag{9.21}$$

The exact time update theorem requires the quantification of the differential $\Delta C_m(t)$ defined as

$$\Delta C_m(t) = C_m(t) - C_m(t-1) = x^T(t)\,\Delta P_m^{\perp}(t)\,x(t-m-1) \quad . \tag{9.22}$$

The quantification of $\Delta C_m(t)$ requires the determination of the *differential projection operator* $\Delta P_m^\perp(t)$. For this purpose, it is necessary to decompose the projection operator $P_m^\perp(t)$ into an "old" projection operator $P_m^\perp(t-1)$ and an "innovation", expressed by the "observable" components (top components) of the residual vectors. To establish such a decomposition, we need to employ a *pinning vector* $\pi(t)$ as follows (note that we have already used this pinning vector in Chap. 4 in conjunction with the fast extraction of prediction errors from the Gentleman and Kung array):

$$\pi(t) = \begin{bmatrix} 1, 0 \ldots 0 \end{bmatrix}^T . \tag{9.23}$$

This pinning vector picks up the *observable component* of a data vector, therefore

$$x(t) = \pi^T(t)\,x(t) \quad . \tag{9.24}$$

Furthermore, we may define a useful projection operator $P_\pi(t)$, which projects signal vectors *orthogonally* onto the pinning vector $\pi(t)$

$$P_\pi(t) = \pi(t) \begin{bmatrix} \pi^T(t)\pi(t) \end{bmatrix}^{-1} \pi^T(t) = \pi(t)\,\pi^T(t) \quad , \tag{9.25a}$$

with the orthogonal complement

$$P_\pi^\perp(t) = I - P_\pi(t) \quad . \tag{9.25b}$$

For example, in the pre-windowed case, the projection operator $P_\pi^\perp(t)$ can be used to express the relationship between subsequent observation vectors as follows:

$$x(t-1) = P_\pi^\perp(t)\,x(t) \quad . \tag{9.26}$$

Furthermore, it can be useful to introduce an *"oblique" projection operator* $P_{s,m}(t)$, which projects signal vectors onto the subspace $X_m(t)$ such that the projection is orthogonal with respect to the *one time step delayed* subspace $X_m(t-1)$. Figure 9.2 illustrates the effect of the oblique projection operator $P_{s,m}(t)$.

From an examination of (9.6 and 10a), it becomes apparent that the orthogonal projection operator $P_m(t)$ corresponds to a transversal predictor parameter set $a^{(m)}(t)$ determined by the actual observation and the actual subspace of past observations via the Normal Equations. On the other hand, the oblique projection operator $P_{s,m}(t)$ corresponds to a parameter set $a^{*(m)}(t)$ determined by the actual observation and the previous (one time step delayed) subspace of past observations as follows:

$$a^{*(m)}(t) = \begin{bmatrix} X_m^T(t-1)X_m(t-1) \end{bmatrix}^{-1} X_m^T(t-1)\,x(t) \quad . \tag{9.27}$$

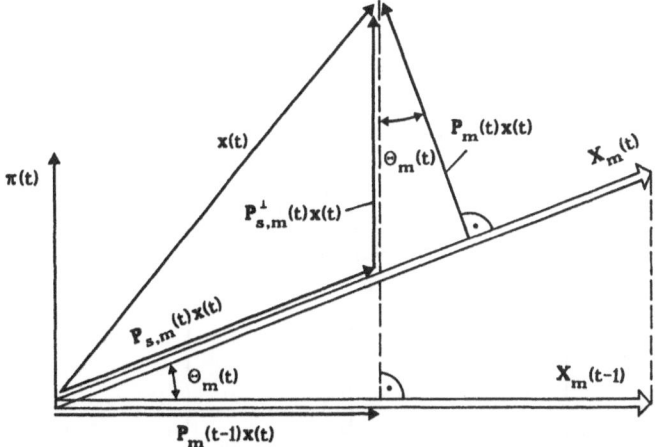

Fig. 9.2. The oblique projection operator $\mathbf{P}_{s,m}(t)$ projects the observation $\mathbf{x}(t)$ onto the subspace $\mathbf{X}_m(t)$. The projection, however, is orthogonal with respect to the *"previous"* subspace $\mathbf{X}_m(t\text{-}1)$

Taking into account that

$$\mathbf{P}_{s,m}^{\perp}(t)\,\mathbf{x}(t) = \mathbf{x}(t) - \mathbf{X}_m(t)\,\mathbf{a}^{*(m)}(t) \quad , \tag{9.28}$$

we obtain

$$\mathbf{P}_{s,m}(t) = \mathbf{X}_m(t)\left[\mathbf{X}_m^{T}(t\text{-}1)\mathbf{X}_m(t\text{-}1)\right]^{-1}\mathbf{X}_m^{T}(t\text{-}1) \quad , \tag{9.29a}$$

$$\mathbf{P}_{s,m}^{\perp}(t) = \mathbf{I} - \mathbf{P}_{s,m}(t) \quad . \tag{9.29b}$$

Figure 9.3 illustrates the successive observations $\mathbf{x}(t)$ and $\mathbf{x}(t\text{-}1)$ and the corresponding subspaces $\mathbf{X}_m(t)$ and $\mathbf{X}_m(t\text{-}1)$, as well as the decomposition of projections at subsequent time steps. In the figure, we have used the following abbreviations, for convenience:

$$\mathbf{a} = \mathbf{P}_{s,m}(t)\mathbf{x}(t) \quad , \tag{9.30a}$$

$$\mathbf{b} = \mathbf{P}_{s,m}^{\perp}(t)\,\mathbf{x}(t) \quad , \tag{9.30b}$$

$$\mathbf{c} = \mathbf{P}_{\pi}(t)\mathbf{P}_{s,m}^{\perp}(t)\,\mathbf{x}(t) \quad , \tag{9.30c}$$

$$\mathbf{d} = \mathbf{P}_m(t)\mathbf{P}_{\pi}(t)\mathbf{P}_{s,m}^{\perp}(t)\,\mathbf{x}(t) \quad , \tag{9.30d}$$

$$\mathbf{e} = \mathbf{P}_m^{\perp}(t)\mathbf{x}(t) \quad , \tag{9.30e}$$

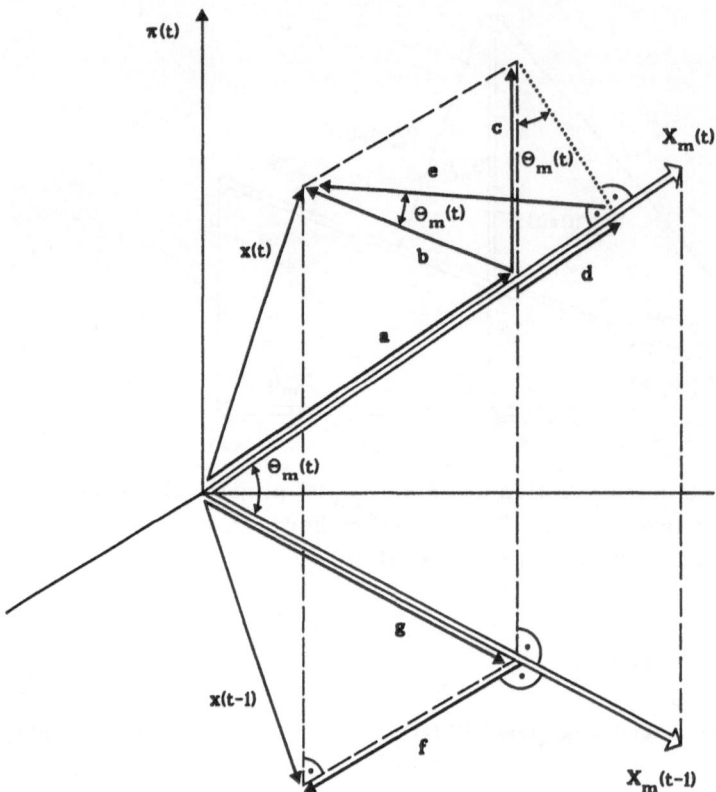

Fig. 9.3. Decomposition of projections of a nonstationary process at subsequent time steps of observation

$$\mathbf{f} = \mathbf{P}_m^\perp(t-1)\mathbf{x}(t-1) \quad , \tag{9.30f}$$

$$\mathbf{g} = \mathbf{P}_m(t-1)\mathbf{x}(t-1) \quad . \tag{9.30g}$$

From Fig. 9.3, one may deduce the following relation:

$$\mathbf{e} = \mathbf{b} - \mathbf{d} \quad , \tag{9.31a}$$

or, equivalently,

$$\mathbf{P}_m^\perp(t)\mathbf{x}(t) = \mathbf{P}_{s,m}^\perp(t)\,\mathbf{x}(t) \; - \; \mathbf{P}_m(t)\mathbf{P}_\pi(t)\mathbf{P}_{s,m}^\perp(t)\,\mathbf{x}(t) \quad . \tag{9.31b}$$

Also,

$$\mathbf{f} = \mathbf{b} - \mathbf{c} \quad , \tag{9.32a}$$

or, explicitly,

$$\mathbf{P}_m^\perp(t-1)\mathbf{x}(t-1) = \mathbf{P}_{s,m}^\perp(t)\mathbf{x}(t) - \mathbf{P}_\pi(t)\mathbf{P}_{s,m}^\perp(t)\mathbf{x}(t) \quad . \tag{9.32b}$$

Substitution of (9.32b) into (9.31b) reveals

$$\mathbf{P}_m^\perp(t)\mathbf{x}(t) = \mathbf{P}_m^\perp(t-1)\mathbf{x}(t-1) + \mathbf{P}_m^\perp(t)\mathbf{P}_\pi(t)\mathbf{P}_{s,m}^\perp(t)\mathbf{x}(t) \quad . \tag{9.33}$$

Note that the desired differential projection operator $\triangle\mathbf{P}_m^\perp(t)$ is given by

$$\triangle\mathbf{P}_m^\perp(t)\,\mathbf{x}(t) = \mathbf{P}_m^\perp(t)\,\mathbf{x}(t) - \mathbf{P}_m^\perp(t-1)\mathbf{x}(t-1)$$

$$= \mathbf{P}_m^\perp(t)\mathbf{P}_\pi(t)\mathbf{P}_{s,m}^\perp(t)\,\mathbf{x}(t) \quad . \tag{9.34}$$

Expressing $\mathbf{P}_\pi(t)$ as the outer product of the pinning vector $\pi(t)$, recall (9.25a), we have

$$\triangle\mathbf{P}_m^\perp(t)\,\mathbf{x}(t) = \mathbf{P}_m^\perp(t)\,\pi(t)\ \pi^T(t)\,\mathbf{P}_{s,m}^\perp(t)\,\mathbf{x}(t) \quad . \tag{9.35}$$

Equation (9.35) is not yet a complete decomposition. It remains to express the oblique projection operator $\mathbf{P}_{s,m}^\perp(t)$ by the orthogonal projection operator $\mathbf{P}_m^\perp(t)$. This can be achieved through the following relation, which can be directly deduced from Fig. 9.3:

$$\cos^2\Theta_m(t) = \frac{\mathbf{e}^T\mathbf{e}}{\mathbf{b}^T\mathbf{b}} = \frac{\mathbf{x}^T(t)\,\mathbf{P}_m^\perp(t)\,\mathbf{x}(t)}{\mathbf{x}^T(t)\,\mathbf{P}_{s,m}^\perp(t)\,\mathbf{x}(t)} \quad , \tag{9.36}$$

or, equivalently,

$$\mathbf{P}_{s,m}^\perp(t)\,\mathbf{x}(t) = \mathbf{P}_m^\perp(t)\,\mathbf{x}(t)\ \frac{1}{\cos^2\Theta_m(t)} \quad . \tag{9.37}$$

Substitution of (9.37) into (9.35) yields the desired differential of the projection operator $\mathbf{P}_m^\perp(t)$:

$$\triangle\mathbf{P}_m^\perp(t) = \mathbf{P}_m^\perp(t)\,\pi(t)\ \pi^T(t)\,\mathbf{P}_m^\perp(t)\ \frac{1}{\cos^2\Theta_m(t)} \quad . \tag{9.38}$$

This relation has become known as the *exact least-squares time update theorem of projection operators*. It constitutes the basis of all fast RLS ladder algorithms. Interestingly, the differential projection operator $\triangle\mathbf{P}_m^\perp(t)$ is proportional to the inverse square of the cosine of the angle between subsequent subspaces. See Fig. 9.3.

Such geometric techniques which make use of the trigonometric functions relating two subspaces have been extensively applied to numerical analysis problems, especially in perturbation analysis. Some representative publications in this area are [9.2-4].

In the area of multivariate statistical analysis, notions of oblique projection and their relation to angles between subspaces have been considered (although in a very different context) by Afriat [9.5], Zassenhaus [9.6], and Greville [9.7].

In the context of signal analysis, the concept of angles between subspaces has been investigated by Landau and Pollak [9.8] and Mazo [9.9]. Finally, the more abstract results in operator theory involving the notion of angles between subspaces have been discussed by Gohberg and Krein [9.10], and more recently by Bart et al. [9.11], and also by Van Dooren and Dewilde [9.12]

These ideas can also be employed for a derivation of fast RLS transversal algorithms. See the work of Cioffi and Kailath [9.13], and the books by Honig and Messerschmitt [9.14] and Alexander [9.15]. In this book, however, we restrict the geometrical considerations to the ladder form only.

9.2 The Algorithm of Lee and Morf

The exact time update theorem of projection operators (9.38) can be directly used to derive the *pre-windowed RLS ladder algorithm* of Lee and Morf [9.1, 16]. This method first accomplished the true least-squares updating of the ladder form reflection coefficients with a computational complexity that grows only *linearly* with the model order.

Substitution of the exact time update theorem of projection operators (9.38) into the definition of the differential covariance $\Delta C_m(t)$ (9.22) yields

$$\Delta C_m(t) = x^T(t) \, \Delta P_m^\perp(t) \, x(t-m-1)$$

$$= x^T(t) \, P_m^\perp(t) \, \pi(t) \, \pi^T(t) \, P_m^\perp(t) \, x(t-m-1) \, \frac{1}{\cos^2 \Theta_m(t)} \, . \tag{9.39}$$

Note that $x^T(t) \, P_m^\perp(t) = e_m^T(t)$ and $P_m^\perp(t) \, x(t-m-1) = r_m(t-1)$. Furthermore, $e_m^T(t) \, \pi(t) = e_m(t)$ and $\pi^T(t) \, r_m(t-1) = r_m(t-1)$. Therefore,

$$\Delta C_m(t) = \frac{e_m(t) \, r_m(t-1)}{\cos^2 \Theta_m(t)} \, , \tag{9.40}$$

which is the desired expression for the differential covariance $\Delta C_m(t)$ in the nonstationary case.

A quick inspection of expression (9.40) shows that the differential covariance requires the *explicit* computation of the residuals $e_m(t)$ and $r_m(t-1)$. This, in fact, is one of the major characteristics of this approach since the explicit quantification of residuals is *always* required, even if we are only interested in the estimation of the ladder reflection coefficients. This is in contrast to other approaches discussed in Chaps. 7 and 8, where ladder algorithms were presented that do not require the explicit computation of the ladder form prediction error filter in the identification case. Recall the structure of PORLA algorithms (Fig. 8.3).

Returning to the algorithm derivation, it remains to note that at this stage the algorithm is not yet complete because of the unknown quantity $\cos^2\Theta_m(t)$. In order to derive a recursion for the updating of this quantity, we consider again the definition of $\Theta_m(t)$ as the angle between the subsequent subspaces $\mathbf{X}_m(t)$ and $\mathbf{X}_m(t-1)$. This relationship is illustrated in Fig. 9.4.

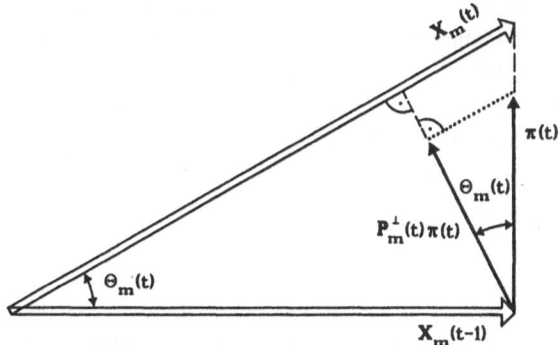

Fig. 9.4. $\Theta_m(t)$ defined as the angle between subsequent subspaces and its computation from a decomposition of projections

Figure 9.4 leads to the relationship

$$\cos^2\Theta_m(t) = \frac{\pi^T(t)\,\mathbf{P}_m^\perp(t)\,\pi(t)}{\pi^T(t)\,\pi(t)} = \pi^T(t)\,\mathbf{P}_m^\perp(t)\,\pi(t) \quad . \tag{9.41}$$

Expression (9.41) together with (9.18a) gives

$$\pi^T(t)\,\mathbf{P}_{m+1}^\perp(t)\,\pi(t) = \pi^T(t)\,\mathbf{P}_m^\perp(t)\,\pi(t)$$

$$- \frac{\pi^T(t)\,\mathbf{r}_m(t-1)\,\mathbf{r}_m^T(t-1)\,\pi(t)}{R_m(t-1)} \quad , \tag{9.42}$$

or, equivalently,

$$\cos^2\Theta_{m+1}(t) = \cos^2\Theta_m(t) - \frac{r_m^2(t-1)}{R_m(t-1)} \quad , \tag{9.43}$$

which is the desired order recursion of the angle between subsequent subspaces. Clearly, an alternative expression to (9.45) can be discovered by combining (9.41) and (9.19a) so that

$$\cos^2\Theta_{m+1}(t+1) = \cos^2\Theta_m(t) - \frac{e_m^2(t)}{E_m(t)} . \qquad (9.44)$$

The order recursions (9.43 and 44) are also based on *explicit* operations on possibly very small and numerically uncertain residuals.

With the shorthand notation

$$\cos^2\Theta_m(t) = \gamma_m(t) , \qquad (9.45)$$

we obtain a closed form recursion for the true least-squares updating of the ladder reflection coefficients at each time step. Note that this procedure is *fast* in terms of a computational complexity of O(m) operations in each recursion for updating the mth-order model. This algorithm is summarized in Table 9.1 and has become known as the fast RLS ladder algorithm of Lee and Morf [9.16].

Interestingly, a quantity like $\gamma_m(t)$ appeared already in the derivation of the RLS transversal algorithm. Recall (5.32) and the subsequent discussion in Chap. 5. There, the quantity $\gamma(t)$ was called the *likelihood variable*, bounded on the interval

$$0 \le \gamma_m(t) \le 1 , \qquad (9.46)$$

as already seen in (5.35). From these considerations, it is obvious that the RLS ladder algorithm of Lee and Morf has an algorithm structure quite similar to the fast RLS transversal algorithms treated in Chap. 5. As a common characteristic, we have also seen the appearance of a likelihood variable in both the RLS transversal and the RLS ladder algorithms. Both types of algorithms require the *explicit computation of residuals* in the adaptation procedure. The resulting algorithm structure may best be named a *"fast Kalman type"* algorithm structure. See Chap. 10 for an illustrative explanation of the RLS problem and how it is related to the Kalman filter approach.

This algorithm structure has the unpleasant effect that it allows the *accumulation of round-off errors in time* at each stage of the "fast" algorithms. Although this error propagation is common to both the transversal and the ladder formulation, the RLS ladder algorithm listed in Table 9.1 was found to have much better numerical properties than its transversal structured counterparts.

The algorithm of Table 9.1 is constructed from the following relations: First, the covariance $C_m(t)$ is updated according to the differential covariance $\Delta C_m(t)$ (9.40) and problem formulation (9.2), where we have reintroduced the exponential weight. The likelihood variable $\gamma_m(t)$ is updated via (9.43) where $\gamma_m(t) = \cos^2\Theta_m(t)$. Note that a counterpart relation exists in the updating of $\gamma_m(t)$ through forward prediction errors (9.44). The remaining recursions in the algorithm of Table 9.1 are relationships well known from the earlier discussion of ladder forms.

Table 9.1. Pre-windowed RLS ladder algorithm of Lee and Morf. λ is an exponential weighting factor. This algorithm involves division. Whenever the divisor is small, set $1/x = x$

FOR   t = 0, 1, 2, . . .

  Input: $x(t)$, $0 \le \lambda \le 1$

  Initialize:

  $e_0(t) = r_0(t) = x(t)$

  $E_0(t) = R_0(t) = \lambda R_0(t-1) + x^2(t)$

  $\gamma_0(t) = 1$

  FOR   m = 0, 1, 2, . . .

$$C_m(t) = \lambda C_m(t-1) + \frac{e_m(t) r_m(t-1)}{\gamma_m(t)} \tag{9.47a}$$

$$\gamma_{m+1}(t) = \gamma_m(t) - \frac{r_m^2(t-1)}{R_m(t-1)} \tag{9.47b}$$

$$K_{m+1}^f(t) = - C_m(t) / R_m(t-1) \tag{9.47c}$$

$$e_{m+1}(t) = e_m(t) + K_{m+1}^f(t)\, r_m(t-1) \tag{9.47d}$$

$$E_{m+1}(t) = E_m(t) + K_{m+1}^f(t)\, C_m(t) \tag{9.47e}$$

$$K_{m+1}^b(t) = - C_m(t) / E_m(t) \tag{9.47f}$$

$$r_{m+1}(t) = r_m(t-1) + K_{m+1}^b(t)\, e_m(t) \tag{9.47g}$$

$$R_{m+1}(t) = R_m(t-1) + K_{m+1}^b(t)\, C_m(t) \tag{9.47h}$$

9.3 Other Forms of Lee's Algorithm

The exact time-update theorem of projection operators allows the derivation of a multitude of fast RLS ladder algorithms. We have already shown the different possibilities of computing the likelihood variable $\gamma_m(t)$ from the forward or, alternatively, from the backward prediction errors. Moreover, other types of Lee's algorithm can be stated where the prediction error energies $E_m(t)$ and $R_m(t)$ are updated recursively in time. Another recently suggested version of Lee's algorithm is based on the *direct updating* of the reflection coefficients without explicit computation of the covariance $C_m(t)$. The idea of formulating such a "direct-update" algorithm can be traced back to the work of Ling et al. [9.17]. These modifications of Lee's algorithm will be discussed next.

9.3.1 A Pure Time Recursive Ladder Algorithm

So far, we have only discussed the time recursion of the covariance $C_m(t)$. Similar time recursions can be stated also in the cases of the prediction error energies $E_m(t)$ and $R_m(t)$.

Utilizing the exact time-update theorem of projection operators (9.38), one can easily establish such recursions. Exploiting the expression (9.10a), we can express the forward prediction error energy $E_m(t)$ as

$$E_m(t) = \mathbf{x}^T(t)\mathbf{P}_m^\perp(t)\mathbf{x}(t) \ . \tag{9.48}$$

Clearly, the differential $\Delta E_m(t)$ satisfies the relation

$$\Delta E_m(t) = \mathbf{x}^T(t) \, \Delta \mathbf{P}_m^\perp(t)\mathbf{x}(t) \ . \tag{9.49}$$

Substituting the exact time update formula of projection operators (9.38) into (9.49), we have

$$\Delta E_m(t) = \mathbf{x}^T(t) \, \mathbf{P}_m^\perp(t)\boldsymbol{\pi}(t) \ \ \boldsymbol{\pi}^T(t) \, \mathbf{P}_m^\perp(t) \, \mathbf{x}(t) \ \frac{1}{\cos^2\Theta_m(t)} \ , \tag{9.50}$$

which gives

$$\Delta E_m(t) = \frac{e_m^2(t)}{\cos^2\Theta_m(t)} \ . \tag{9.51}$$

Therefore, the time recursion of $E_m(t)$ becomes

$$E_m(t) = \lambda E_m(t-1) + \frac{e_m^2(t)}{\cos^2\Theta_m(t)} \ , \tag{9.52}$$

where λ is an exponential decay factor.

Following this line of thought, one can also derive a time recursion of $R_m(t)$ by noting that, according to (9.13b),

$$\Delta R_m(t-1) = \mathbf{x}^T(t-m-1) \, \Delta \mathbf{P}_m^\perp(t) \, \mathbf{x}(t-m-1) \tag{9.53}$$

and, by substituting the exact time update formula into (9.53), we finally obtain the time recursion

$$R_m(t-1) = \lambda R_m(t-2) + \frac{r_m^2(t-1)}{\cos^2\Theta_m(t)} \ . \tag{9.54}$$

Table 9.2 summarizes this pure time recursive RLS ladder algorithm in the pre-windowed case with exponential decay. In this pure time recursive RLS ladder algorithm, the covariance $C_m(t)$ is again computed via the key relation (9.47a). In contrast to the algorithm of Table 9.1, the forward prediction error energy $E_m(t)$ and the backward prediction error energy $R_m(t-1)$ are now updated in time according to (9.52) and (9.54). Again, we compute the likelihood variable $\gamma_{m+1}(t)$ according to (9.47b). The remaining recursions are already known from the earlier dicussion of ladder forms.

Table 9.2. Pre-windowed pure time recursive RLS ladder algorithm. λ is an exponential weighting factor. This algorithm involves division. Whenever the divisor is small, set $1/x = x$

FOR $t = 0, 1, 2, \ldots$

  Input: $x(t)$, $0 \le \lambda \le 1$

  Initialize:

  $e_0(t) = r_0(t) = x(t)$

  $\gamma_0(t) = 1$

  FOR $m = 0, 1, 2, \ldots$

$$C_m(t) = \lambda C_m(t-1) + \frac{e_m(t) r_m(t-1)}{\gamma_m(t)} \tag{9.55a}$$

$$E_m(t) = \lambda E_m(t-1) + \frac{e_m^2(t)}{\gamma_m(t)} \tag{9.55b}$$

$$R_m(t-1) = \lambda R_m(t-2) + \frac{r_m^2(t-1)}{\gamma_m(t)} \tag{9.55c}$$

$$\gamma_{m+1}(t) = \gamma_m(t) - \frac{r_m^2(t-1)}{R_m(t-1)} \tag{9.55d}$$

$$K_{m+1}^f(t) = - C_m(t)/R_m(t-1) \tag{9.55e}$$

$$K_{m+1}^b(t) = - C_m(t)/E_m(t) \tag{9.55f}$$

$$e_{m+1}(t) = e_m(t) + K_{m+1}^f(t) r_m(t-1) \tag{9.55g}$$

$$r_{m+1}(t) = r_m(t-1) + K_{m+1}^b(t) e_m(t) \tag{9.55h}$$

9.3.2 Direct Updating of Reflection Coefficients

Recent investigations by Ling et al. [9.17] have shown that another form of Lee's algorithm is possible where the reflection coefficients are *directly updated in time* without explicit computation of the error covariance $C_m(t)$. Such types of algorithms have slightly better numerical properties than the original counterparts of Tables 9.1 and 9.2. We show the derivation of such an algorithm and start by expressing the covariance $C_m(t)$ in terms of reflection coefficients as follows:

$$C_m(t) = - K^f_{m+1}(t) R_m(t-1) \quad , \tag{9.56a}$$

$$C_m(t) = - K^b_{m+1}(t) E_m(t) \quad . \tag{9.56b}$$

Substitution of (9.56a,b) into the time recursion of the covariance $C_m(t)$ (9.55a) yields a time recursion of reflection coefficients:

$$K^f_{m+1}(t) = \frac{K^f_{m+1}(t-1) R_m(t-2)}{R_m(t-1)} - \frac{e_m(t) r_m(t-1)}{\gamma_m(t) R_m(t-1)} \quad , \tag{9.57a}$$

$$K^b_{m+1}(t) = \frac{K^b_{m+1}(t-1) E_m(t-1)}{E_m(t)} - \frac{e_m(t) r_m(t-1)}{\gamma_m(t) E_m(t)} \quad . \tag{9.57b}$$

Table 9.3 is a summary of this fast RLS ladder algorithm with direct updating of the ladder reflection coefficients. In the algorithm of Table 9.3, one first computes the residual energies $E_m(t)$ and $R_m(t-1)$ according to (9.55b,c). The ladder reflection coefficients $K^f_{m+1}(t)$ and $K^b_{m+1}(t)$ are then updated *directly* via (9.57a,b), hereby avoiding the explicit quantification of the covariance $C_m(t)$. ·

9.4 Gradient Adaptive Ladder Algorithms

Gradient adaptive ladder algorithms were developed long before the exact time update theorem of projection operators was discovered. In fact, the absence of an exact least-squares time update theorem was the *major motivation* for the development of gradient ladder algorithms. In Chap. 5, we discussed the relationships between the true least-squares RLS transversal algorithms and the LMS algorithm. It has been demonstrated that the LMS algorithm can be interpreted as a special simplified case of the RLS algorithm where the modified Kalman gain vector $k^*(t)$ has been replaced by the state vector $z(t)$ multiplied by a constant stepsize μ. Recall (5.242).

Table 9.3. Pre-windowed RLS ladder algorithm with direct updating of the ladder reflection coefficients. λ is an exponential weighting factor. This algorithm involves division. Whenever the divisor is small, set $1/x = x$

$K_m^f(-1) = K_m^b(-1) = 0$

FOR $t = 0, 1, 2, \ldots$

> *Input:* $x(t),\ 0 \le \lambda \le 1$
>
> *Initialize:*
>
> $e_0(t) = r_0(t) = x(t)$
>
> $\gamma_0(t) = 1$
>
> FOR $m = 0, 1, 2, \ldots$
>
> > $$E_m(t) = \lambda E_m(t-1) + \frac{e_m^2(t)}{\gamma_m(t)} \tag{9.58a}$$
> >
> > $$R_m(t-1) = \lambda R_m(t-2) + \frac{r_m^2(t-1)}{\gamma_m(t)} \tag{9.58b}$$
> >
> > $$K_{m+1}^f(t) = \frac{K_{m+1}^f(t-1) R_m(t-2)}{R_m(t-1)} - \frac{e_m(t) r_m(t-1)}{\gamma_m(t) R_m(t-1)} \tag{9.58c}$$
> >
> > $$K_{m+1}^b(t) = \frac{K_{m+1}^b(t-1) E_m(t-1)}{E_m(t)} - \frac{e_m(t) r_m(t-1)}{\gamma_m(t)\, E_m(t)} \tag{9.58d}$$
> >
> > $$\gamma_{m+1}(t) = \gamma_m(t) - \frac{r_m^2(t-1)}{R_m(t-1)} \tag{9.58e}$$
> >
> > $$e_{m+1}(t) = e_m(t) + K_{m+1}^f(t)\, r_m(t-1) \tag{9.58f}$$
> >
> > $$r_{m+1}(t) = r_m(t-1) + K_{m+1}^b(t)\, e_m(t) \tag{9.58g}$$

Although the development of gradient ladder algorithms can be traced back to the early work of Griffiths [9.18,19], Makhoul [9.20] and others, who investigated descent methods for adapting ladder form prediction error filters long before the true least-squares algorithms were developed, it seems to be very useful to introduce the gradient ladder algorithms in a similar way to the LMS algorithm, namely, as special *simplified cases* of the true least-squares solution.

As a major difference to the transversal case where the LMS algorithm exhibited a much lower computational and structural complexity than the exact RLS transversal counterparts, the ladder algorithms in the simplified (gradient descent) case do *not* result in considerable computational savings.

9.4.1 Gradient Adaptive Ladder Algorithm GAL 2

We may start the dicussion with a gradient ladder algorithm for the ladder form with different reflection coefficients in the forward and backward predictors. This algorithm will be derived from Lee's exact approach as listed in Table 9.2.

According to (9.55b,c), the quotient of prediction error energies in subsequent time steps becomes

$$\frac{E_m(t-1)}{E_m(t)} = 1 - \frac{e_m^2(t)}{\gamma_m(t) E_m(t)} \quad , \tag{9.59a}$$

$$\frac{R_m(t-2)}{R_m(t-1)} = 1 - \frac{r_m^2(t-1)}{\gamma_m(t) R_m(t-1)} \quad . \tag{9.59b}$$

Taking the ladder equations for the instantaneous errors at the "previous" time step

$$e_{m+1}(t-1) = e_m(t-1) + K_{m+1}^f(t-1) r_m(t-2) \quad , \tag{9.60a}$$

$$r_{m+1}(t-1) = r_m(t-2) + K_{m+1}^b(t-1) e_m(t-1) \quad , \tag{9.60b}$$

and solving (9.60a,b) for the reflection coefficients, we have

$$K_{m+1}^f(t-1) = \frac{e_{m+1}(t-1) - e_m(t-1)}{r_m(t-2)} \quad , \tag{9.61a}$$

$$K_{m+1}^b(t-1) = \frac{r_{m+1}(t-1) - r_m(t-2)}{e_m(t-1)} \quad . \tag{9.61b}$$

Now we replace the energy quotients in (9.58c,d) by the expressions (9.59a,b), to obtain

$$K_{m+1}^f(t) = K_{m+1}^f(t-1)\left(1 - \frac{r_m^2(t-1)}{\gamma_m(t) R_m(t-1)}\right) - \frac{e_m(t) r_m(t-1)}{\gamma_m(t) R_m(t-1)} \quad , \tag{9.62a}$$

$$K_{m+1}^b(t) = K_{m+1}^b(t-1)\left(1 - \frac{e_m^2(t)}{\gamma_m(t) E_m(t)}\right) - \frac{e_m(t) r_m(t-1)}{\gamma_m(t) E_m(t)} \quad . \tag{9.62b}$$

Note that by using (9.61a,b) one finds

$$K^f_{m+1}(t-1) \frac{r^2_m(t-1)}{\gamma_m(t) R_m(t-1)} = \frac{r^2_m(t-1) e_{m+1}(t-1) - r^2_m(t-1) e_m(t-1)}{\gamma_m(t) R_m(t-1) r_m(t-2)} \quad , \qquad (9.63a)$$

$$K^b_{m+1}(t-1) \frac{e^2_m(t)}{\gamma_m(t) E_m(t)} = \frac{e^2_m(t) r_{m+1}(t-1) - e^2_m(t) r_m(t-2)}{\gamma_m(t) E_m(t) e_m(t-1)} \quad . \qquad (9.63b)$$

Using these expressions, together with (9.62a,b), a new set of direct coefficient update equations can be stated as

$$K^f_{m+1}(t) = K^f_{m+1}(t-1)$$

$$- \frac{r^2_m(t-1) e_{m+1}(t-1) - r^2_m(t-1) e_m(t-1) + e_m(t) r_m(t-1) r_m(t-2)}{\gamma_m(t) R_m(t-1) r_m(t-2)} \quad , \qquad (9.64a)$$

$$K^b_{m+1}(t) = K^b_{m+1}(t-1)$$

$$- \frac{e^2_m(t) r_{m+1}(t-1) - e^2_m(t) r_m(t-2) + e_m(t) r_m(t-1) e_m(t-1)}{\gamma_m(t) E_m(t) e_m(t-1)} \quad . \qquad (9.64b)$$

Note that these recursions can be viewed as alternative schemes for a direct updating of the reflection coefficients in the true least-squares case. Recall Table 9.3.

We may introduce some obvious approximations for the stationary case. Clearly, in the case of stationary data, the angle between subsequent subspaces is zero, and hence we may set

$$\gamma_m(t) = 1 \quad . \qquad (9.65)$$

Furthermore, we assume that

$$e_m(t) r_m(t-1) = e_m(t-1) r_m(t-2) \qquad (9.66)$$

and, similarly,

$$\frac{e^2_m(t)}{E_m(t)} = \frac{e^2_m(t-1)}{E_m(t-1)} \quad , \qquad (9.67a)$$

$$r^2_m(t-1) = r^2_m(t-2) \quad , \qquad (9.67b)$$

$$\frac{e_m(t) r_m(t-1)}{E_m(t)} = \frac{e_m(t-1) r_m(t-2)}{E_m(t-1)} \quad . \qquad (9.67c)$$

With these approximations, the exact recursions (9.64a,b) attain the following approximate (gradient) forms, which already determine the gradient ladder algorithm GAL 2, where μ is a constant stepsize:

$$K^f_{m+1}(t) = K^f_{m+1}(t-1) - \mu \frac{e_{m+1}(t-1)\, r_m(t-2)}{R_m(t-1)} \quad , \tag{9.68a}$$

$$K^b_{m+1}(t) = K^b_{m+1}(t-1) - \mu \frac{e_m(t-1)\, r_{m+1}(t-1)}{E_m(t-1)} \quad . \tag{9.68b}$$

The algorithm GAL 2 was orignally proposed by Griffiths [9.18] as the result of quite different considerations. The algorithm is listed in Table 9.4 and consists basically of the direct gradient update of the reflection coefficients (9.68a,b). Furthermore, simple update recursions of the forward and backward residual energies $E_m(t)$ and $R_m(t)$ are obtained from (9.59a,b), by taking the "zero angle" assumption (9.65) into account.

A more detailed discussion of gradient ladder algorithms can be found in the original work of Griffiths [9.18] and the tutorial paper of Friedlander [9.21], where, in particular, the appropriate choice of the parameters μ and σ is discussed more in detail.

9.4.2 Gradient Adaptive Ladder Algorithm GAL 1

Gradient adaptive ladder algorithms may also be useful for the adaptation of PARCOR ladder forms. We shall report an early approach also suggested by Griffiths. Schemes different from the one reported here have been suggested by Makhoul [9.20] and others. The algorithm GAL 1 is easily deduced from GAL 2 simply by introducing the assumptions

$$K^f_m(t) = K^b_m(t) = K_m(t) \tag{9.70}$$

and

$$D_m(t) = 2\, E_m(t) = 2\, R_m(t) = E_m(t) + R_m(t) \quad . \tag{9.71}$$

Using these expressions, the recursions of GAL 2 may be further simplified in that

$$2K_{m+1}(t) = 2K^f_{m+1}(t) = 2K^b_{m+1}(t) = K^f_{m+1}(t) + K^b_{m+1}(t)$$

$$= K_{m+1}(t-1) - \frac{e_{m+1}(t-1)r_m(t-2)}{1/2\, D_m(t-1)}$$

$$+ K_{m+1}(t-1) - \frac{r_{m+1}(t-1)e_m(t-1)}{1/2\, D_m(t-1)} \quad , \tag{9.72}$$

Table 9.4. Gradient adaptive ladder algorithm GAL 2. The parameter μ is a constant stepsize and σ is an initial guess for the expected prediction error energy. This algorithm involves division. Whenever the divisor is small, set $1/x = x$

$E_m(-1) = R_m(-1) = \sigma$

$K_m^f(-1) = K_m^b(-1) = 0$

FOR t = 0, 1, 2, . . .

   Input: $x(t)$, $0 \le \lambda \le 1$

   Initialize:

   $e_0(t) = r_0(t) = x(t)$

   FOR m = 0, 1, 2, . . .

$$K_{m+1}^f(t) = K_{m+1}^f(t-1) - \mu\,\frac{e_{m+1}(t-1)\,r_m(t-2)}{R_m(t-1)} \tag{9.69a}$$

$$K_{m+1}^b(t) = K_{m+1}^b(t-1) - \mu\,\frac{e_m(t-1)\,r_{m+1}(t-1)}{E_m(t-1)} \tag{9.69b}$$

$$E_m(t) = \lambda E_m(t-1) + e_m^2(t) \tag{9.69c}$$

$$R_m(t) = \lambda R_m(t-1) + r_m^2(t) \tag{9.69d}$$

$$e_{m+1}(t) = e_m(t) + K_{m+1}^f(t)\,r_m(t-1) \tag{9.69e}$$

$$r_{m+1}(t) = r_m(t-1) + K_{m+1}^b(t)\,e_m(t) \tag{9.69f}$$

---

which finally gives

$$K_{m+1}(t) = K_{m+1}(t-1) - \frac{e_{m+1}(t-1)r_m(t-2) + r_{m+1}(t-1)e_m(t-1)}{D_m(t-1)} \;. \tag{9.73}$$

Table 9.5 summarizes this gradient adaptive ladder algorithm for PARCOR ladder forms, which has become known as GAL 1.

Table 9.5. Gradient adaptive ladder algorithm GAL 1. The parameter μ is a constant stepsize and σ is the sum of the expected forward/backward residual energies. This algorithm involves division. Whenever the divisor is small, set $1/x = x$

$$D_m(-1) = \sigma$$

$$K_m(-1) = 0$$

FOR   t = 0, 1, 2, . . .

> *Input:* $x(t)$, $0 \le \lambda \le 1$
>
> *Initialize:*
>
> $e_0(t) = r_0(t) = x(t)$
>
> FOR   m = 0, 1, 2, . . .
>
>> $$K_{m+1}(t) = K_{m+1}(t-1)$$
>>
>> $$- \mu \, \frac{e_{m+1}(t-1)r_m(t-2) + r_{m+1}(t-1)e_m(t-1)}{D_m(t-1)} \qquad (9.74a)$$
>>
>> $$D_m(t) = \lambda D_m(t-1) + \left[e_m^2(t) + r_m^2(t-1) \right] \qquad (9.74b)$$
>>
>> $$e_{m+1}(t) = e_m(t) + K_{m+1}(t)\, r_m(t-1) \qquad (9.74c)$$
>>
>> $$r_{m+1}(t) = r_m(t-1) + K_{m+1}(t)\, e_m(t) \qquad (9.74d)$$

9.5 Lee's Normalized RLS Ladder Algorithm

The objective of deriving normalized algorithms is to obtain recursions on *bounded* variables, where the dynamic range requirements are fixed, and independent of the variance and characteristics of the observed process. Normalized recursions have already been discussed in the case of Potter's square-root normalized RLS transversal algorithm. Recall Sect. 5.2. Another example is the LeRoux-Gueguen algorithm, which operates on normalized recursions initialized from the (power-normalized) auto-correlation coefficients.

We will next develop the normalized recursions of Lee's RLS ladder algorithm. Interestingly, the normalization is carried out in *two* steps. This kind of normalization was, in fact, another fundamental result in Lee's Ph.D. thesis [9.1]. First, we have a conventional *square-root normalization*, and second, an *angle normalization* is applied to the angle between subsequent subspaces.

Normalized RLS ladder algorithms have been the topic of intensive research. The numerical behavior of RLS ladder algorithms was investigated by Samson and Reddy [9.22]. Porat et al. [9.23] presented normalized RLS ladder algorithms with a sliding rectangular window and Porat and Kailath [9.24] extended the normalized RLS ladder algorithm to FIR system identification.

9.5.1 Power Normalization

First, the power normalized residual vectors $\mathbf{v}_m(t)$ and $\mathbf{w}_m(t)$ are defined:

$$\mathbf{v}_m(t) = \mathbf{e}_m(t)\left[\mathbf{e}_m^T(t)\mathbf{e}_m(t)\right]^{-1/2} = \mathbf{e}_m(t)\,E_m^{-1/2}(t) \quad , \tag{9.75a}$$

$$\mathbf{w}_m(t) = \mathbf{r}_m(t)\left[\mathbf{r}_m^T(t)\mathbf{r}_m(t)\right]^{-1/2} = \mathbf{r}_m(t)\,R_m^{-1/2}(t) \quad . \tag{9.75b}$$

Clearly, the energy of $\mathbf{v}_m(t)$ and $\mathbf{w}_m(t)$ is normalized so that

$$\mathbf{v}_m^T(t)\,\mathbf{v}_m(t) = 1 \quad , \tag{9.76a}$$

$$\mathbf{w}_m^T(t)\,\mathbf{w}_m(t) = 1 \quad . \tag{9.76b}$$

The normalized counterparts to $E_m(t)$ and $R_m(t)$ are equal to *unity*, and therefore they do *not* appear in the further discussion. This leads to a considerable simplification of the normalized RLS ladder algorithm compared to its unnormalized counterparts. Only the *normalized covariance* $\rho_{m+1}(t)$ will be required. This quantity is defined as

$$\rho_{m+1}(t) = \mathbf{v}_m^T(t)\,\mathbf{w}_m(t-1) = C_m(t)\,E_m^{-1/2}(t)\,R_m^{-1/2}(t-1) \quad . \tag{9.77}$$

Note that in the unnormalized case the reflection coefficients were defined by

$$K_m^f(t) = -\,C_{m-1}(t)\big/R_{m-1}(t-1) \quad , \tag{9.78a}$$

$$K_m^b(t) = -\,C_{m-1}(t)\big/E_{m-1}(t) \quad . \tag{9.78b}$$

A comparison of (9.77) with (9.78a,b) reveals that the normalized forward/backward reflection coefficients have identical values, which are equal to the normalized covariance $\rho_{m+1}(t)$, hence

$$K_m^f(t),\ K_m^b(t) \ \xrightarrow{\ \textit{Normalization}\ }\ \rho_{m+1}(t) \quad . \tag{9.79}$$

Substitution of (9.75a,b, 76a,b and 77) into the unnormalized ladder vector order recursions

$$e_{m+1}(t) = e_m(t) + K_{m+1}^f(t) \, r_m(t-1) \quad , \tag{9.80a}$$

$$r_{m+1}(t) = r_m(t-1) + K_{m+1}^b(t) \, e_m(t) \tag{9.80b}$$

gives

$$v_m(t) E_m^{1/2}(t) E_{m-1}^{-1/2}(t) = v_{m-1}(t) - \rho_m(t) w_{m-1}(t-1) \quad , \tag{9.81a}$$

$$w_m(t) R_m^{1/2}(t) R_{m-1}^{-1/2}(t-1) = w_{m-1}(t-1) - \rho_m(t) v_{m-1}(t) \quad . \tag{9.81b}$$

Taking the squared value of both sides of (9.81a) yields the expression

$$v_m^T(t) \, v_m(t) \, E_m(t) \, E_{m-1}^{-1}(t) = v_{m-1}^T(t) \, v_{m-1}(t)$$
$$- 2 \rho_m(t) \, v_{m-1}^T(t) \, w_{m-1}(t-1) + \rho_m^2(t) \, w_{m-1}^T(t-1) \, w_{m-1}(t-1) \quad . \tag{9.82}$$

Taking into account (9.76a,b and 77), it is easy to see that (9.81a) reduces to

$$E_m(t) \, E_{m-1}^{-1}(t) = 1 - \rho_m^2(t) \quad . \tag{9.83a}$$

A similar procedure can be applied in the case of (9.81b) to give

$$R_m(t) \, R_{m-1}^{-1}(t-1) = 1 - \rho_m^2(t) \quad . \tag{9.83b}$$

Interestingly, it turns out that

$$E_m(t) \, E_{m-1}^{-1}(t) = R_m(t) \, R_{m-1}^{-1}(t-1) \quad . \tag{9.84}$$

The desired ladder vector order recursions in the normalized case are obtained by substituting (9.83a,b) into (9.81a,b):

$$v_m(t) = \left[v_{m-1}(t) - \rho_m(t) \, w_{m-1}(t-1) \right] \left[1 - \rho_m^2(t) \right]^{-1/2} \quad , \tag{9.85a}$$

$$w_m(t) = \left[w_{m-1}(t-1) - \rho_m(t) \, v_{m-1}(t) \right] \left[1 - \rho_m^2(t) \right]^{-1/2} \quad . \tag{9.85b}$$

Although at this stage of the discussion the normalization process is not yet complete, we may visualize the power normalized vector order recursions (9.85a,b) as a ladder structure with *identical* reflection coefficients

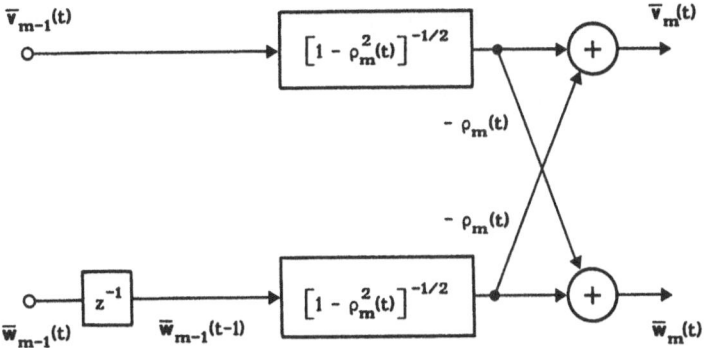

Fig. 9.5. Section m of a power normalized ladder form

in the forward and backward predictors. Note, however, that this is not a PARCOR ladder form since it requires, at each stage of the recursion, a *normalization* of the error vectors. See Fig. 9.5.

9.5.2 Angle Normalization

A second normalization of the vector order recursions (9.85a,b) might be of advantage in that the power normalized forward/backward residual vectors $\mathbf{v}_m(t)$ and $\mathbf{w}_m(t)$ are normalized with respect to the *cosine* of the angle $\Theta_m(t)$ between subsequent subspaces at stage m as follows:

$$\overline{\mathbf{v}}_m(t) = \mathbf{v}_m(t) \frac{1}{\cos \Theta_m(t)} \quad , \tag{9.86a}$$

$$\overline{\mathbf{w}}_m(t-1) = \mathbf{w}_m(t-1) \frac{1}{\cos \Theta_m(t)} \quad . \tag{9.86b}$$

The relationships between the unnormalized residual vectors $\mathbf{e}_m(t)$ and $\mathbf{r}_m(t-1)$ and their angle and power normalized counterparts are then given by

$$\mathbf{e}_m(t) = E_m^{1/2}(t) \cos \Theta_m(t) \, \overline{\mathbf{v}}_m(t) \quad , \tag{9.87a}$$

$$\mathbf{r}_m(t-1) = R_m^{1/2}(t-1) \cos \Theta_m(t) \, \overline{\mathbf{w}}_m(t-1) \quad . \tag{9.87b}$$

Now recall the time recursion of forward and backward residual energies (9.52 and 54). These recursions may be expressed in terms of residual vectors and the pinning vector $\pi(t)$:

$$\lambda E_m(t-1) = E_m(t) - \mathbf{e}_m^T(t)\pi(t) \, \pi^T(t)\mathbf{e}_m(t) \frac{1}{\cos^2 \Theta_m(t)} \quad , \tag{9.88a}$$

$$\lambda R_m(t-2) = R_m(t-1) - r_m^T(t-1)\,\pi(t)\,\pi^T(t)\,r_m(t-1)\,\frac{1}{\cos^2\Theta_m(t)}\quad. \qquad (9.88b)$$

Substitution of (9.87a,b) into (9.88a,b) gives

$$\lambda E_m(t-1) = E_m(t) - E_m(t)\,\overline{v}_m^T(t)\,\pi(t)\,\pi^T(t)\,\overline{v}_m(t)\quad, \qquad (9.89a)$$

$$\lambda R_m(t-2) = R_m(t-1) - R_m(t-1)\,\overline{w}_m^T(t-1)\,\pi(t)\,\pi^T(t)\,\overline{w}_m(t-1)\quad, \qquad (9.89b)$$

or, equivalently,

$$E_m^{1/2}(t-1)E_m^{-1/2}(t) = \lambda^{-1/2}\left[1 - \overline{v}_m^T(t)\,\pi(t)\,\pi^T(t)\,\overline{v}_m(t)\right]^{1/2}, \qquad (9.90a)$$

$$R_m^{1/2}(t-2)R_m^{-1/2}(t-1) = \lambda^{-1/2}\left[1 - \overline{w}_m^T(t-1)\,\pi(t)\,\pi^T(t)\,\overline{w}_m(t-1)\right]^{1/2}. \qquad (9.90b)$$

A similar procedure may be applied also to the unnormalized covariance $C_m(t)$,

$$C_m(t) = \lambda C_m(t-1) + e_m^T(t)\,\pi(t)\,\pi^T(t)\,r_m(t-1)\,\frac{1}{\cos^2\Theta_m(t)}\quad. \qquad (9.91)$$

Using (9.77), together with (9.87a,b), we obtain the normalized covariance as a function of the twofold normalized forward and backward residual vectors:

$$\rho_{m+1}(t) = \lambda E_m^{1/2}(t-1)E_m^{-1/2}(t)R_m^{1/2}(t-2)R_m^{-1/2}(t-1)\rho_{m+1}(t-1)$$

$$+ \overline{v}_m^T(t)\,\pi(t)\,\pi^T(t)\,\overline{w}_m(t-1)\quad. \qquad (9.92)$$

Clearly, according to (9.90a,b), the quotients of residual energies in (9.92) can be expressed in terms of normalized residual vectors, giving

$$\rho_{m+1}(t) = \left[1 - \overline{v}_m^T(t)\,\pi(t)\,\pi^T(t)\,\overline{v}_m(t)\right]^{1/2}$$

$$\times \left[1 - \overline{w}_m^T(t-1)\,\pi(t)\,\pi^T(t)\,\overline{w}_m(t-1)\right]^{1/2}\rho_{m+1}(t-1) \qquad (9.93)$$

$$+ \overline{v}_m^T(t)\,\pi(t)\,\pi^T(t)\,\overline{w}_m(t-1)$$

or, more compactly,

$$\rho_{m+1}(t) = \left[1 - \overline{v}_m^2(t)\right]^{1/2}\left[1 - \overline{w}_m^2(t-1)\right]^{1/2}\rho_{m+1}(t-1)$$

$$+ \overline{v}_m(t)\,\overline{w}_m(t-1)\quad. \qquad (9.94)$$

Equation (9.94) is the final time recursion of the normalized covariance $\rho_{m+1}(t)$. We see that this recursion contains *hyperbolic rotations* of the normalized residual signals. This is a typical situation where CORDIC computations can be of great interest. Recall the early discussion of the CORDIC computation technique in Sect. 4.4.

Another important point is that the exponential weighting factor λ has completely *disappeared* in the recursion (9.94). Additionally, we recognize that the gain variable $\cos^2\Theta_m(t)$ required in the unnormalized recursion has now been *omitted* in the time recursion of the normalized covariance. This was in fact the main reason for the introduction of the angle normalization. We end up with an algorithm that is based on just *three recursion variables:* $\rho_{m+1}(t)$, $\overline{v}_m(t)$ and $\overline{w}_m(t)$. The exponential weighting factor appears only in the first stage of the algorithm. In subsequent stages, the quantities $\overline{v}_m(t)$ and $\overline{w}_m(t)$ also carry the weighting information.

Next we determine the recursions of the twofold normalized forward and backward residual signals $\overline{v}_m(t)$ and $\overline{w}_m(t)$. A simple substitution of (9.86a,b) into (9.85a,b) yields the expressions

$$\overline{v}_m(t)\,\frac{\cos\Theta_m(t)}{\cos\Theta_{m-1}(t)} = \left[\overline{v}_{m-1}(t) - \rho_m(t)\,\overline{w}_{m-1}(t-1)\right]$$
$$\times\left[1 - \rho_m^2(t)\right]^{-1/2}, \tag{9.95a}$$

$$\overline{w}_m(t)\,\frac{\cos\Theta_m(t+1)}{\cos\Theta_{m-1}(t)} = \left[\overline{w}_{m-1}(t-1) - \rho_m(t)\,\overline{v}_{m-1}(t)\right]$$
$$\times\left[1 - \rho_m^2(t)\right]^{-1/2}. \tag{9.95b}$$

Note that

$$\cos^2\Theta_{m+1}(t) = \cos^2\Theta_m(t)$$
$$- \pi^T(t)r_m(t-1)r_m^T(t-1)\,\pi(t)\,R_m^{-1}(t-1) \tag{9.96}$$

or, equivalently,

$$\frac{\cos\Theta_m(t)}{\cos\Theta_{m-1}(t)} = \left[1 - \frac{\pi^T(t)r_{m-1}(t-1)r_{m-1}^T(t-1)\,\pi(t)}{\cos^2\Theta_{m-1}(t)\,R_{m-1}(t-1)}\right]^{1/2} \tag{9.97}$$

or, in terms of twofold normalized residual vectors,

$$\frac{\cos\Theta_m(t)}{\cos\Theta_{m-1}(t)} = \left[1 - \pi^T(t)\,\overline{w}_{m-1}(t-1)\,\overline{w}_{m-1}^T(t-1)\pi(t)\right]^{1/2}, \tag{9.98}$$

which results in the final relationship between the angle $\Theta_m(t)$ and the normalized backward residual signal:

$$\frac{\cos\Theta_m(t)}{\cos\Theta_{m-1}(t)} = \left[1 - \bar{w}_{m-1}^2(t-1)\right]^{1/2} . \tag{9.99}$$

Similarly, we know from (9.44) that

$$\cos^2\Theta_m(t+1) = \cos^2\Theta_{m-1}(t)$$

$$- \pi^T(t)\,\mathbf{e}_{m-1}(t)\,\mathbf{e}_{m-1}^T(t)\,\pi(t)\,E_{m-1}^{-1}(t) \tag{9.100}$$

and therefore

$$\frac{\cos\Theta_m(t+1)}{\cos\Theta_{m-1}(t)} = \left[1 - \frac{\pi^T(t)\,\mathbf{e}_{m-1}(t)\mathbf{e}_{m-1}^T(t)\,\pi(t)}{\cos^2\Theta_{m-1}(t)\,E_{m-1}(t)}\right]^{1/2} . \tag{9.101}$$

Using the twofold normalized residual vector $\bar{\mathbf{v}}_{m-1}(t)$ one obtains

$$\frac{\cos\Theta_m(t+1)}{\cos\Theta_{m-1}(t)} = \left[1 - \pi^T(t)\,\bar{\mathbf{v}}_{m-1}(t)\,\bar{\mathbf{v}}_{m-1}^T(t)\,\pi(t)\right]^{1/2} \tag{9.102}$$

or, in terms of the top component,

$$\frac{\cos\Theta_m(t+1)}{\cos\Theta_{m-1}(t)} = \left[1 - \bar{v}_{m-1}^2(t)\right]^{1/2} . \tag{9.103}$$

This is the desired relationship between the subspace angle $\Theta_m(t)$ and the normalized forward residual signal $\bar{v}_{m-1}(t)$.

Substitution of (9.99 and 103) into (9.95a,b) yields the final recursions

$$\bar{\mathbf{v}}_m(t) = \left[\bar{\mathbf{v}}_{m-1}(t) - \rho_m(t)\,\bar{\mathbf{w}}_{m-1}(t-1)\right]\left[1 - \rho_m^2(t)\right]^{-1/2}$$

$$\times \left[1 - \bar{w}_{m-1}^2(t-1)\right]^{-1/2} , \tag{9.104a}$$

$$\bar{\mathbf{w}}_m(t) = \left[\bar{\mathbf{w}}_{m-1}(t-1) - \rho_m(t)\,\bar{\mathbf{v}}_{m-1}(t)\right]\left[1 - \rho_m^2(t)\right]^{-1/2}$$

$$\times \left[1 - \bar{v}_{m-1}^2(t)\right]^{-1/2} , \tag{9.104b}$$

where

$$\rho_m(t) = \left[1 - \bar{v}_{m-1}^2(t)\right]^{1/2}\left[1 - \bar{w}_{m-1}^2(t-1)\right]^{1/2}\rho_m(t-1)$$

$$+ \bar{v}_{m-1}(t)\,\bar{w}_{m-1}(t-1) \tag{9.105}$$

is the recursive computation of the reflection coefficient (or normalized covariance) $\rho_m(t)$. With this step, we have reduced the recursive least-squares ladder problem to only *three* recursions, namely, the filter recursions (9.104a,b) and the covariance recursion (9.105). Table 9.6 summarizes this exponentially weighted, square-root normalized RLS ladder algorithm. Figure 9.6 shows one section of the underlying normalized feed-forward ladder form.

Note that the normalized RLS ladder algorithm presented in Table 9.6 involves a *nonlinear* filter operation on the data sequence. The non-linearity arises from the *data-dependent normalization* that appeared through the angle normalization process. When these ladder recursions are used for data analysis or modeling, this normalization can be very useful. It guarantees that all internal variables in the algorithm are of magnitude less than one.

Table 9.6. Lee's exponentially weighted normalized RLS ladder algorithm. λ is an exponential weighting factor. This algorithm involves division. Whenever the divisor is small, set $1/x = 1$

FOR t = 0, 1, 2, . . .

> *Input:* $x(t)$, $0 \le \lambda \le 1$
>
> $s(t) = \lambda s(t-1) + x^2(t)$
>
> *Initialize:*
>
> $\overline{v}_0(t) = \overline{w}_0(t) = s^{-1/2}(t)x(t)$
>
> FOR m = 1, 2, 3, . . .
>
> > $\rho_m(t) = \left[1 - \overline{v}_{m-1}^2(t) \right]^{1/2} \left[1 - \overline{w}_{m-1}^2(t-1) \right]^{1/2} \rho_m(t-1)$
> >
> > $\qquad + \overline{v}_{m-1}(t)\overline{w}_{m-1}(t-1)$
> >
> > $\overline{v}_m(t) = \left[\overline{v}_{m-1}(t) - \rho_m(t)\overline{w}_{m-1}(t-1) \right] \left[1 - \rho_m^2(t) \right]^{-1/2}$
> >
> > $\qquad \times \left[1 - \overline{w}_{m-1}^2(t-1) \right]^{-1/2}$
> >
> > $\overline{w}_m(t) = \left[\overline{w}_{m-1}(t-1) - \rho_m(t)\overline{v}_{m-1}(t) \right] \left[1 - \rho_m^2(t) \right]^{-1/2}$
> >
> > $\qquad \times \left[1 - \overline{v}_{m-1}^2(t) \right]^{-1/2}$

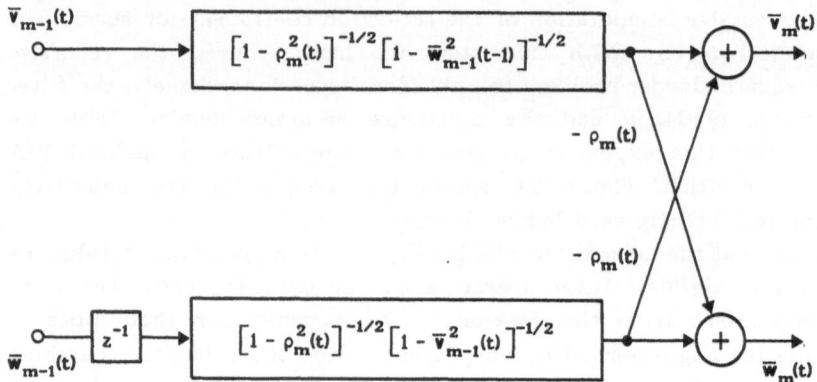

Fig. 9.6. Section m of a power and angle normalized feed-forward ladder form

9.6 Chapter Summary

In this chapter, we have scanned the principles of fast RLS ladder algorithms. This class of algorithms is based on the exact time-update theorem of projection operators introduced by Lee [9.1] in 1980. The theory and application of this class of ladder algorithms has been discussed extensively in the literature. See also the excellent review paper of Friedlander [9.21]. Several different forms of the fast RLS ladder algorithms are possible. We have discussed a pure time recursive version and an algorithm with direct updating of the reflection coefficients without explicit computation of the error covariance. Moreover, the fast RLS ladder algorithm has been extended to ARMA modeling [9.25].

Normalization plays a central role in RLS ladder algorithms. By employing both a variance (square-root) and a subsequent angle normalization, we obtained an algorithm that is based on only three recursion variables, which are all less than one in magnitude. Interestingly, in this normalized algorithm, the forward and backward reflection coefficients attain *identical* values. Moreover, this common reflection coefficient is identical with the normalized error covariance. A data-dependent gain appeared in the direct path of the forward and backward predictors, causing the same adaptive filtering effect on nonstationary data as separate forward/backward reflection coefficients in the unnormalized case. The normalized ladder recursions can actually be expressed as a sequence of circular and hyperbolic rotations. Efficient computation of these rotations can be achieved by the use of COordinate Rotation DIgital Computer (CORDIC) and other bit-recursive algorithms; see, e.g., [9.26, 27] and the discussion of the CORDIC technique in Chap. 4.

Gradient ladder algorithms have been discussed as special cases of the more general exact least-squares case in just the same way as we introduced the LMS algorithm as a special case of the RLS algorithm when the gain vector is set equal to the attenuated state vector. Note, however, that gradient ladder algorithms require much more computation than the LMS algorithm, and hence they are less attractive. Their gain in speed over the exact RLS ladder algorithms is negligible.

Although very attractive due to their low computational costs, the fast RLS ladder algorithms suffer from some inherent problems, such as:

(1) The time-recursion allows round-off errors to accumulate in the recursion variables, which causes *biased estimates* when the algorithms are operated in numerically uncertain environments.

(2) We have observed a dramatic *loss* in convergence speed when the algorithms are implemented with fixed-point arithmetic and a short word length. In this case, the exact least-squares algorithms performed not significantly better than their gradient-based counterparts.

(3) Another drawback of fast RLS ladder algorithms is the limitation of the data observation to rectangular and exponential windows or simple modifications thereof. The incorporation of a rectangular window even *doubles* the number of recursions and the total operations count. The time recursion in the algorithm does not allow the incorporation of higher-order windows, as seen in the case of the covariance ladder algorithms of the PORLA type (recall Chap. 8).

10. Special Signal Models and Extensions

Until now, we have only considered the problem of predicting a process from its own subspace of past observations. The algorithms obtained for this simple case can, however, be extended to more involved problems. Assume that one needs to predict a process from the subspace of a *related (correlated)* process. This case is commonly referred to as the *"joint-process"* case of linear prediction. A second case of interest is *system identification*, where we assume not the simple AR process model, but possibly an MA (all-zero) process model, or even a more general ARMA (pole-zero) process model. This leads directly to the most general one-dimensional problem, namely, the identification of a *multichannel (vector-autoregressive) process.* In fact, it turns out that we can handle the MA (FIR) system identification problem with the joint-process approach, whereas the ARMA system identification problem can be embedded in a two-channel vector autoregressive process model.

In some cases, one is interested in even higher-dimensional process models for, e.g., image processing applications. Unfortunately, the nice shift-invariance properties of the observation matrix does not hold in these extended cases, and hence there are no simple extensions of the discussed algorithms to higher-dimensional cases. This fact, and the tremendously increasing computational complexity has limited the interest in multidimensional linear prediction models. The true multidimensional case is sometimes approximated by *separable models* where the objective is to approximate a true multidimensional model by two or more one-dimensional models.

10.1 Joint Process Estimation

In many applications, such as noise cancelling, one is frequently faced with the problem of modeling a process $y(t)$ given a related process $x(t)$. The problem can be solved by the joint process estimator [10.1] that simply performs the orthogonal projection of $y(t)$ onto the subspace spanned by $\mathbf{X}_m(t)$. Figure 10.1 shows the orthogonal projection of a process $y(t)$ onto the subspace of past observations $\mathbf{X}_m(t)$ of a related process $x(t)$.

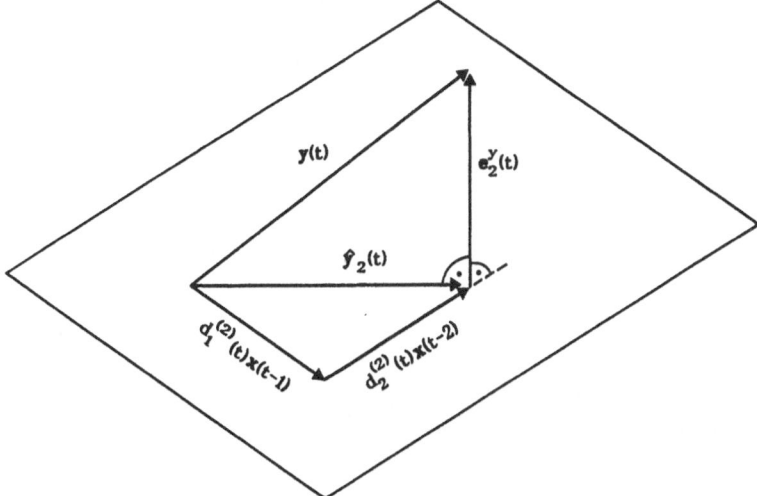

Fig. 10.1. Orthogonal projection of a process $y(t)$ onto the subspace spanned by the observations $x(t-1)$ and $x(t-2)$

The component of $y(t)$ that lies in the subspace $X_m(t)$ is obtained via the projection

$$\hat{y}_m(t) = P_m^x(t)y(t) \quad , \tag{10.1}$$

while the prediction error is given by

$$e_m^y(t) = \left[I - P_m^x(t) \right] y(t) = P_m^{x\perp}(t) y(t) \quad , \tag{10.2}$$

where

$$P_m^x(t) = X_m(t)\left[X_m^T(t)X_m(t) \right]^{-1} X_m^T(t) \quad . \tag{10.3}$$

Clearly,

$$y(t) = e_m^y(t) + \hat{y}_m(t) \tag{10.4}$$

and the parameter set $d^{(m)}(t)$ of the transversal joint process predictor is determined by the *joint process Normal Equations*

$$d^{(m)}(t) = \left[X_m^T(t)X_m(t) \right]^{-1} X_m^T(t)y(t) \quad . \tag{10.5}$$

All the algorithms that have been developed for the computation of the AR parameters $a^{(m)}(t)$ can be used with minor modifications to solve the joint process estimation problem (10.5).

10.1.1 Ladder Formulation of the Joint Process Problem

From the previous considerations of the joint process estimation problem, we may directly deduce its ladder formulation. According to Fig. 10.1 and relation (10.1), it is necessary to project $y(t)$ orthogonally onto the subspace $X_m(t)$. This can be accomplished by a recursive orthogonalization procedure as follows:

$$e_0^y(t) = y(t) \quad , \tag{10.6a}$$

$$e_1^y(t) = y(t)\langle x(t-1)\rangle \quad , \tag{10.6b}$$

$$e_2^y(t) = y(t)\langle x(t-1), x(t-2)\rangle \quad , \tag{10.6c}$$

$$\vdots \qquad \vdots \qquad \vdots$$

$$e_{m-1}^y(t) = y(t)\langle X_{m-1}(t)\rangle \quad , \tag{10.6d}$$

$$e_m^y(t) = y(t)\langle X_{m-1}(t), x(t-m)\rangle \quad . \tag{10.6e}$$

Note the similarity between (10.6a–e) and (6.12a–e). This leads to the joint process vector order recursions

$$e_0^y(t) = y(t) \quad , \tag{10.7a}$$

$$e_m^y(t) = e_{m-1}^y(t) + H_m(t)r_{m-1}(t-1) \quad . \tag{10.7b}$$

The scalar $H_m(t)$ is the *joint ladder reflection coefficient* determined by

$$H_m(t) = -\frac{e_{m-1}^{y\,T}(t)r_{m-1}(t-1)}{r_{m-1}^T(t-1)\,r_{m-1}(t-1)} \quad . \tag{10.7c}$$

The joint recursion (10.7a–c), together with the AR-part ladder recursions (6.15a,b), constitutes the *joint process ladder form* shown in Fig. 10.2.

10.1.2 The Joint Process Model

Next, we are interested in the relationships between the processes $x(t)$ and $y(t)$ allowing a perfect modeling by the joint process estimator. In other words, we are seeking the underlying *model assumptions* in the discussed joint process estimator. From the *transversal* realization of the joint process prediction error filter

$$e_m^y(t) = y(t) - X_m(t)d^{(m)}(t) \quad , \tag{10.8}$$

it is clear that this must be a model of the type

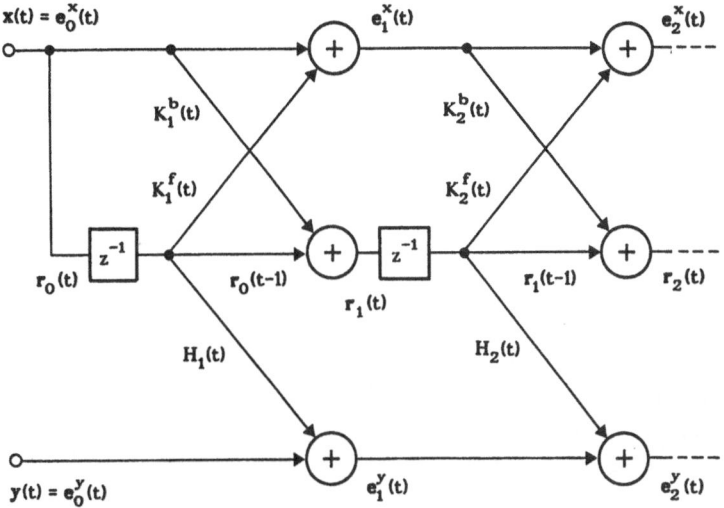

Fig. 10.2. Joint process ladder form

$$y(t) = X_m(t) d^{*(m)}(t) + \varepsilon_m(t) \ , \tag{10.9}$$

where $\varepsilon_m(t)$ is a process that is already *orthogonal* with respect to the subspace $X_m(t)$, hence

$$\varepsilon_m^T(t) x(t-j) = 0 \ ; \qquad 1 \le j \le m \ . \tag{10.10}$$

This model (10.9) generates a process that is [except for its orthogonal component $\varepsilon_m(t)$] *perfectly predictable* by the joint process prediction error filter (10.8). Figure 10.3 illustrates this joint process model.

The model shown in Fig. 10.3 is useful in those cases where two processes are *"coupled"* by systems that are best modeled by a FIR filter. The model can be further extended in that we can have *more than one related channel* $y_1(t)$, $y_2(t)$, . . . , $y_q(t)$.

10.1.3 FIR System Identification

A quick inspection of the joint process model of Fig. 10.3 reveals that we can use the joint process estimator for *FIR system identification* [10.2-4]. In the system identification case, both the input and the output of the unknown system are accessible. The structure of the corresponding FIR system identification model is shown in Fig. 10.4. The unknown system is modeled by a FIR filter with a sufficiently large order m.

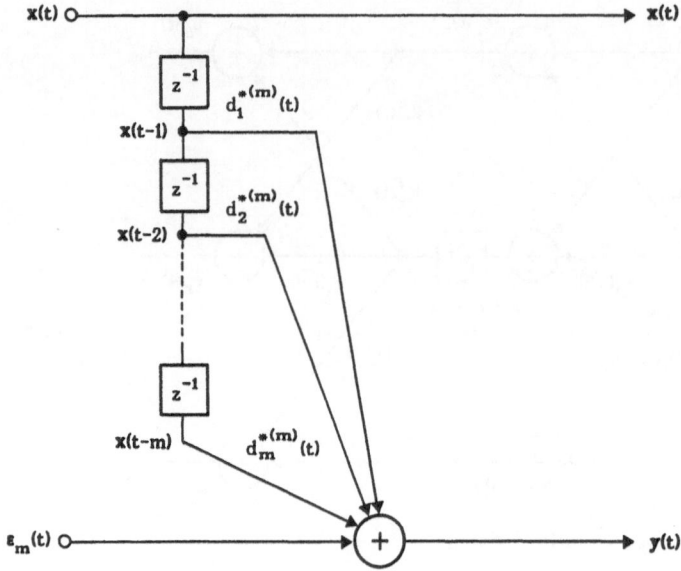

Fig. 10.3. Joint process model of order m

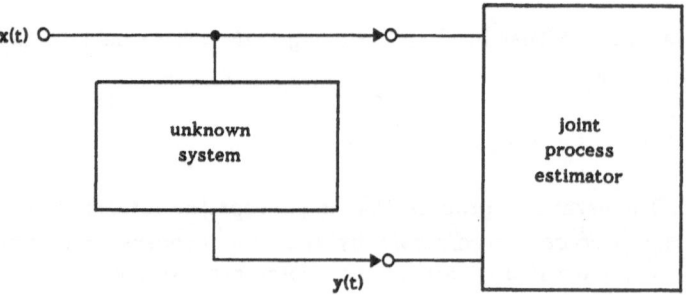

Fig. 10.4. FIR system identification model

10.1.4 Noise Cancelling

Another important application of joint process estimation is *noise cancelling* [10.5]. In noise cancelling, a process of interest that is corrupted by a disturbing process is to be *reconstructed*, i.e., the disturbing process components are to be removed. In the case that the disturbing process is observable, the reconstructed process is simply the *orthogonal component* of the corrupted process with respect to the sub-space of past observations of the disturbing process. This useful geo-metrical interpretation emerges directly from our previous discussions of joint process estimation. Clearly, the projection the corrupted process onto the subspace of past observations of the disturbing process can be accomplished by the joint process prediction error filter, as discussed.

In this application, the disturbing process is connected to the x-input of the joint process estimator and the corrupted process is connected to the y-input (joint-input) of the joint process prediction error filter. The reconstructed process is obtained at the y-output (joint-output) of the joint process prediction error filter.

The underlying model assumption in this case of noise-cancelling becomes apparent if we recall the structure of the joint process model (Fig. 10.3). According to this model, we assume that the process of interest is $\varepsilon_m(t)$. We further assume that the corrupted process $y(t)$ is generated by *linear superposition* of *weighted, time-shifted* versions of the disturbing process $x(t)$ onto $\varepsilon_m(t)$, as illustrated in Fig. 10.3.

We note that the observation of the disturbing process presents a problem in many applications. The interested reader is referred to the special applications-oriented literature cited in Chap. 11 for a more detailed discussion of the different practical aspects of noise cancelling.

10.2 ARMA System Identification

Another case of interest arises when the unknown system is to be modeled by a more general ARMA (pole-zero) filter. Fig. 10.5 is a block diagram of the ARMA system identification model. Both the input and the output of the unknown system are accessible. No restricting pre-assumptions are made about the input/output processes $x(t)$ and $u(t)$. It is desired to model the unknown system as an ARMA filter [10.6-10] with a sufficiently large order m .

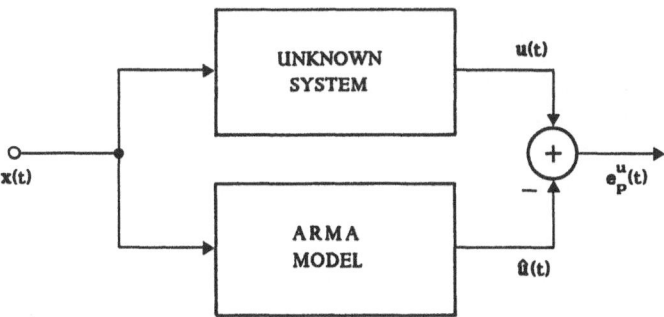

Fig. 10.5. ARMA system identification model

10.2.1 The ARMA Normal Equations

The approach used here is to determine the ARMA parameters $\mathbf{b}^{(m)}(t)$ and $\mathbf{a}^{(m)}(t)$,

$$\mathbf{b}^{(m)}(t) = \left[b_1^{(m)}(t), b_2^{(m)}(t), \ldots, b_m^{(m)}(t) \right]^T , \tag{10.11a}$$

$$\mathbf{a}^{(m)}(t) = \left[a_1^{(m)}(t), a_2^{(m)}(t), \ldots, a_m^{(m)}(t) \right]^T , \tag{10.11b}$$

in the least-squares sense, hence

$$E^{uu}(t) = \mathbf{e}_m^{u^T}(t)\mathbf{e}_m^u(t) \overset{!}{=} \min . \tag{10.12}$$

We assume that the covariance matrices $\Phi^{xx}(t)$ and $\Phi^{uu}(t)$, as well as the cross covariance matrix $\Phi^{xu}(t)$,

$$\Phi_{i,j}^{xx}(t) = \mathbf{x}^T(t-i)\mathbf{x}(t-j) , \tag{10.13a}$$

$$\Phi_{i,j}^{uu}(t) = \mathbf{u}^T(t-i)\mathbf{u}(t-j) , \tag{10.13b}$$

$$\Phi_{i,j}^{xu}(t) = \mathbf{x}^T(t-i)\mathbf{u}(t-j) , \tag{10.13c}$$

are known and positive definite. The vectors $\mathbf{x}(t)$ and $\mathbf{u}(t)$ are the observable input/output processes of the unknown system. Suppose now that the unknown system can be modeled by a time-varying ARMA filter as shown in Fig. 10.6, where $\mathbf{X}_m(t)$ and $\mathbf{U}_m(t)$ are the matrices of the joint input/output subspace of past observations, and $\mathbf{a}^{(m)}(t)$ and $\mathbf{b}^{(m)}(t)$ are the ARMA coefficient vectors:

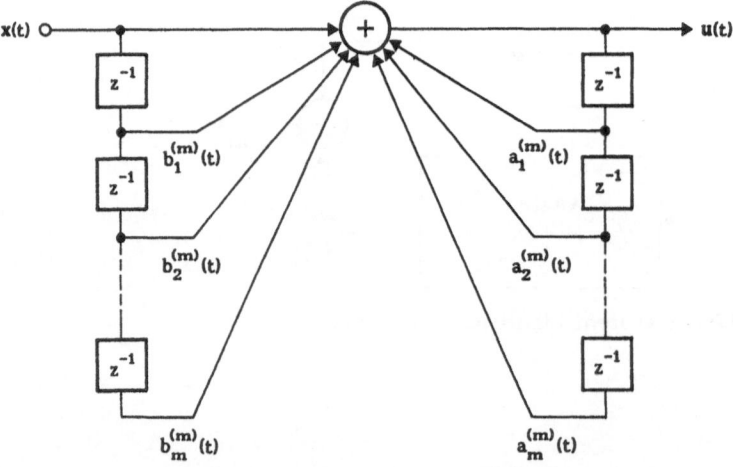

Fig. 10.6. ARMA representation of the unknown system in Fig. 10.5

$$\mathbf{X}_m(t) = \left[\mathbf{x}(t-1),\ \mathbf{x}(t-2),\ \dots\ ,\ \mathbf{x}(t-m)\right]\quad, \tag{10.14a}$$

$$\mathbf{U}_m(t) = \left[\mathbf{u}(t-1),\ \mathbf{u}(t-2),\ \dots\ ,\ \mathbf{u}(t-m)\right]\quad. \tag{10.14b}$$

The predicted process vector $\hat{\mathbf{u}}(t)$ is then given by

$$\hat{\mathbf{u}}(t) = \mathbf{x}(t) + \mathbf{X}_m(t)\,\mathbf{b}^{(m)}(t) + \mathbf{U}_m(t)\,\mathbf{a}^{(m)}(t) \tag{10.15}$$

and the prediction error $\mathbf{e}_m^u(t)$ is expressed as

$$\mathbf{e}_m^u(t) = \mathbf{u}(t) - \mathbf{x}(t) - \mathbf{X}_m(t)\,\mathbf{b}^{(m)}(t) - \mathbf{U}_m(t)\,\mathbf{a}^{(m)}(t)\quad. \tag{10.16}$$

It is desired to minimize the prediction error energy $E^{uu}(t)$. This can be accomplished by setting the partial derivatives of $E^{uu}(t)$ with respect to the parameter vectors $\mathbf{a}^{(m)}(t)$ and $\mathbf{b}^{(m)}(t)$ to zero

$$\frac{\partial E^{uu}}{\partial \mathbf{a}^{(m)T}} = -\,\mathbf{U}_m^T\mathbf{u} + \mathbf{U}_m^T\mathbf{x} + \mathbf{U}_m^T\mathbf{X}_m\mathbf{b}^{(m)} + \mathbf{U}_m^T\mathbf{U}_m\mathbf{a}^{(m)}$$

$$= \left[0 \dots 0\right]^T\ , \tag{10.17a}$$

$$\frac{\partial E^{uu}}{\partial \mathbf{b}^{(m)T}} = -\,\mathbf{X}_m^T\mathbf{u} + \mathbf{X}_m^T\mathbf{x} + \mathbf{X}_m^T\mathbf{X}_m\mathbf{b}^{(m)} + \mathbf{X}_m^T\mathbf{U}_m\mathbf{a}^{(m)}$$

$$= \left[0 \dots 0\right]^T\ . \tag{10.17b}$$

The relations (10.17a,b) may be rewritten in a more convenient matrix/vector notation, where the time index has been omitted for convenience:

$$\begin{bmatrix} \mathbf{U}_m^T\mathbf{U}_m & \mathbf{U}_m^T\mathbf{X}_m \\[4pt] \mathbf{X}_m^T\mathbf{U}_m & \mathbf{X}_m^T\mathbf{X}_m \end{bmatrix} \begin{bmatrix} \mathbf{a}^{(m)} \\[4pt] \mathbf{b}^{(m)} \end{bmatrix} = \begin{bmatrix} \mathbf{U}_m^T(\mathbf{u} - \mathbf{x}) \\[4pt] \mathbf{X}_m^T(\mathbf{u} - \mathbf{x}) \end{bmatrix}\quad. \tag{10.18}$$

The system of equations (10.18) has been termed the *Normal Equations of ARMA system identification*. Recalling the definitions of the input/output covariance matrix (10.13a,b) and the input/output cross covariance matrix (10.13c), we may verify that the ARMA Normal Equations can be constructed from the elements of the following partitioning schemes:

$$\boldsymbol{\Phi}^{xx}(t) = \begin{bmatrix} \mathbf{x}^T\mathbf{x} & \mathbf{x}^T\mathbf{X}_m \\[4pt] \mathbf{X}_m^T\mathbf{x} & \mathbf{X}_m^T\mathbf{X}_m \end{bmatrix}\quad, \tag{10.19a}$$

$$\boldsymbol{\Phi}^{uu}(t) = \begin{bmatrix} \mathbf{u}^T\mathbf{u} & \mathbf{u}^T\mathbf{U}_m \\[4pt] \mathbf{U}_m^T\mathbf{u} & \mathbf{U}_m^T\mathbf{U}_m \end{bmatrix}\quad, \tag{10.19b}$$

$$\Phi^{xu}(t) = \begin{bmatrix} \mathbf{x}^T \mathbf{u} & \mathbf{x}^T \mathbf{U}_m \\ \mathbf{X}_m^T \mathbf{u} & \mathbf{X}_m^T \mathbf{U}_m \end{bmatrix} . \qquad (10.19c)$$

10.2.2 ARMA Embedding

Instead of solving the ARMA Normal Equations (10.18) directly, we introduce a useful *embedding* approach, which allows the ARMA system identification problem to be embedded in a more general two-channel AR estimation framework. For this purpose, consider the two-channel transversal prediction error filter of the form

$$\begin{bmatrix} \mathbf{e}_m^{x^T}(t) \\ \mathbf{e}_m^{y^T}(t) \end{bmatrix} = \begin{bmatrix} \mathbf{x}^T(t) \\ \mathbf{y}^T(t) \end{bmatrix} + \sum_{j=1}^{m} \begin{bmatrix} a_{11}^{(j)}(t) & a_{21}^{(j)}(t) \\ a_{12}^{(j)}(t) & a_{22}^{(j)}(t) \end{bmatrix} \begin{bmatrix} \mathbf{x}^T(t-j) \\ \mathbf{y}^T(t-j) \end{bmatrix} , \qquad (10.20)$$

where $\mathbf{x}(t)$ and $\mathbf{y}(t)$ are the observed processes, $\mathbf{e}_m^x(t)$ and $\mathbf{e}_m^y(t)$ are the prediction error vectors and $a_{**}^{(j)}(t)$ are the time-varying *two-channel AR parameters* of stage j.

Rewriting the lower part of (10.20) gives

$$\mathbf{e}_m^y(t) = \mathbf{y}(t) + \sum_{j=1}^{m} a_{12}^{(j)}(t)\, \mathbf{x}(t-j) + \sum_{j=1}^{m} a_{22}^{(j)}(t)\, \mathbf{y}(t-j) . \qquad (10.21)$$

Now let the input process $\mathbf{y}(t)$ of the two-channel prediction error filter (10.20) be the *"reduced"* output signal of the unknown system as follows:

$$\mathbf{y}(t-j) = \mathbf{u}(t-j) - \mathbf{x}(t-j) ; \qquad 0 \le j \le m , \qquad (10.22)$$

and let the input signal $\mathbf{x}(t)$ of the two-channel prediction error filter (10.20) be connected with the input of the unknown system. Then, by substituting (10.22) into (10.21), we obtain

$$\mathbf{e}_m^y(t) = \mathbf{u}(t) - \mathbf{x}(t) - \sum_{j=1}^{m} \left[a_{22}^{(j)}(t) - a_{12}^{(j)}(t) \right] \mathbf{x}(t-j)$$

$$- \sum_{j=1}^{m} \left[-a_{22}^{(j)}(t) \right] \mathbf{u}(t-j) . \qquad (10.23)$$

Comparing (10.23) and the ARMA prediction error filter (10.16), we conclude that ARMA system identification can be embedded into a two-channel AR estimation framework, as defined by (10.20). Note that in this case, the reduced output signal $y(t)$ of the unknown system, namely, the output signal $u(t)$ minus the excitation $x(t)$, must be fed to the y-input of the two-channel all-zero prediction error filter, whereas the excitation signal $x(t)$ must be connected to the x-input. Then, by comparing (10.23) and (10.16), it is easy to see that the ARMA parameters can be calculated from the two-channel AR parameters via the relations

$$a_j^{(m)}(t) = -a_{22}^{(j)}(t) \quad , \qquad\qquad\qquad\qquad (10.24a)$$

$$\left.\begin{array}{l} \\ \\ \end{array}\right\} \quad 1 \le j \le m \quad .$$

$$b_j^{(m)}(t) = a_{22}^{(j)}(t) - a_{12}^{(j)}(t) \quad , \qquad\qquad\qquad (10.24b)$$

10.2.3 Ladder Formulation of the ARMA System Identification Problem

Until now, we have only considered the "fixed-order" formulation of the ARMA system identification problem based on the Normal Equations. In the following, we will develop the "growing-order" or ladder formulation of the ARMA system identification problem. From a geometrical viewpoint, the two-channel prediction error filter (10.20) provides the projection of the reduced output signal $y(t)$ onto the *joint subspace* spanned by the observations $X_m(t)$ and $Y_m(t)$. The optimal parameter sets $a^{(m)}(t)$ and $b^{(m)}(t)$ ensure that this projection is *orthogonal*, i.e., $y(t)$ is *decomposed* into one (in-space) component $\hat{y}_m(t)$, which lies in the joint subspace, and into another component $e_m^y(t)$ which is orthogonal with respect to the joint subspace of past observations. The residual vector $e_m^y(t)$ thus obtained is of least Euclidean length and therefore satisfies the least-squares cost function (10.12).

From these considerations, it becomes apparent that an alternative way of constructing $e_m^y(t)$ exists as follows. First, construct a set of orthogonal basis vectors of the joint subspace, and second, project the vector $y(t)$ successively onto the orthogonal basis vectors. As known from previous considerations in this book, such a procedure is termed the ladder formulation of a least-squares problem.

Similarly to the scalar (single-channel) case treated in Chap. 6, we may state the following *two-channel Gram-Schmidt orthogonalization procedure*:

$$e_0^y(t) = y(t) \quad , \qquad\qquad\qquad\qquad\qquad\qquad (10.25a)$$

$$e_1^y(t) = y(t)\langle x(t-1); y(t-1)\rangle \quad , \qquad\qquad\qquad (10.25b)$$

$$e_2^y(t) = y(t)\langle x(t-1), x(t-2); y(t-1), y(t-2)\rangle \quad , \qquad (10.25c)$$

$$\vdots$$

$$e_{m-1}^y(t) = y(t)\langle X_{m-1}(t); Y_{m-1}(t)\rangle \quad , \qquad\qquad (10.25d)$$

$$e_m^y(t) = y(t)\langle X_m(t); Y_m(t)\rangle \quad , \qquad\qquad\qquad (10.25e)$$

$$e_m^y(t) = y(t)\langle X_{m-1}(t); Y_{m-1}(t)\rangle$$

$$+ K_m^{fyx}(t)x(t-m)\langle X_{m-1}(t); Y_{m-1}(t)\rangle$$

$$+ K_m^{fyy}(t)y(t-m)\langle X_{m-1}(t); Y_{m-1}(t)\rangle \quad . \tag{10.26}$$

Similarly, we have

$$e_0^x(t) = x(t) \quad , \tag{10.27a}$$

$$e_1^x(t) = x(t)\langle x(t-1); y(t-1)\rangle \quad , \tag{10.27b}$$

$$e_2^x(t) = x(t)\langle x(t-1), x(t-2); y(t-1), y(t-2)\rangle \quad , \tag{10.27c}$$

$$\vdots$$

$$e_{m-1}^x(t) = x(t)\langle X_{m-1}(t); Y_{m-1}(t)\rangle \quad , \tag{10.27d}$$

$$e_m^x(t) = x(t)\langle X_m(t); Y_m(t)\rangle \quad , \tag{10.27e}$$

$$e_m^x(t) = x(t)\langle X_{m-1}(t); Y_{m-1}(t)\rangle \quad ,$$

$$+ K_m^{fxx}(t)x(t-m)\langle X_{m-1}(t); Y_{m-1}(t)\rangle$$

$$+ K_m^{fxy}(t)y(t-m)\langle X_{m-1}(t); Y_{m-1}(t)\rangle \quad . \tag{10.28}$$

Likewise, we may construct the orthogonal basis vectors $r_m^x(t)$ and $r_m^y(t)$ spanning the joint subspace of past observations:

$$r_0^y(t) = y(t) \quad , \tag{10.29a}$$

$$r_1^y(t) = y(t-1)\langle x(t); y(t)\rangle \quad , \tag{10.29b}$$

$$r_2^y(t) = y(t-2)\langle x(t), x(t-1); y(t), y(t-1)\rangle \quad , \tag{10.29c}$$

$$\vdots$$

$$r_{m-1}^y(t) = y(t-m+1)\langle X_{m-1}(t+1); Y_{m-1}(t+1)\rangle \quad , \tag{10.29d}$$

$$r_m^y(t) = y(t-m)\langle X_m(t+1); Y_m(t+1)\rangle \quad , \tag{10.29e}$$

$$r_m^y(t) = y(t-m)\langle X_{m-1}(t); Y_{m-1}(t)\rangle$$

$$+ K_m^{byx}(t)x(t)\langle X_{m-1}(t); Y_{m-1}(t)\rangle$$

$$+ K_m^{byy}(t)y(t)\langle X_{m-1}(t); Y_{m-1}(t)\rangle \tag{10.30}$$

and, consequently,

$$r_0^x(t) = x(t) \quad , \tag{10.31a}$$

$$r_1^x(t) = x(t-1)\langle x(t); y(t)\rangle \quad , \tag{10.31b}$$

$$r_2^x(t) = x(t-2)\langle x(t), x(t-1); y(t), y(t-1)\rangle \quad , \tag{10.31c}$$

$$\vdots$$

$$r_{m-1}^x(t) = x(t-m+1)\langle X_{m-1}(t+1); Y_{m-1}(t+1)\rangle \quad , \tag{10.31d}$$

$$r_m^x(t) = x(t-m)\langle X_m(t+1); Y_m(t+1)\rangle \quad , \tag{10.31e}$$

$$r_m^x(t) = x(t-m)\langle X_{m-1}(t); Y_{m-1}(t)\rangle$$

$$+ K_m^{bxx}(t)x(t)\langle X_{m-1}(t); Y_{m-1}(t)\rangle$$

$$+ K_m^{bxy}(t)y(t)\langle X_{m-1}(t); Y_{m-1}(t)\rangle \quad . \tag{10.32}$$

Exploiting these expressions, we may note that the following vector order recursions hold

$$e_m^x(t) = e_{m-1}^x(t) + K_m^{fxx}(t)r_{m-1}^x(t-1) + K_m^{fxy}(t)r_{m-1}^y(t-1) \quad , \tag{10.33a}$$

$$e_m^y(t) = e_{m-1}^y(t) + K_m^{fyx}(t)r_{m-1}^x(t-1) + K_m^{fyy}(t)r_{m-1}^y(t-1) \quad , \tag{10.33b}$$

$$r_m^x(t) = r_{m-1}^x(t-1) + K_m^{bxx}(t)e_{m-1}^x(t) + K_m^{bxy}(t)e_{m-1}^y(t) \quad , \tag{10.34a}$$

$$r_m^y(t) = r_{m-1}^y(t-1) + K_m^{byx}(t)e_{m-1}^x(t) + K_m^{byy}(t)e_{m-1}^y(t) \quad , \tag{10.34b}$$

where

$$e_0^x(t) = r_0^x(t) = x(t) \quad , \tag{10.35a}$$

$$e_0^y(t) = r_0^y(t) = y(t) \quad . \tag{10.35b}$$

The relations (10.33a,b and 34a,b), together with (10.35a,b), constitute the desired vector order recursions for recursive orthogonalization of the observation $y(t)$ with respect to the joint subspace of past observations. The orthogonal basis of the joint subspace is then spanned by the set of orthogonal basis vectors $\{r_j^x(t), r_j^y(t); 1 \leq j \leq m\}$. They may be termed the *backward residual vectors* of stage j at time step t. The recursion vectors $\{e_j^x(t), e_j^y(t); 1 \leq j \leq m\}$ are known as the *partially orthogonalized forward residual vectors*. The scalar quantities $\{K_j^{f**}(t), K_j^{b**}(t); 1 \leq j \leq m\}$ are

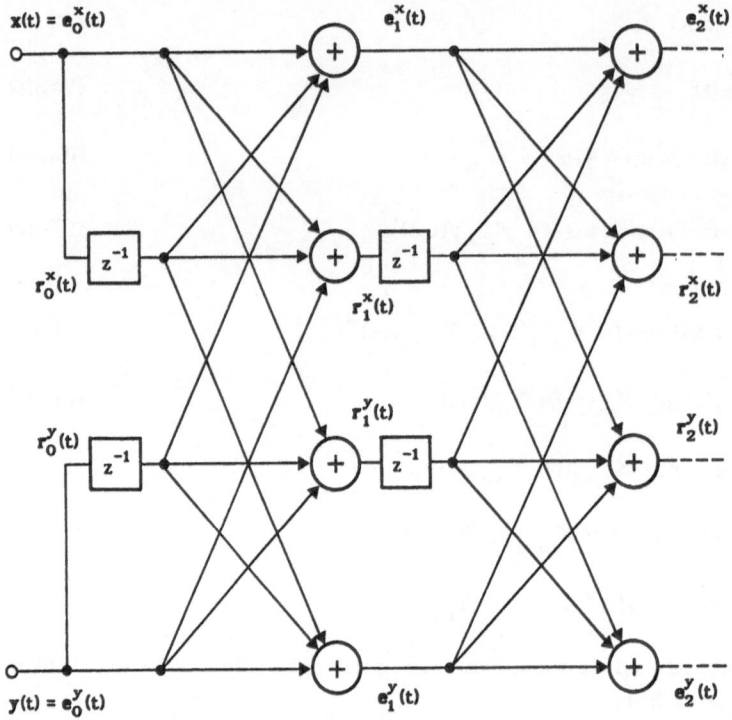

Fig. 10.7. Canonical ARMA ladder form

the (time–varying) ARMA ladder reflection coefficients. The vector order recursions (10.33a,b and 34a,b), together with the initialization scheme (10.35a,b) can be interpreted as a filter structure, the so-called *ARMA ladder form* displayed in Fig. 10.7.

The ARMA ladder recursions in their original formulation (10.33a,b and 34a,b) are fairly complex. They may be rewritten in a more compact matrix/vector notation as

$$\mathbf{e}_m(t) = \mathbf{e}_{m-1}(t) + \mathbf{r}_{m-1}(t-1)\mathbf{K}_m^f(t) \quad , \tag{10.36a}$$

$$\mathbf{r}_m(t) = \mathbf{r}_{m-1}(t-1) + \mathbf{e}_{m-1}(t)\,\mathbf{K}_m^b(t) \; . \tag{10.36b}$$

with the initial condition

$$\mathbf{e}_0(t) = \mathbf{r}_0(t) = \Big[\mathbf{x}(t),\, \mathbf{y}(t)\Big] \quad , \tag{10.37}$$

where

$$\mathbf{e}_m(t) = \Big[\mathbf{e}_m^x(t),\, \mathbf{e}_m^y(t)\Big] \quad , \tag{10.38a}$$

$$\mathbf{r}_m(t) = \Big[\mathbf{r}_m^x(t),\, \mathbf{r}_m^y(t)\Big] \tag{10.38b}$$

and

$$\mathbf{K}_m^f(t) = \begin{bmatrix} K_m^{fxx}(t) & K_m^{fyx}(t) \\ K_m^{fxy}(t) & K_m^{fyy}(t) \end{bmatrix} \quad , \tag{10.39a}$$

$$\mathbf{K}_m^b(t) = \begin{bmatrix} K_m^{bxx}(t) & K_m^{byx}(t) \\ K_m^{bxy}(t) & K_m^{byy}(t) \end{bmatrix} \quad . \tag{10.39b}$$

We are interested in determining the ARMA ladder reflection matrices $\mathbf{K}_m^f(t)$ and $\mathbf{K}_m^b(t)$. A minimization of the growing order least-squares cost functions

$$\mathbf{e}_m^T(t)\,\mathbf{e}_m(t) \stackrel{!}{=} \min \quad , \tag{10.40a}$$

$$\mathbf{r}_m^T(t)\,\mathbf{r}_m(t) \stackrel{!}{=} \min \quad , \tag{10.40b}$$

yields the desired rule for adjusting the ARMA ladder reflection matrices in the least-squares sense. Note that (10.40a,b) implies a minimization of each element of the respective energy matrix. After some algebra, one finds the following rule for adjusting the ladder reflection matrices in the least-squares sense:

$$\mathbf{K}_m^f(t) = -\mathbf{R}_{m-1}^{-1}(t-1)\,\mathbf{C}_{m-1}^T(t) \quad , \tag{10.41a}$$

$$\mathbf{K}_m^b(t) = -\mathbf{E}_{m-1}^{-1}(t)\,\mathbf{C}_{m-1}(t) \quad , \tag{10.41b}$$

where

$$\mathbf{E}_{m-1}(t) = \mathbf{e}_{m-1}^T(t)\,\mathbf{e}_{m-1}(t) \quad , \tag{10.42a}$$

$$\mathbf{R}_{m-1}(t) = \mathbf{r}_{m-1}^T(t)\,\mathbf{r}_{m-1}(t) \quad , \tag{10.42b}$$

$$\mathbf{C}_{m-1}(t) = \mathbf{e}_{m-1}^T(t)\,\mathbf{r}_{m-1}(t-1) \quad , \tag{10.42c}$$

are the residual energy matrices at stage m−1. Note at this point that a commutation of matrices in (10.41a,b) is generally *not* permitted. Furthermore, we may note that the covariance matrix $\mathbf{C}_{m-1}(t)$ appears *transposed* in (10.41a).

This completes the discussion of the two-channel ARMA ladder form. With these considerations, it is only a small step to extend the algorithms that were presented for the scalar (single-channel) case to the more general two-channel (ARMA) case.

10.2.4 The PORLA Method for ARMA System Identification

We present an example of how a scalar algorithm can be extended to the two-channel (ARMA) case. We consider the extension of the PORLA method, as introduced in Chap. 8, to the ARMA case [10.7]. In a first step, one has to *expand* the residual energy matrices (10.42a-c) to *block marix forms* as follows:

$$\mathbf{E}_{m,i,j}(t) = \mathbf{e}_m^T(t-i)\mathbf{e}_m(t-j) \quad , \tag{10.43a}$$

$$\mathbf{R}_{m,i,j}(t) = \mathbf{r}_m^T(t-1-i)\mathbf{r}_m(t-1-j) \quad , \tag{10.43b}$$

$$\mathbf{C}_{m,i,j}(t) = \mathbf{e}_m^T(t-i)\,\mathbf{r}_m(t-1-j) \quad . \tag{10.43c}$$

These block matrix forms of residual energies have been termed *"block generalized residual energies"* (BGREs). More precisely, we can rewrite the element $\mathbf{E}_{m,i,j}(t)$ in terms of the error processes $\mathbf{e}_m^x(t)$ and $\mathbf{e}_m^y(t)$,

$$\mathbf{E}_{m,i,j}(t) = \begin{bmatrix} \mathbf{e}_m^{x^T}(t-i)\mathbf{e}_m^x(t-j) & \mathbf{e}_m^{x^T}(t-i)\,\mathbf{e}_m^y(t-j) \\ \mathbf{e}_m^{y^T}(t-j)\mathbf{e}_m^x(t-j) & \mathbf{e}_m^{y^T}(t-j)\,\mathbf{e}_m^y(t-j) \end{bmatrix} \quad , \tag{10.44}$$

while the entire BGRE $\mathbf{E}_m(t)$ is a *block matrix* of the form

$$\mathbf{E}_m(t) = \begin{bmatrix} E_{m,0,0}(t) & E_{m,0,1}(t) \cdots\cdots E_{m,0,k}(t) \\ E_{m,1,0}(t) & E_{m,1,1}(t) \cdots\cdots E_{m,1,k}(t) \\ \vdots & \vdots \quad \ddots \quad \vdots \\ E_{m,k,0}(t) & E_{m,k,1}(t) \cdots\cdots E_{m,k,k}(t) \end{bmatrix} \quad , \tag{10.45a}$$

where

$$k = p - m - 1 \quad . \tag{10.45b}$$

The BGREs $\mathbf{R}_m(t)$ with elements $\{\mathbf{R}_{m,i,j}(t)\,,\ 0 \le i,j \le p - m - 1\}$ and $\mathbf{C}_m(t)$ with elements $\{\mathbf{C}_{m,i,j}(t)\,,\ 0 \le i,j \le p - m - 1\}$ have analogous structures.

In the next step, one substitutes the order recursion of the time-shifted residual matrices, as obtained from the ARMA ladder recursions (10.36a,b), into the definition of the BGREs (10.43a-c) to obtain *block matrix order recursions* of the type

$$\mathbf{E}_{m,i,j}(t) = \mathbf{E}_{m-1,i,j}(t) + \mathbf{K}_m^{f^T}(t-i)\mathbf{C}_{m-1,j,i}^T(t) + \mathbf{C}_{m-1,i,j}(t)\mathbf{K}_m^f(t-j)$$

$$+ \mathbf{K}_m^{f^T}(t-i)\,\mathbf{R}_{m-1,i,j}(t)\,\mathbf{K}_m^f(t-j) \quad . \tag{10.46}$$

Analogous recursions can be stated for the order recursive updating of $C_{m,i,j}(t)$ and $R_{m,i,j}(t)$. Note that these procedures are very similar to the scalar case, except that scalar quantities in the scalar recursions can be commuted whereas this is, in general, not true for the matrix-valued recursions. Moreover, it can be expected that the BGREs will also obey similar symmetric and shift invariance properties as discovered in the scalar case. In the ARMA case, it turns out that

$$E_{m,i,j}(t) = E_{m,j,i}^T(t) \quad , \tag{10.47a}$$

$$R_{m,i,j}(t) = R_{m,j,i}^T(t) \quad , \tag{10.47b}$$

which is the property analogous to the symmetric property in the scalar case (8.48a,b).

In this way, a *shift invariance property* similar to (8.53a–c) can be stated as

$$E_{m,i,j}(t) = E_{m,i+1,j+1}(t-1) \quad , \tag{10.48a}$$

$$R_{m,i,j}(t) = R_{m,i+1,j+1}(t-1) \quad , \tag{10.48b}$$

$$C_{m,i,j}(t) = C_{m,i+1,j+1}(t-1) \quad . \tag{10.48c}$$

Exploiting these expressions, the recursion (10.46) is completely determined by the *block vector order recursion*

$$E_{m,0,j}(t) = E_{m-1,0,j}(t) + K_m^{f^T}(t) C_{m-1,j,0}^T(t) + C_{m-1,0,j}(t) K_m^f(t-j)$$

$$+ K_m^{f^T}(t) R_{m-1,0,j}(t) K_m^f(t-j) \quad , \tag{10.49}$$

and similar recursions can be found in the cases of $R_{m,0,j}(t)$, $C_{m,0,j}(t)$ and $C_{m,j,0}(t)$. The recursions are initialized from the *covariance block matrix* of the observed process as follows:

$$E_{0,i,j}(t) = \Phi_{i,j}(t) \quad , \tag{10.50a}$$

$$R_{0,i,j}(t) = \Phi_{i+1,j+1}(t) \quad , \tag{10.50b}$$

$$C_{0,i,j}(t) = \Phi_{i,j+1}(t) \quad , \tag{10.50c}$$

where

$$\Phi_{i,j}(t) = \begin{bmatrix} x^T(t-i)x(t-j) & x^T(t-i)\,y(t-j) \\ y^T(t-i)x(t-j) & y^T(t-i)y(t-j) \end{bmatrix} \quad . \tag{10.51}$$

A complete algorithm of this type can be found in [10.7]. Nevertheless, we should note again at this point that all other algorithms discussed in the earlier chapters can be extended to the ARMA case in quite a similar manner.

10.2.5 Computing the ARMA Parameters from the Ladder Reflection Matrices

After constructing the true least–squares ARMA ladder form, it is some-times of interest to determine the transversal-form ARMA parameter estimates $\mathbf{a}^{(m)}(t)$ and $\mathbf{b}^{(m)}(t)$ from the ARMA ladder reflection matrices. In the stationary case, these parameters can be easily obtained as the impulse response of the ARMA ladder form (with time-invariant para-meters). In the more general nonstationary case, however, such a simple procedure does not hold. A *two-channel Levinson type recursion* is required to transform the (time-varying) ARMA ladder reflection matrices into the transversal-form two-channel AR parameters. This useful pro-cedure is listed in Table 10.1. It can be interpreted as the two-channel version of the Levinson type recursion given in Table 6.3 for the scalar case.

Table 10.1. Levinson type recursion for computation of the two-channel AR transversal predictor parameters from the ARMA ladder reflection matrices. $\mathbf{I}$ is the identity matrix

FOR $t = 0, 1, 2, \ldots$

Initialize: $\quad \mathbf{a}_0^{(1)}(t) = -\mathbf{I} \qquad\qquad \mathbf{a}_1^{(1)}(t) = -\mathbf{K}_1^f(t)$

$\qquad\qquad\quad \mathbf{b}_0^{(1)}(t) = -\mathbf{K}_1^b(t) \qquad \mathbf{b}_1^{(1)}(t) = -\mathbf{I}$

FOR $m = 2, 3, \ldots, p$

$\quad$ FOR $j = 1, 2, \ldots, m-1$

$\qquad \mathbf{b}_j^{(m)}(t) = \mathbf{b}_{j-1}^{(m-1)}(t-1) + \mathbf{a}_j^{(m-1)}(t)\,\mathbf{K}_m^b(t)$

$\qquad \mathbf{a}_j^{(m)}(t) = \mathbf{a}_j^{(m-1)}(t) + \mathbf{b}_{j-1}^{(m-1)}(t-1)\,\mathbf{K}_m^f(t)$

$\quad \mathbf{a}_0^{(m)}(t) = -\mathbf{I} \qquad\qquad\qquad \mathbf{a}_m^{(m)}(t) = -\mathbf{K}_m^f(t)$

$\quad \mathbf{b}_0^{(m)}(t) = -\mathbf{K}_m^b(t) \qquad\qquad \mathbf{b}_m^{(m)}(t) = -\mathbf{I}$

FOR $j = 1, 2, \ldots, p$

$\quad \mathbf{a}_j(t) = \mathbf{a}_j^{(p)}(t) \qquad\qquad\qquad \mathbf{b}_j(t) = \mathbf{b}_j^{(p)}(t)$

10.3 Identification of Vector Autoregressive Processes

After illustrating the ARMA case, it is now only a small step to the most general nonstationary vector autoregressive (multichannel) case, where a multichannel process model must be fitted onto a set of q nonstationary time series. In the vector autoregressive (VAR) case, the *input signal matrix* $\mathbf{S}(t)$ is of the form

$$\mathbf{S}(t) = \left[\mathbf{x}_1(t),\ \mathbf{x}_2(t),\ \ldots,\ \mathbf{x}_q(t) \right] \ . \tag{10.52}$$

Note the input signal matrix $\mathbf{S}(t)$ is just the counterpart to the signal vector $\mathbf{x}(t)$ in the scalar (single-channel) case. The vector autoregressive process is completely determined by its covariance block matrix

$$\Phi_{i,j}(t) = \mathbf{S}^T(t-i)\ \mathbf{S}(t-j) \tag{10.53}$$

with the property

$$\Phi_{i,j}(t) = \Phi_{j,i}^T(t) \ . \tag{10.54}$$

It is now an easy task to extend the variables appearing in the two-channel case to the multichannel case. For example, the forward reflection matrix of the multichannel ladder form (q channels) has the structure

$$\mathbf{K}_m^f(t) = \begin{bmatrix} K_m^{f11}(t) & K_m^{f21}(t) & \cdots\cdots & K_m^{fq1}(t) \\ K_m^{f12}(t) & K_m^{f22}(t) & \cdots\cdots & K_m^{fq2}(t) \\ \vdots & \vdots & \ddots & \vdots \\ K_m^{f1q}(t) & K_m^{f2q}(t) & \cdots\cdots & K_m^{fqq}(t) \end{bmatrix} \ . \tag{10.55}$$

The remaining variables of the algorithms are obtained analogously. See also Appendix A.3 for a discussion of two recently developed algorithms for parameter estimation in the vector autoregressive case.

In the strictly stationary case, a vector autoregressive process is completely determined by the block-Toeplitz structured autocorrelation matrix. Solutions to this problem were suggested by Whittle [10.11] and Wiggins and Robinson [10.12].

10.4 Parametric Spectral Estimation

In various signal processing applications, the frequency domain description of a wide sense stationary time series $\mathbf{x}(t)$ is sought. Generally, for this class of problems, one seeks to determine the underlying *power spectral density function* $S_x(\omega)$ as formally defined by

$$S_x(\omega) = \sum_{n = -\infty}^{\infty} \Phi_{0,n}\, e^{-j\omega n} \quad . \tag{10.56}$$

The power spectral density is recognized as being the *Fourier transform* of the unnormalized process autocorrelation sequence

$$\Phi_{0,n} = \mathbf{x}^T(t)\mathbf{x}(t-n) \quad . \tag{10.57}$$

The determination of the power spectral density is seen to entail complete knowledge of the generally infinite extent autocorrelation sequence. In practical spectral estimation, one seeks to estimate the power spectral density from a *finite set* of time series observations.

Clearly, there exists a basic incompatibility between this *finite set of sample observations* upon which the spectral estimate is to be made and the definition of the power spectral density, which is dependent on the *infinite length* autocorrelation sequence. A convenient way out of this incompatibility lies in the *parametrization* of the observed process.

The most general parametrization is the *rational* (ARMA) model discussed earlier in this chapter. In this approach, the desired power spectral density is simply approximated by the squared absolute value of the ARMA transfer function evaluated at the unit circle, hence

$$S_x(\omega) = \left| \frac{b_0 + b_1 e^{-j\omega} + b_2 e^{-2j\omega} + \ldots + b_q e^{-qj\omega}}{1 - a_1 e^{-j\omega} - a_2 e^{-2j\omega} - \ldots - a_p e^{-qj\omega}} \right|^2 \quad . \tag{10.58}$$

A model of this type can be justified on the basis that *any* continuous power spectral density can be approximated arbitrarily closely by an ARMA model provided the numerator and denominator order pair (q,p) attains sufficiently large values.

Clearly, this approach reduces the spectral estimation problem merely to a *signal identification problem* where it is desired to determine the ARMA parameters from a given set of data. The data set is then assumed to be the output of an unknown system with white Gaussian excitation modeled by a (time-invariant) ARMA filter. See e.g. [10.13-15], for a more detailed discussion of ARMA spectral estimation. It remains to mention that - besides the general case of ARMA spectral estimation - there exist simplified approaches which are distinguished by the particular model being involved. For instance, we may use simpler AR or MA models as a substitute for the general ARMA model.

10.5 Relationships Between Parameter Estimation and Kalman Filter Theory

The fields of parameter estimation and Kalman filter theory have many basic ideas in common [10.16,17]. An investigation of the relationships between these two fields can therefore be instructive.

For this purpose, we may introduce the Kalman filter estimator of a *state vector* $\mathbf{a}(t)$ as shown in Fig. 10.8. The system model of Fig. 10.8 is a first-order recursive model with *state transition matrix* $\mathbf{A}(t)$ and *system output matrix* $\mathbf{C}(t)$. The system is excited by the *input process* $\mathbf{u}(t)$, which is allowed to be further transformed by $\mathbf{B}(t)$. Zero-mean white Gaussian noise processes $\mathbf{v}(t)$ and $\mathbf{w}(t)$ are *superimposed* at the input and the output of the modeled system. The corresponding Kalman filter has, in general, *the same* first-order recursive structure as the system model.

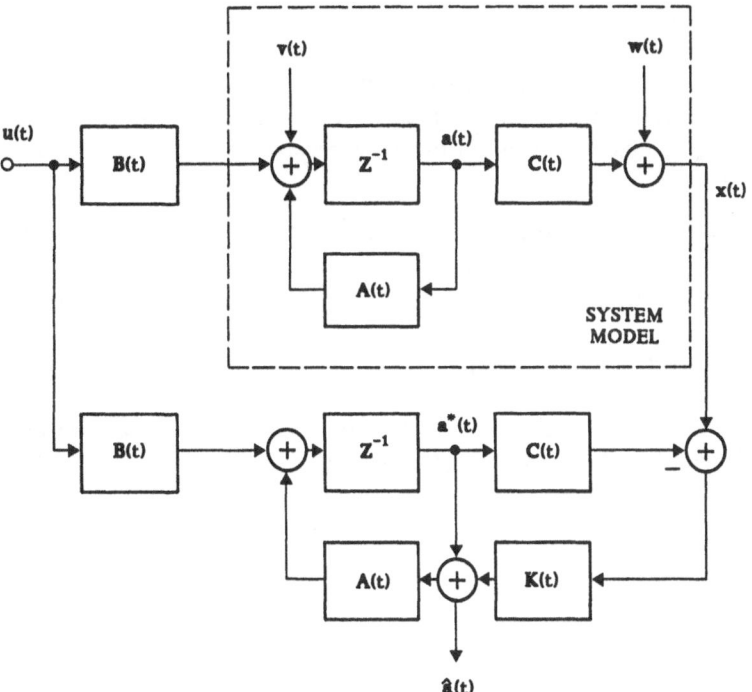

Fig. 10.8. State–space model of a linear system and the corresponding Kalman filter estimator for computation of the state vector estimate $\hat{\mathbf{a}}(t)$. $\mathbf{Z}^{-1}$ is the unit-delay matrix

Note that the Kalman filter does not employ "inverse" modeling as appeared in the linear prediction approach (recall Chap. 2). The Kalman filter observes the same excitation as the unknown (modeled) system. The output of the Kalman filter is now compared with the output of the unknown system and the so-obtained error process is used as an *additional excitation* of the first-order recursive loop model inside the Kalman filter. This excitation in terms of the error process is additionally controlled by the *Kalman gain matrix* $\mathbf{K}(t)$. The *estimated state vector* $\hat{\mathbf{a}}(t)$ can be observed at the "input" of the state transition matrix $\mathbf{A}(t)$. See Fig. 10.8.

The state-space model of the linear system in Fig. 10.8 is given by

$$\mathbf{a}(t+1) = \mathbf{A}(t)\mathbf{a}(t) + \mathbf{B}(t)\mathbf{u}(t) + \mathbf{v}(t) \quad , \tag{10.59a}$$

$$\mathbf{x}(t) = \mathbf{C}(t)\mathbf{a}(t) + \mathbf{w}(t) \quad , \tag{10.59b}$$

while the corresponding Kalman filter obeys the relations

$$\hat{\mathbf{a}}(t) = \mathbf{a}^*(t) + \mathbf{K}(t)\Big[\mathbf{x}(t) - \mathbf{C}(t)\mathbf{a}^*(t)\Big] \quad , \tag{10.60a}$$

$$\mathbf{a}^*(t+1) = \mathbf{A}(t)\hat{\mathbf{a}}(t) + \mathbf{B}(t)\mathbf{u}(t) \quad . \tag{10.60b}$$

Note that the corresponding estimator of the Kalman gain matrix $\mathbf{K}(t)$ that minimizes the estimation error term $\mathbf{x}(t) - \mathbf{C}(t)\mathbf{a}^*(t)$ in the least-squares sense is given by the following recursive relations (see [10.16] for details):

$$\mathbf{K}(t) = \mathbf{P}^*(t)\mathbf{C}^T(t)\Big[\mathbf{C}(t)\mathbf{P}^*(t)\mathbf{C}^T(t) + \mathbf{R}(t)\Big]^{-1} , \tag{10.61a}$$

$$\overline{\mathbf{P}}(t) = \mathbf{P}^*(t) - \mathbf{K}(t)\mathbf{C}(t)\mathbf{P}^*(t) \quad , \tag{10.61b}$$

$$\mathbf{P}^*(t+1) = \mathbf{A}(t)\overline{\mathbf{P}}(t)\mathbf{A}^T(t) + \mathbf{Q}(t) \quad , \tag{10.61c}$$

where $\mathbf{Q}(t)$ and $\mathbf{R}(t)$ are the *covariance matrices* of the vector-valued, zero-mean Gaussian processes $\{\mathbf{v}(t)\}$ and $\{\mathbf{w}(t)\}$, hence

$$\mathbf{Q}(t) = \mathbf{V}^T(t)\mathbf{V}(t) \quad , \tag{10.62a}$$

$$\mathbf{R}(t) = \mathbf{W}^T(t)\mathbf{W}(t) \quad , \tag{10.62b}$$

where

$$\mathbf{V}^T(t) = \Big[\mathbf{v}(t), \mathbf{v}(t-1), \ldots , \mathbf{v}(1), \mathbf{v}(0)\Big] \quad , \tag{10.63a}$$

$$\mathbf{W}^T(t) = \Big[\mathbf{w}(t), \mathbf{w}(t-1), \ldots , \mathbf{w}(1), \mathbf{w}(0)\Big] \quad . \tag{10.63b}$$

The noise processes $\{v(t)\}$ and $\{w(t)\}$ are *decoupled* and *orthogonal* with respect to the sequence of state vectors $\{a(t)\}$, hence

$$\mathbf{v}^T(t)\,\mathbf{W}(t) = \mathbf{0} \quad , \tag{10.64a}$$

$$\mathbf{a}^T(t)\,\mathbf{V}(t) = \mathbf{0} \quad , \tag{10.64b}$$

$$\mathbf{a}^T(t)\,\mathbf{W}(t) = \mathbf{0} \quad , \tag{10.64c}$$

where

$$\mathbf{a}^T(t) = \Big[\,a(t),\ a(t-1),\ \dots\ ,\ a(1),\ a(0)\,\Big] \tag{10.65}$$

is the sequence of state vectors and

$$\mathbf{P}^*(t+1) = \mathbf{a}^{*T}(t+1)\mathbf{a}^*(t+1) \tag{10.66}$$

is the covariance matrix of the sequence of "extrapolated" state vectors. Now consider the *simplified* state-space model

$$a(t+1) = a(t) \quad , \tag{10.67a}$$

$$x(t) = \mathbf{z}^T(t)\,a(t) + w(t) \quad . \tag{10.67b}$$

Clearly, this state-space model may be deduced from the general case (10.59a,b) simply by setting $v(t)$ equal to the zero process. The matrix $A(t)$ is replaced by the identity matrix and $B(t)$ vanishes completely (zero matrix). The excitation $u(t)$ is the zero vector and the output matrix $C(t)$ is replaced by the state vector $\mathbf{z}^T(t)$ of the transversal prediction error filter [recall (5.10)]. Obviously, relation (10.67b) can be interpreted as an autoregressive signal model with parameter vector $a(t)$, state vector $z(t)$ and white Gaussian excitation $w(t)$. Then, $x(t)$ is a sample of the generated autoregressive process $\{x(t)\}$. Note that the terminology "state vector $z(t)$" in this context denotes the state vector of the transversal prediction error filter and should not be confused with the state vector $a(t)$ of the observed linear system in the Kalman approach, which is, in fact, identical with the *parameter vector* of the autoregressive signal model. This clever approach enables the AR parameter estimation problem to be *embedded* in the more general state vector estimation problem. Figure 10.9 illustrates the underlying state-space model for AR parameter estimation using the Kalman filter approach. Interestingly, (10.67a) indicates that the state-space model thus obtained is a *"free"* model in the sense that the excitation signal $u(t)$ is the zero process. This "free" state-space model is illustrated in Fig. 10.9.

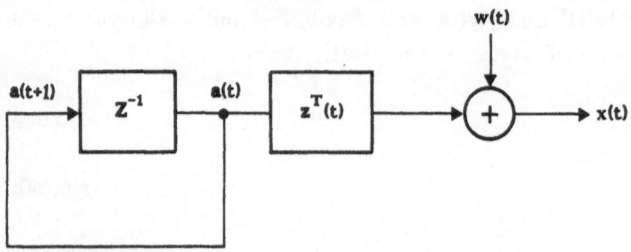

Fig. 10.9. "Free" state-space model as associated with AR parameter estimation

Applying the same ideas of simplification to the general Kalman estimator equations, we may obtain the *reduced* Kalman estimator equations from the general recursions (10.60a,b and 61a-c) as follows:

$$\hat{a}(t) = a^*(t) + k^*(t)\left[x(t) - z^T(t)\,a^*(t)\right] \quad , \tag{10.68a}$$

$$a^*(t+1) = \hat{a}(t) \quad , \tag{10.68b}$$

$$k^*(t) = P^*(t)\,z(t)\left[z^T(t)\,P^*(t)\,z(t) + R(t)\right]^{-1} \quad , \tag{10.69a}$$

$$\overline{P}(t) = P^*(t) - k^*(t)\,z^T(t)\,P^*(t) \quad , \tag{10.69b}$$

$$P^*(t+1) = \overline{P}(t) \quad . \tag{10.69c}$$

Note that the Kalman gain matrix has now changed into a *vector-valued* quantity. Moreover, {w(t)} is a scalar process with unit variance, hence the covariance matrix $R(t)$ turns into the scalar quantity 1 and therefore *vanishes* from the further discussion:

$$R(t) \longrightarrow 1 \quad . \tag{10.70}$$

With these simplifications, the recursions (10.68a,b and 69a-c) may be condensed to the *three-term recurrence relation* for the estimate $\hat{a}(t)$.

$$\hat{a}(t) = \hat{a}(t-1) + k^*(t)\left[x(t) - z^T(t)\,\hat{a}(t-1)\right] \quad , \tag{10.71a}$$

$$k^*(t) = \overline{P}(t-1)\,z(t)\left[z^T(t)\,\overline{P}(t-1)\,z(t) + 1\right]^{-1} \quad , \tag{10.71b}$$

$$\overline{P}(t) = \overline{P}(t-1) - k^*(t)\,z^T(t)\,\overline{P}(t-1) \quad . \tag{10.71c}$$

These recursions are in fact identical representations of the RLS algorithm of Table 5.1 when we only assume that $\overline{P}(t) = A^{-1}(t)$ is the inverse

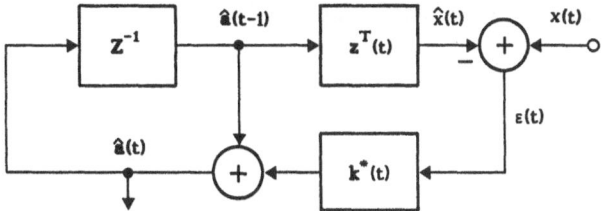

Fig. 10.10. RLS algorithm as the Kalman filter of the "free" state-space model (Fig. 10.9)

covariance matrix of the AR process $\{x(t)\}$ in the serialized data case. The vector $\mathbf{k}^*(t)$ is then easily recognized as the modified Kalman gain vector of the RLS algorithm of Table 5.1. Note also that a growing window ($\lambda = 1$) was assumed throughout our considerations on AR parameter estimation using Kalman's algorithm. Figure 10.10 illustrates this interpretation of the RLS algorithm as the Kalman filter of the "free" state-space model (Fig. 10.9).

10.6 Chapter Summary

The purpose of this chapter was to introduce several useful extensions of the simple scalar AR process model, such as the joint process model, the ARMA model, and the vector autoregressive (VAR) process model. It was demonstrated how these models can be used for MA and ARMA system identification, where the input process of the unknown system was assumed to be accessible. The problem solutions were formulated employing a Normal Equation framework, or, alternatively, a ladder formulation framework. We refer to Appendix A.3, where the VAR case of two recently discovered recursive covariance ladder algorithms is presented in detail (with algorithm summaries).

Besides the extension of signal models and algorithms, we discussed the relationships between system identification in the stationary case and spectral estimation. As an interesting exercise, we showed how the AR estimation problem can be embedded in the Kalman state vector estimation problem. This approach constitutes an alternative way of deriving the RLS algorithm, which was obtained much earlier in the book by using the Sherman-Morrison identity (matrix inversion lemma). The approach shown in this chapter justifies the terminology fast "Kalman" algorithm which is sometimes used for the fast O(p) versions of the RLS algorithm.

11. Concluding Remarks and Applications

Linear prediction theory and algorithms have found numerous applications. Conceptual approaches that clearly derive from linear prediction theory can be found in such diverse fields as radar, sonar, communications, control, seismology, image processing, pattern recognition and large-scale artificial neural networks and associative memory. In all these applications, the common problem is to *adapt* some predetermined filter structure or model parameters effectively in an unknown environment and also to *track variations* of the observed process or model parameters as soon as the statistics of these inputs change.

Most of these essential adaptation algorithms and structures derive from the fundamental theory of least squares discovered by the German mathematician Carl Friedrich Gauss in the early 19th century. The success of least squares stems from two very clear and fundamental observations.

1. The method of least squares lends itself to mathematically tractable problem formulations and efficient algorithms for their solution.
2. The Gauss statistics is a good approximation for many real-world processes and the first- and second-order optimization underlying the least-squares method therefore performs often very close to the optimal, but computationally burdensome, maximum likelihood method.

The majority of algorithms presented in this monograph have been derived for the simplest case of a scalar process in order to make the derivations more accessible for the rader unfamiliar with them. Once one knows how to derive a specific algorithm in the scalar case, it is only a small step to apply the same ideas to more complicated process models.

Linear prediction algorithms can be distinguished in many different ways. They are based on the *autocorrelation* (Toeplitz) formulation or on the *covariance* formulation of the data observation problem. They can be classified into *recursive* and *nonrecursive* approaches. Furthermore, they may be based on different structures such as the *(fixed-order)* transversal filter, the *(growing-order)* ladder form or even an approach that does not assume a filter structure at all, such as "least-squares frequency transform" algorithms based on the recursive QR decomposition. Finally, the algorithms may be divided into those methods which require the explicit computation of residuals and other techniques based merely on pure inner product recursions, as seen in the case of the pure order recursive ladder algorithm (PORLA).

Although many of these schemes provide identical solutions to the same problem under the theoretical assumption of infinite precision arithmetic, the situation can change dramatically as soon as the algorithms are operated under real-world finite precision arithmetic. In this case, the algorithms may be divided in two basic categories associated with the two basic problems of finite precision implementations, namely, the problem of *numerical accuracy* and the problem of *numerical stability*. An algorithm is said to have a good numerical accuracy if the error in the estimated parameters is small when compared to an infinite precision realization of the same algorithm. An algorithm is called numerically stable if it can be operated over infinite time with a *bounded error* in the estimated parameters. The problems of numerical accuracy and stability have a great impact on the implementation of recursive least-squares algorithms. It is emphasized that, in principle, these two problems can be completely *independent*. For example, an algorithm that performs very accurately on short data records could become unstable in the long run, i.e., the error in the estimated parameters grows unbounded with time until it destroys the normal operation of the algorithm. On the other hand, an algorithm that performs very inaccurately could exhibit the property that it runs stably over infinite time, i.e., the error in the estimated parameters may be considerable, but *bounded*. It is clear from these considerations that numerical stability is an *indispensable require-ment* for real-time or "on-line" applications of recursive least-squares algorithms. Unfortunately, several of the sophisticated "fast" recursive least-squares algorithms have proved to be *intrinsically unstable*. For example, the update equations of the fast Kalman and related algorithms can be interpreted as a recursive system with the poles at or outside the unit circle. This fact finds its underlying cause in the appearance of *hyperbolic rotations* in the backward predictor update equations. The unpleasant practical effect frequently observed with this class of algo-rithms is that the intrinsically unstable control mechanism "explodes" as soon as a certain quantity of round-off error is induced at a certain state of operation. This effect has been a severe problem which has almost blocked the real-time application of fast RLS algorithms during the last decade. Another inherent problem associated with the "fast" schemes is that they are implicitly based on the update of the *inverse* system matrix of the Normal Equations. Hence these schemes implicitly assume that the system matrix will remain nonsingular throughout all recursions and for all time. This, in fact, is an assumption that seldom holds in practice.

Another reason for instability can be associated with the special *arithmetic* involved in the computations. For example, it can be shown that a recursive loop system with a pole at the unit circle (a theoretically wide sense stable system) can be operated stably over infinite time when *fixed-point arithmetic* is used. The same system will be numerically unstable when *floating-point arithmetic* is used. This surprising statement is true independent of how large the mantissa of the floating-point format might grow. This effect stems from the fact that floating-point arithmetic involves rounding in both the multiplication and addition operations, whereas fixed-point arithmetic involves rounding in only the multiplication operation. The importance of these considerations becomes very obvious when we discover that all sliding window algorithms involve recursive computation structures with poles located exactly on the unit circle. Such wide-sense stable adaptation algorithms cannot be implemented using conventional floating-point signal processors when an "on-line" application is an issue; a fact that seems to be not commonly known among the wider family of algorithm users. However, sliding windows are sometimes essential for fast parameter tracking. For example, the fact that sliding window RLS algorithms cannot be implemented with floating-point arithmetic again makes it interesting to investigate the numerical accuracy of RLS algorithms when fixed-point arithmetic is used. Here, the general observation is that a smaller complexity in terms of arithmetic operations is often paid for with a worse numerical accuracy. For example, the $O(p^2)$ PORLA computation schemes have a much better numerical accuracy than the fast $O(p)$ recursive least-squares ladder algorithms. These considerations, which have motivated the investigation of PORLA computation techniques, have also initiated the development of the recursive QRLS algorithms. See Appendix A.2 for a comparison of recently developed PORLA methods and the QRLS algorithms.

A basic knowledge of these facts and the behavior of the algorithms seems indispensable for the proper choice of an appropriate algorithm for a specific application. The applications may be grouped according to four basic information processing operations.

Parameter estimation, where one is interested in extracting the parameters of an underlying process model from an observed process.

Adaptive filtering, which means the extraction of information about quantities of interest at time t by using data measurements of the same or related processes in a certain interval of past observations.

Smoothing, which differs from filtering in that the quantity of interest is already available and is included in the information extraction process. This means that in the case of smoothing there is a delay in producing the result of interest. This also means that the smoothing process is, in some sense, more accurate than the filtering process since it is more "complete" in terms of the available information at a certain time step.

Prediction or "forecasting", where the aim is to predict a process sample at some future time t+τ from the "subspace of past observations" spanned by process vectors consisting of past data samples of the observed process itself or the observation of a related process.

In principle, the filters used for these basic processing operations are *linear filters* as long as their parameters are *time-invariant*. As soon as we use *adaptive* techniques, however, the filter parameters become dependent to some extent on the data, and hence the simple linearity assumption may no longer hold for these adaptive forms.

Besides these principal considerations, it is a purpose of this chapter to provide a link to the applications-oriented literature. Specific applications of linear prediction theory have been extensively discussed. An excellent overview of this topic can be found in the book by Widrow and Stearns [11.1] and the review papers by Bellanger [11.2] and Claasen and Mecklenbräuker [11.3]. Applications concerning the **system identification** problem are provided in [11.4], the fundamental paper of Mullis and Roberts [11.5] and the work of Lee et al. [11.6] and Strobach [11.7]. **Adaptive equalization and noise cancelling** are treated in [11.8-14].

Speech analysis, synthesis and coding methods based on the linear prediction approach have been initiated by the fundamental papers of Atal and Hanauer [11.15] and Schroeder [11.16]. An excellent overview paper about linear predictive coding of speech signals was provided by Schroeder [11.17]. Other papers dealing with the topic of linear predictive coding of speech are [11.18-20]. In this context, one should also mention the early work about linear prediction of speech by Markel and Gray [11.21]. Finally, Ahmed et al. have suggested a VLSI speech analysis chip set [11.22].

Finite precision effects are of particular interest as soon as on-line applications of linear prediction algorithms are an issue. Concerning the LMS algorithm, the influences of round-off error have been studied by Alexander [11.23] and Caraiscos and Liu [11.24]. The fixed-point round-off error behavior of the RLS algorithm was investigated in [11.25 and 26] whereas implementation of fast Kalman algorithms was considered in [11.27]. Finite precision effects were also studied in the case of the ladder algorithms. Results concerning this topic can be found in [11.7, 28-32]. The fixed-point implementation of some gradient-based algorithms was considered in [11.33]. Pepe and Rogers have written an interesting and particularly useful paper about the "simulation of fixed-point operations with high level languages" [11.34]. Linear prediction algorithms in **failure detection and adaptive segmentation** systems were the topic of [11.4, 35, 36]. Azenkot and Gertner [11.37] have suggested a linear prediction based approach for **time delay estimation**. Another important field of parametric linear predictive modeling is **parametric spectral estimation**. See the papers of Cadzow [11.38] and Kay [11.39].

Implementations of linear prediction algorithms were treated in [11.40 and 41]. The design of special architectures for implementation of adaptive algorithms was discussed in [11.42]. Chan et al. have suggested a linear prediction based approach to the **estimation of frequencies of sinusoids** [11.43] and Janssen et al. have shown how autoregressive modeling can be used for an **adaptive interpolation of discrete time signals** [11.44]. A new field for the application of linear prediction algorithms is adaptive two-dimensional (image) processing. Here, we may find problems and approaches similar to those in the one-dimensional (time-series) case. For example the **estimation of blur parameters** [11.45] can be viewed as the analog problem to equalization in the one-dimensional case. General considerations about two-dimensional linear prediction and adaptive filtering were made in [11.46-51].

Two-dimensional object detection by means of linear prediction algorithms was the topic of [11.52 and 53], while an algorithm for **linear prediction image coding** was described in [11.54]. Strobach showed in [11.55] how linear prediction models can be embedded in the quadtree segmentation structure for two-dimensional object detection and image sequence coding. Linear prediction algorithms were also extended to the **training of associative memories and artificial neural networks**. This emerging field of higher-order adaptive structures for perception and pattern recognition is closely related to adaptive filtering. A major difference to known models from linear prediction theory is that some artificial neural networks involve *nonlinear threshold elements* at the output of linear combiners, which complicates the application of adaptation algorithms known from purely linear models. The appearance of these nonlinearities can be justified from the analogy with the biological neuron. This, however, is only an unsatisfactory explanation for the observation that the incorporation of such nonlinearities enables the structure to "learn" certain information in the sense of an analog associative memory [11.56]. A deeper theoretical explanantion of this effect can be that the input layer of a neural network performs a *partitioning* of the input space into a number of cells formed by intersecting hyperplanes. The sole function of succeeding layers is then to perform a *"grouping"* of these cells in order to approximate a desired *decision region* in the input space [11.57]. The classical replicas of artificial neural nets that are currently in use are commonly trained by a procedure called the "back propagation algorithm" [11.58, 59], which is an extended formulation of Widrow's LMS algorithm. A novel approach by Singhal and Wu [11.60] uses the extended Kalman algorithm [11.61] for the purpose of training of a nonlinear artificial neural network. A good description of the current state of research into artificial neural networks is given by the papers [11.62-68].

Concluding this chapter, we note that our discussion about applications of linear prediction algorithms was mainly limited to the fields of communications, system identification and pattern recognition. We have not discussed the diverse applications of linear prediction algorithms in the forecasting of economic processes and seismic processing.

Appendix

A.1 Summary of the Most Important Forward/Backward Linear Prediction Relationships

This section presents a summary of the most important formulas and relationships between forward and backward linear prediction. First, the *forward observation vector* $\mathbf{x}(t)$ and the *backward observation vector* $\mathbf{x}(t-m)$ are defined:

$$\mathbf{x}(t) = \left[x(t), x(t-1), \ldots, x(t-L+1) \right]^{T} , \tag{A.1a}$$

$$\mathbf{x}(t-m) = \left[x(t-m), x(t-m-1), \ldots, x(t-m-L+1) \right]^{T} , \tag{A.1b}$$

where m denotes the model order and L is a window (record) length. The corresponding *subspace of past observations* $\mathbf{X}_m(t)$ and the *subspace of previous observations* $\mathbf{X}_m(t+1)$ are defined as

$$\mathbf{X}_m(t) = \left[\mathbf{x}(t-1), \mathbf{x}(t-2), \ldots, \mathbf{x}(t-m) \right] , \tag{A.2a}$$

$$\mathbf{X}_m(t+1) = \left[\mathbf{x}(t), \mathbf{x}(t-1), \ldots, \mathbf{x}(t-m+1) \right] . \tag{A.2b}$$

The *forward prediction error vector* $\mathbf{e}_m(t)$ is defined as the complement of the orthogonal projection of $\mathbf{x}(t)$ onto the subspace spanned by $\mathbf{X}_m(t)$, whereas the *backward prediction error vector* $\mathbf{r}_m(t)$ is the complement of the orthogonal projection of $\mathbf{x}(t-m)$ onto the subspace spanned by $\mathbf{X}_m(t+1)$:

$$\mathbf{e}_m(t) = \mathbf{x}(t) \langle \mathbf{X}_m(t) \rangle , \tag{A.3a}$$

$$\mathbf{r}_m(t) = \mathbf{x}(t-m) \langle \mathbf{X}_m(t+1) \rangle . \tag{A.3b}$$

The *forward predicted process vector* $\hat{\mathbf{x}}_m^f(t)$ is the orthogonal projection of $\mathbf{x}(t)$ onto the subspace spanned by $\mathbf{X}_m(t)$, whereas the *backward predicted process vector* $\hat{\mathbf{x}}_m^b(t)$ is the orthogonal projection of $\mathbf{x}(t-m)$ onto the subspace spanned by $\mathbf{X}_m(t+1)$, hence

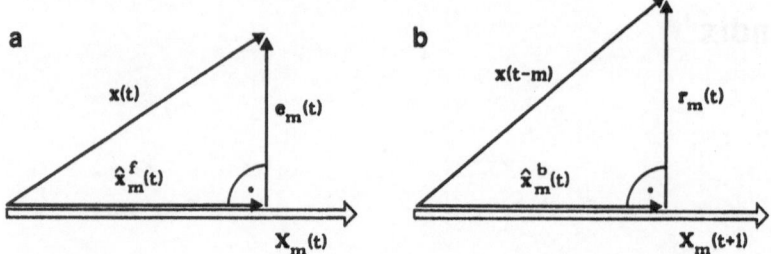

Fig. A.1. Decomposition of observations. (a) Forward linear prediction. (b) Backward linear prediction

$$\mathbf{e}_m(t) = \mathbf{x}(t) - \hat{\mathbf{x}}_m^f(t) \quad , \tag{A.4a}$$

$$\mathbf{r}_m(t) = \mathbf{x}(t-m) - \hat{\mathbf{x}}_m^b(t) \quad . \tag{A.4b}$$

Fig. A.1 illustrates these relationships.

The forward predicted process vector $\hat{\mathbf{x}}_m^f(t)$ is obtained as the weighted linear combination of oblique basis vectors of the subspace $\mathbf{X}_m(t)$, where $\mathbf{a}^{(m)}(t)$ is the set of weighting coefficients or the *forward transversal predictor parameter set*. The backward predicted process vector $\hat{\mathbf{x}}_m^b(t)$ is obtained as the weighted linear combination of oblique basis vectors of the subspace $\mathbf{X}_m(t+1)$, where $\mathbf{b}^{(m)}(t)$ is the associated set of weighting coefficients or the *backward transversal predictor parameter set*:

$$\hat{\mathbf{x}}_m^f(t) = \mathbf{X}_m(t)\mathbf{a}^{(m)}(t) \quad , \tag{A.5a}$$

$$\hat{\mathbf{x}}_m^b(t) = \mathbf{X}_m(t+1)\mathbf{b}^{(m)}(t) \quad , \tag{A.5b}$$

$$\mathbf{a}^{(m)}(t) = \left[a_1^{(m)}(t), a_2^{(m)}(t), \ldots , a_m^{(m)}(t) \right]^T \quad , \tag{A.6a}$$

$$\mathbf{b}^{(m)}(t) = \left[b_1^{(m)}(t), b_2^{(m)}(t), \ldots , b_m^{(m)}(t) \right]^T \quad . \tag{A.6b}$$

Clearly,

$$\hat{\mathbf{x}}_m^f(t) \langle \mathbf{X}_m(t) \rangle = \left[0 \ldots 0 \right]^T , \tag{A.7a}$$

$$\hat{\mathbf{x}}_m^b(t) \langle \mathbf{X}_m(t+1) \rangle = \left[0 \ldots 0 \right]^T , \tag{A.7b}$$

$$\mathbf{e}_m^T(t)\hat{\mathbf{x}}_m^f(t) = 0 \quad , \tag{A.8a}$$

$$\mathbf{r}_m^T(t)\hat{\mathbf{x}}_m^b(t) = 0 \quad , \tag{A.8b}$$

$$x^T(t)e_m(t) \; = \; e_m^T(t)e_m(t) \quad , \tag{A.9a}$$

$$x^T(t-m)r_m(t) = r_m^T(t)r_m(t) \quad , \tag{A.9b}$$

$$x^T(t-m-1)e_m(t) = e_m^T(t)r_m(t-1) \quad , \tag{A.10a}$$

$$x^T(t)r_m(t-1) \; = \; e_m^T(t)r_m(t-1) \quad . \tag{A.10b}$$

The *forward transversal prediction error filter* is defined by (A.11a) and the *backward transversal prediction error filter* by (A.11b):

$$e_m(t) = x(t) - X_m(t)a^{(m)}(t) \quad , \tag{A.11a}$$

$$r_m(t) = x(t-m) - X_m(t+1)b^{(m)}(t) \quad . \tag{A.11b}$$

The *forward residual energy* $E_m(t)$, the *backward residual energy* $R_m(t)$, and the *covariance* $C_m(t)$ are defined as follows:

$$E_m(t) = e_m^T(t)e_m(t) \quad , \tag{A.12a}$$

$$R_m(t) = r_m^T(t)r_m(t) \quad , \tag{A.12b}$$

$$C_m(t) = e_m^T(t)r_m(t-1) \quad . \tag{A.12c}$$

The forward residual energy and the backward residual energy can be written as *quadratic functions* of the forward and backward transversal predictor parameter sets ,

$$
\begin{aligned}
E_m(t, \, a^{(m)}(t)) = \; & x^T(t)x(t) - a^{(m)T}(t)X_m^T(t)x(t) \\
& - x^T(t)X_m(t)a^{(m)}(t) \\
& + a^{(m)T}(t)X_m^T(t)X_m(t)a^{(m)}(t) \overset{!}{=} min \quad ,
\end{aligned}
\tag{A.13a}
$$

$$
\begin{aligned}
R_m(t, \, b^{(m)}(t)) = \; & x^T(t-m)x(t-m) - b^{(m)T}(t)X_m^T(t+1)x(t-m) \\
& - x^T(t-m)X_m(t+1)b^{(m)}(t) \\
& + b^{(m)T}(t)X_m^T(t+1)X_m(t+1)b^{(m)}(t) \overset{!}{=} min \quad .
\end{aligned}
\tag{A.13b}
$$

Taking the gradients of the energy functions E and R and setting them to zero yields the *Normal Equations of forward linear prediction* (A.14a) and the corresponding *Normal Equations of backward linear prediction* (A.14b):

$$\frac{\partial E_m(t,\, \mathbf{a}^{(m)}(t))}{\partial \mathbf{a}^{(m)^T}(t)} = -\, 2\mathbf{X}_m^T(t)\mathbf{x}(t)$$

$$+\, 2\mathbf{X}_m^T(t)\mathbf{X}_m(t)\mathbf{a}^{(m)}(t) = \begin{bmatrix} 0 \ldots 0 \end{bmatrix}^T ,\qquad \text{(A.14a)}$$

$$\frac{\partial R_m(t,\, \mathbf{b}^{(m)}(t))}{\partial \mathbf{b}^{(m)^T}(t)} = -\, 2\mathbf{X}_m^T(t+1)\mathbf{x}(t-m)$$

$$+\, 2\mathbf{X}_m^T(t+1)\mathbf{X}_m(t+1)\mathbf{b}^{(m)}(t) = \begin{bmatrix} 0 \ldots 0 \end{bmatrix}^T ,\quad \text{(A.14b)}$$

$$\mathbf{X}_m^T(t)\mathbf{X}_m(t)\mathbf{a}^{(m)}(t) = \mathbf{X}_m^T(t)\mathbf{x}(t) \quad , \tag{A.15a}$$

$$\mathbf{X}_m^T(t+1)\mathbf{X}_m(t+1)\mathbf{b}^{(m)}(t) = \mathbf{X}_m^T(t+1)\mathbf{x}(t-m) \quad . \tag{A.15b}$$

Defining the *covariance matrix of input data* $\mathbf{\Phi}^{(m)}(t)$ for the *serialized data case* $\mathbf{x}(t), \mathbf{x}(t-1), \ldots, \mathbf{x}(t-m)$

$$\mathbf{\Phi}^{(m)}(t) = \begin{bmatrix} \Phi_{i,j}^{(m)}(t) \end{bmatrix} \quad ;$$

$$\Phi_{i,j}^{(m)}(t) = \mathbf{x}^T(t-i)\mathbf{x}(t-j) \quad ; \qquad 0 \le i,j \le m , \tag{A.16a}$$

with the shift-invariance property

$$\Phi_{i,j}^{(m)}(t) = \Phi_{i-1,j-1}^{(m)}(t-1) \quad , \tag{A.16b}$$

we may interpret the *forward autocorrelation sequence* $c_i^f(t)$ as the upper row vector and the *backward autocorrelation sequence* $c_i^b(t)$ as the reversely ordered lower row vector of the covariance matrix:

$$c_i^f(t) = \mathbf{x}^T(t)\mathbf{x}(t-i) \quad ; \qquad\qquad 0 \le i \le m , \tag{A.17a}$$

$$c_i^b(t) = \mathbf{x}^T(t-m)\mathbf{x}(t-m+i) \quad ; \qquad 0 \le i \le m . \tag{A.17b}$$

The *ladder form prediction error filter* is defined by the equations

$$\mathbf{e}_m(t) = \mathbf{e}_{m-1}(t) + K_m^f(t)\mathbf{r}_{m-1}(t-1) \quad , \tag{A.18a}$$

$$\mathbf{r}_m(t) = \mathbf{r}_{m-1}(t-1) + K_m^b(t)\mathbf{e}_{m-1}(t) \quad , \tag{A.18b}$$

where the scalar quantities $K_m^f(t)$ and $K_m^b(t)$ are termed the *forward ladder reflection coefficient* and the *backward ladder reflection coefficient,* respectively:

$$K_m^f(t) = - C_{m-1}(t) / R_{m-1}(t-1) \quad , \tag{A.19a}$$

$$K_m^b(t) = - C_{m-1}(t) / E_{m-1}(t) \quad . \tag{A.19b}$$

The forward residual energy $E_m(t)$ and the backward residual energy $R_m(t)$ obey the following recursive relationships:

$$E_m(t) = E_{m-1}(t) + K_m^f(t) C_{m-1}(t) \quad , \tag{A.20a}$$

$$R_m(t) = R_{m-1}(t-1) + K_m^b(t) C_{m-1}(t) \quad , \tag{A.20b}$$

$$E_m(t) = E_{m-1}(t) \left[1 - K_m^f(t) K_m^b(t) \right] \quad , \tag{A.21a}$$

$$R_m(t) = R_{m-1}(t-1) \left[1 - K_m^f(t) K_m^b(t) \right] \quad . \tag{A.21b}$$

The covariance matrix can be partitioned as

$$\Phi^{(m)}(t) = \left[x(t) \ X_m(t) \right]^T \left[x(t) \ X_m(t) \right] \quad , \tag{A.22a}$$

$$\Phi^{(m)}(t) = \left[X_m(t+1) \ x(t-m) \right]^T \left[X_m(t+1) \ x(t-m) \right] \quad , \tag{A.22b}$$

or, explicitly,

$$\Phi^{(m)}(t) = \begin{bmatrix} x^T(t) x(t) & x^T(t) X_m(t) \\ X_m^T(t) x(t) & X_m^T(t) X_m(t) \end{bmatrix} \quad , \tag{A.23a}$$

$$\Phi^{(m)}(t) = \begin{bmatrix} X_m^T(t+1) X_m(t+1) & X_m^T(t+1) x(t-m) \\ x^T(t-m) X_m(t+1) & x^T(t-m) x(t-m) \end{bmatrix} \quad . \tag{A.23b}$$

The parameter sets $a^{(m)}(t)$ and $b^{(m)}(t)$ obtained as the solutions of the Normal Equations (A.15a and b) determine the *minimal residual energies*

$$E_m(t) = x^T(t) x(t) - x^T(t) X_m(t) a^{(m)}(t) \quad , \tag{A.24a}$$

$$R_m(t) = x^T(t-m) x(t-m) - x^T(t-m) X_m(t+1) b^{(m)}(t) \quad . \tag{A.24b}$$

Inspecting these relations, one discovers that

$$\Phi^{(m)}(t) \begin{bmatrix} -1 \\ a^{(m)}(t) \end{bmatrix} = \begin{bmatrix} -E_m(t) \\ 0 \\ \vdots \\ 0 \end{bmatrix} \quad , \tag{A.25a}$$

$$
\Phi^{(m)}(t) \begin{bmatrix} \mathbf{b}^{(m)}(t) \\ \vdots \\ -1 \end{bmatrix} = \begin{bmatrix} 0 \\ \vdots \\ 0 \\ -R_m(t) \end{bmatrix} .
\tag{A.25b}
$$

The transversal form parameter sets are related to the ladder reflection coefficients by the following *Levinson type recursions*:

$$
\begin{bmatrix} -1 \\ \mathbf{a}^{(m)}(t) \end{bmatrix} = \begin{bmatrix} -1 \\ \mathbf{a}^{(m-1)}(t) \\ 0 \end{bmatrix} + K_m^f(t) \begin{bmatrix} 0 \\ \mathbf{b}^{(m-1)}(t-1) \\ -1 \end{bmatrix} ,
\tag{A.26a}
$$

$$
\begin{bmatrix} \mathbf{b}^{(m)}(t) \\ -1 \end{bmatrix} = \begin{bmatrix} 0 \\ \mathbf{b}^{(m-1)}(t-1) \\ -1 \end{bmatrix} + K_m^b(t) \begin{bmatrix} -1 \\ \mathbf{a}^{(m-1)}(t) \\ 0 \end{bmatrix} .
\tag{A.26b}
$$

The forward/backward prediction error filters can then be expressed as

$$
\mathbf{e}_m(t) = - \begin{bmatrix} \mathbf{x}(t) & \mathbf{X}_m(t) \end{bmatrix} \begin{bmatrix} -1 \\ \mathbf{a}^{(m)}(t) \end{bmatrix} ,
\tag{A.27a}
$$

$$
\mathbf{r}_m(t-1) = - \begin{bmatrix} \mathbf{X}_m(t) & \mathbf{x}(t-m-1) \end{bmatrix} \begin{bmatrix} \mathbf{b}^{(m)}(t-1) \\ -1 \end{bmatrix} .
\tag{A.27b}
$$

The covariance $C_m(t)$ can be expressed as

$$
C_m(t) = \begin{bmatrix} -1 & \mathbf{a}^{(m)T}(t) \end{bmatrix}
$$
$$
\times \begin{bmatrix} \mathbf{x}^T(t)\mathbf{X}_m(t) & \mathbf{x}^T(t)\mathbf{x}(t-m-1) \\ \mathbf{X}_m^T(t)\mathbf{X}_m(t) & \mathbf{X}_m^T(t)\mathbf{x}(t-m-1) \end{bmatrix} \begin{bmatrix} \mathbf{b}^{(m)}(t-1) \\ -1 \end{bmatrix} ,
\tag{A.28}
$$

or, equivalently,

$$
C_m(t) = \begin{bmatrix} \mathbf{X}_m^T(t)\mathbf{X}_m(t)\mathbf{a}^{(m)}(t) - \mathbf{X}_m^T(t)\mathbf{x}(t) \\ \mathbf{a}^{(m)T}(t)\mathbf{X}_m^T(t)\mathbf{x}(t-m-1) - \mathbf{x}^T(t)\mathbf{x}(t-m-1) \end{bmatrix}^T \begin{bmatrix} \mathbf{b}^{(m)}(t-1) \\ -1 \end{bmatrix} .
\tag{A.29a}
$$

From (A.5a, 4a, 10a and 29a), one sees that

$$
C_m(t) = \mathbf{x}^T(t)\mathbf{x}(t-m-1) - \mathbf{a}^{(m)T}(t)\mathbf{X}_m^T(t)\mathbf{x}(t-m-1) ,
\tag{A.29b}
$$

or, using (A.17b),

$$C_m(t) = - \left[c_{m+1}^b(t-1),\ c_m^b(t-1),\ \ldots,\ c_1^b(t-1) \right] \begin{bmatrix} -1 \\ \mathbf{a}^{(m)}(t) \end{bmatrix} . \qquad \text{(A.29c)}$$

On the other hand

$$C_m(t) = \left[-1 \quad \mathbf{a}^{(m)^T}(t) \right]$$

$$\times \begin{bmatrix} \mathbf{x}^T(t)\,\mathbf{X}_m(t)\,\mathbf{b}^{(m)}(t-1)\ -\ \mathbf{x}^T(t)\mathbf{x}(t-m-1) \\ \mathbf{X}_m^T(t)\,\mathbf{X}_m(t)\,\mathbf{b}^{(m)}(t-1)\ -\ \mathbf{X}_m^T(t)\mathbf{x}(t-m-1) \end{bmatrix} . \qquad \text{(A.30a)}$$

From (A.5b, 4b, 10b and 30a), one sees that

$$C_m(t) = \mathbf{x}^T(t)\mathbf{x}(t-m-1)\ -\ \mathbf{x}^T(t)\,\mathbf{X}_m(t)\,\mathbf{b}^{(m)}(t-1)\ , \qquad \text{(A.30b)}$$

or, using (A.17a),

$$C_m(t) = - \left[c_1^f(t),\ c_2^f(t),\ \ldots,\ c_{m+1}^f(t) \right] \begin{bmatrix} \mathbf{b}^{(m)}(t-1) \\ -1 \end{bmatrix} . \qquad \text{(A.30c)}$$

From (A.29a and 30a) one finds

$$\mathbf{a}^{(m)^T}(t)\,\mathbf{X}_m^T(t)\,\mathbf{X}_m(t)\mathbf{b}^{(m)}(t-1) = \mathbf{b}^{(m)^T}(t-1)\mathbf{X}_m^T(t)\mathbf{x}^T(t)\ , \qquad \text{(A.31a)}$$

$$\mathbf{a}^{(m)^T}(t)\,\mathbf{X}_m^T(t)\,\mathbf{X}_m(t)\mathbf{b}^{(m)}(t-1) = \mathbf{a}^{(m)^T}(t)\mathbf{X}_m^T(t)\mathbf{x}(t-m-1)\ . \qquad \text{(A.31b)}$$

We note at this point that the Levinson type ladder algorithm presented in Table 7.5 can be speeded up to $1.5p^2 + O(p)$ computations per recursion if (7.102g) is replaced by either (A.29c) or (A.30c). These recursions make a better use of the shift-invariance properties of a recursively updated covariance matrix. When (A.30c) is used, the recursive least-squares Levinson algorithm of Table 7.5 is initialized from the time-recursively updated sequence of forward autocorrelation coefficients. In the case of (A.29c), the time-recursively updated sequence of backward autocorrelation coefficients is used as input to the recursive least-squares Levinson algorithm of Table 7.5 at each time step.

We may discover some other interesting partitioning schemes. From (A.29b, 30b and 23a,b), as well as (A.15a,b), it follows that

$$
\begin{bmatrix} \mathbf{x}^T(t)\mathbf{x}(t-m-1) & \mathbf{x}^T(t-m-1)\mathbf{X}_m(t) \\ \mathbf{X}_m^T(t)\mathbf{x}(t) & \mathbf{X}_m^T(t)\mathbf{X}_m(t) \end{bmatrix} \begin{bmatrix} -1 \\ \mathbf{a}^{(m)}(t) \end{bmatrix} = \begin{bmatrix} -C_m(t) \\ 0 \\ 0 \end{bmatrix}, \quad \text{(A.32a)}
$$

$$
\begin{bmatrix} \mathbf{X}_m^T(t)\mathbf{X}_m(t) & \mathbf{X}_m^T(t)\mathbf{x}(t-m-1) \\ \mathbf{x}^T(t)\mathbf{X}_m(t) & \mathbf{x}^T(t)\mathbf{x}(t-m-1) \end{bmatrix} \begin{bmatrix} \mathbf{b}^{(m)}(t-1) \\ -1 \end{bmatrix} = \begin{bmatrix} 0 \\ 0 \\ -C_m(t) \end{bmatrix}, \quad \text{(A.32b)}
$$

and from (A.25a and 29b), it follows that

$$
\Phi^{(m)}(t) \begin{bmatrix} -1 \\ \mathbf{a}^{(m-1)}(t) \\ 0 \end{bmatrix} = \begin{bmatrix} -E_{m-1}(t) \\ 0 \\ \vdots \\ 0 \\ -C_{m-1}(t) \end{bmatrix}. \quad \text{(A.33a)}
$$

Similarly, (A.25b and 29b) give

$$
\Phi^{(m)}(t) \begin{bmatrix} 0 \\ \mathbf{b}^{(m-1)}(t-1) \\ -1 \end{bmatrix} = \begin{bmatrix} -C_{m-1}(t) \\ 0 \\ \vdots \\ 0 \\ -R_{m-1}(t-1) \end{bmatrix}. \quad \text{(A.33b)}
$$

We find a last type of partitioning scheme in that

$$
\Phi^{(m)}(t) \begin{bmatrix} 0 \\ -1 \\ \mathbf{a}^{(m-1)}(t-1) \end{bmatrix} = \begin{bmatrix} U_{m-1}(t) \\ -E_{m-1}(t-1) \\ 0 \\ \vdots \\ 0 \end{bmatrix}, \quad \text{(A.34a)}
$$

where

$$
U_{m-1}(t) = \begin{bmatrix} c_1^f(t), & c_2^f(t), & \ldots, & c_m^f(t) \end{bmatrix} \begin{bmatrix} -1 \\ \mathbf{a}^{(m-1)}(t-1) \end{bmatrix} \quad \text{(A.34b)}
$$

and, analogously,

$$
\Phi^{(m)}(t) \begin{bmatrix} \mathbf{b}^{(m-1)}(t) \\ -1 \\ 0 \end{bmatrix} = \begin{bmatrix} 0 \\ \vdots \\ 0 \\ -R_{m-1}(t) \\ V_{m-1}(t) \end{bmatrix}, \quad \text{(A.35a)}
$$

where

$$V_{m-1}(t) = \left[c_m^b(t), \ c_{m-1}^b(t), \ \ldots, \ c_1^b(t) \right] \begin{bmatrix} \mathbf{b}^{(m-1)}(t) \\ \\ -1 \end{bmatrix} , \qquad (A.35b)$$

which concludes our discussion of useful partitioning schemes.

A.2 New PORLA Algorithms and Their Systolic Array Implementation

Two new and previously unpublished PORLA algorithms [A.1] for implementation on systolic arrays are presented in this section. The first algorithm is just a speeded-up version of the Levinson type least-squares ladder algorithm listed in Table 7.5. As already mentioned in Sect. A.1, a computationally attractive version of this algorithm can be obtained by using either (A.29c) or (A.30c) for a fast computation of the covariance $C_m(t)$. This algorithm, named ARRAYLAD 1, computes literally everything: transversal predictor parameters, ladder reflection coefficients, and residual energies. The prediction errors themselves can be obtained from either a transversal or a ladder form prediction error filter operated with the estimated coefficient sets.

In many cases, it is sufficient to compute only the ladder reflection coefficients and the residual energies. In this case, one can define an intermediate recursion variable, namely, the so-called *generalized covariance*, which circumvents the direct computation of transversal predictor parameters and leads to an algorithm that is particularly well suited to a systolic array implementation. This algorithm was named ARRAYLAD 2. It is shown that in the stationary case ARRAYLAD 2 converges to an algorithm that is very similar to the LeRoux-Gueguen algorithm [A.2], but differs from its classical counterpart in that the inner loop recursion in the new algorithm requires access to only two recursion variables, whereas the LeRoux-Gueguen algorithm requires access to four recursion variables. Hence, the new scheme might be of advantage in those cases where the memory access is the limitation.

A.2.1 Triangular Array Ladder Algorithm ARRAYLAD 1

The recursions of ARRAYLAD 1 may be grouped as follows. First compute the forward and backward residual energies via (A.20a,b). Next use either (A.29c) or (A.30c) for the computation of the covariance $C_m(t)$. When (A.30c) is used, the algorithm is initialized from the time-recursively updated forward autocorrelation coefficients $\{c_i^f(t), \ 0 \leq i \leq p\}$, where p is the maximum order of the algorithm. When (A.29c) is used, the time-recursively updated backward autocorrelation coefficients $\{c_i^b(t),$

$0 \leq i \leq p\}$ may apply for initialization. After the covariance has been computed, the forward/backward transversal predictor parameter sets are computed order recursively using the Levinson type recursions (A.26a,b). The computation of the ladder reflection coefficients via (A.19a,b) concludes the algorithm. Table A.1 provides a summary of this algorithm named ARRAYLAD 1. In this example, relationship (A.30c) was used for the computation of the covariance $C_m(t)$.

We note that ARRAYLAD 1 has quite a similar structure to the Levinson–Durbin algorithm. In fact, it can be viewed as a *"time-recursive"* formulation of the Levinson–Durbin algorithm [A.3, 4] in that it computes both the transversal predictor parameters and the ladder reflection coefficients at a computational complexity of $1.5p^2$ multiplications and additions per recursion. An inspection of ARRAYLAD 1 furthermore reveals that two Levinson type recursions [A.3] are required. The structure of these Levinson type recursions indicates that forward and backward linear prediction are two strongly interrelated problems, which cannot be separated as long as the observed process is *nonstationary*, i.e., as long as the correlation coefficients change with time. In the stationary case, however, the backward autocorrelation sequence becomes equal to the forward autocorrelation sequence and the transversal backward predictor coefficient set is just the reversely ordered forward transversal predictor coefficient set. In this case, both Levinson recursions in ARRAYLAD 1 are identical and the algorithm simply turns into the classical Levinson–Durbin algorithm [A.3, 4].

A.2.2 Triangular Array Ladder Algorithm ARRAYLAD 2

In some cases, knowledge about the transversal predictor parameters is not required and it is sufficient to compute only the ladder reflection coefficients. In this case, a simplified algorithm may be derived in that one first defines a *generalized forward covariance* $C_{m,j}^f(t)$ and an associated *generalized backward covariance* $C_{m,j}^b(t)$ as follows:

$$C_{m,j}^f(t) = -\left[\Phi_{j,0}(t), \Phi_{j,1}(t), \ldots, \Phi_{j,m}(t)\right] \begin{bmatrix} -1 \\ \mathbf{a}^{(m)}(t) \end{bmatrix}, \qquad \text{(A.37a)}$$

$$C_{m,j}^b(t) = -\left[\Phi_{j,0}(t), \Phi_{j,1}(t), \ldots, \Phi_{j,m}(t)\right] \begin{bmatrix} \mathbf{b}^{(m)}(t) \\ -1 \end{bmatrix}. \qquad \text{(A.37b)}$$

A simple comparison of (A.37a) with the definition of the covariance (A.29c), recalling the definition of the covariance matrix (A.16) and (A.17a,b), gives

Table A.1. Triangular array ladder algorithm ARRAYLAD 1. The algorithm is initialized from the time-recursively updated autocorrelation coefficients. See Chap. 2 for a summary of the most important procedures for the time-recursive computation of autocorrelation coefficients using higher-order windows. This algorithm involves division. Whenever the divisor is small and positive, set $1/x = x$. When the divisor is negative, set $1/x = 0$

FOR $t = 0, 1, 2, \ldots$

Input: forward autocorrelation sequence $c_0^f(t), c_1^f(t), \ldots, c_p^f(t)$

Initialize: $E_0(t) = R_0(t) = c_0^f(t)$

$$C_0(t) = c_1^f(t)$$

$$K_1^f(t) = - c_1^f(t) \big/ c_0^f(t-1)$$

$$K_1^b(t) = - c_1^f(t) \big/ c_0^f(t)$$

FOR $m = 1, 2, \ldots, p - 1$ *compute:*

$$E_m(t) = E_{m-1}(t) + K_m^f(t)\, C_{m-1}(t) \tag{A.36a}$$

$$R_m(t) = R_{m-1}(t-1) + K_m^b(t)\, C_{m-1}(t) \tag{A.36b}$$

$$C_m(t) = - \left[c_1^f(t),\, c_2^f(t),\, \ldots,\, c_{m+1}^f(t) \right] \begin{bmatrix} b^{(m)}(t-1) \\ -1 \end{bmatrix} \tag{A.36c}$$

$$\begin{bmatrix} -1 \\ a^{(m)}(t) \end{bmatrix} = \begin{bmatrix} -1 \\ a^{(m-1)}(t) \\ 0 \end{bmatrix} + K_m^f(t) \begin{bmatrix} 0 \\ b^{(m-1)}(t-1) \\ -1 \end{bmatrix} \tag{A.36d}$$

$$\begin{bmatrix} b^{(m)}(t) \\ -1 \end{bmatrix} = \begin{bmatrix} 0 \\ b^{(m-1)}(t-1) \\ -1 \end{bmatrix} + K_m^b(t) \begin{bmatrix} -1 \\ a^{(m-1)}(t) \\ 0 \end{bmatrix} \tag{A.36e}$$

$$K_{m+1}^f(t) = - C_m(t) \big/ R_m(t-1) \tag{A.36f}$$

$$K_{m+1}^b(t) = - C_m(t) \big/ E_m(t) \tag{A.36g}$$

$$C^f_{m,m+1}(t) = C_m(t) , \tag{A.38}$$

where we have also used the shift-invariance property of the covariance matrix (A.16b). Note the "generalized covariance" intermediate variables have very many similarities to the generalized residual energies used by Cumani and Strobach in the context of the PORLA technique.

Clearly, we may substitute the Levinson type recursions (A.26a,b) into the definition of the generalized forward/backward covariance (A.37a,b) to obtain

$$C^f_{m,j}(t) = - \left[\Phi_{j,0}(t), \Phi_{j,1}(t), \ldots, \Phi_{j,m}(t) \right] \begin{bmatrix} -1 \\ a^{(m-1)}(t) \\ 0 \end{bmatrix}$$

$$- K^f_m(t) \left[\Phi_{j,0}(t), \Phi_{j,1}(t), \ldots, \Phi_{j,m}(t) \right] \begin{bmatrix} 0 \\ b^{(m-1)}(t-1) \\ -1 \end{bmatrix} , \tag{A.39a}$$

$$C^b_{m,j}(t) = - \left[\Phi_{j,0}(t), \Phi_{j,1}(t), \ldots, \Phi_{j,m}(t) \right] \begin{bmatrix} 0 \\ b^{(m-1)}(t-1) \\ -1 \end{bmatrix}$$

$$- K^b_m(t) \left[\Phi_{j,0}(t), \Phi_{j,1}(t), \ldots, \Phi_{j,m}(t) \right] \begin{bmatrix} -1 \\ a^{(m-1)}(t) \\ 0 \end{bmatrix} . \tag{A.39b}$$

Exploiting the shift-invariance property of the covariance matrix (A.16b), we may express the right sides of (A.39a,b) in terms of the generalized forward/backward error covariance of order (m-1) as follows:

$$C^f_{m,j}(t) = C^f_{m-1,j}(t) + K^f_m(t) C^b_{m-1,j-1}(t-1) , \tag{A.40a}$$

$$C^b_{m,j}(t) = C^b_{m-1,j-1}(t-1) + K^b_m(t) C^f_{m-1,j}(t) , \tag{A.40b}$$

and hence, one of the three $0.5p^2$ loops in ARRAYLAD 1 has been annihilated; the price paid is that the transversal predictor parameters are no longer computed. The resulting algorithm was named ARRAYLAD 2 and is listed in Table A.2.

Table A.2. Triangular array ladder algorithm ARRAYLAD 2. The algorithm is initialized from the time-recursively updated sequence of autocorrelation coefficients. This algorithm involves division. Whenever the divisor is small and positive, set $1/x = x$. When the divisor is negative, set $1/x = 0$

FOR t = 0, 1, 2, . . .

Input: forward autocorrelation sequence $c_0^f(t), c_1^f(t), \ldots, c_p^f(t)$

Initialize: $E_0(t) = R_0(t) = c_0^f(t)$

$$K_1^f(t) = -c_1^f(t)\big/c_0^f(t-1)$$

$$K_1^b(t) = -c_1^f(t)\big/c_0^f(t)$$

FOR j = 1, 2, 3, . . . , p

$$\left[\; C_{0,j}^f(t) = c_j^f(t)\; ; \qquad C_{0,j}^b(t) = c_j^f(t)\right.$$

FOR m = 1, 2, . . . , p − 1 *compute:*

$$E_m(t) = E_{m-1}(t) + K_m^f(t)\,C_{m-1,m}^f(t) \tag{A.41a}$$

$$R_m(t) = R_{m-1}(t-1) + K_m^b(t)\,C_{m-1,m}^f(t) \tag{A.41b}$$

FOR j = m+1, m+2, . . . , p *compute:*

$$\left[\; C_{m,j}^f(t) = C_{m-1,j}^f(t) + K_m^f(t)\,C_{m-1,j-1}^b(t-1)\right. \tag{A.41c}$$

$$\left.\; C_{m,j}^b(t) = C_{m-1,j-1}^b(t-1) + K_m^b(t)\,C_{m-1,j}^f(t)\right. \tag{A.41d}$$

$$K_{m+1}^f(t) = -C_{m,m+1}^f(t)\big/R_m(t-1) \tag{A.41e}$$

$$K_{m+1}^b(t) = -C_{m,m+1}^f(t)\big/E_m(t) \tag{A.41f}$$

A.2.3 Systolic Array Implementation

This section presents the triangular array structures [A.5, 6] of ARRAYLAD 1 and ARRAYLAD 2. The structures are easily derived from the algorithms listed in Tables A.1 and A.2. It turns out that one requires a common boundary cell for both algorithms. This boundary cell is shown in Fig. A.2. The internal (or "rotational") cell of ARRAYLAD 1 is shown

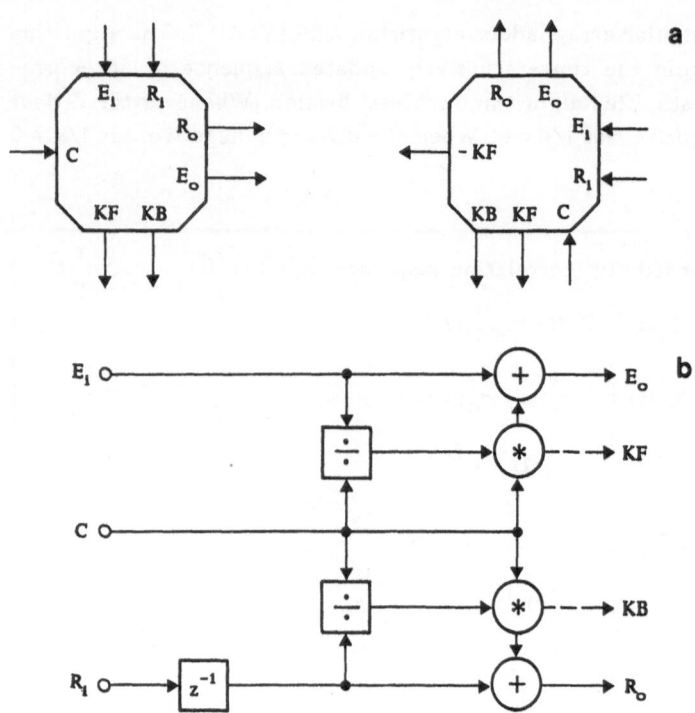

Fig. A.2. Boundary cell of ARRAYLAD 1 and ARRAYLAD 2. (a) Block symbol. (b) Internal structure

in Fig. A.3, whereas Fig. A.4 is the internal cell of ARRAYLAD 2. Figure A.5 is the structure of the triangular array of ARRAYLAD 1 using the boundary cell of Fig. A.2 and the internal cell of Fig. A.3. Finally, the triangular systolic array of ARRAYLAD 2 is shown in Fig. A.6.

A.2.4 Comparison of Ladder and Givens Rotors

The triangular array structures of ARRAYLAD 1 and ARRAYLAD 2 require further discussion. Clearly, the main component in both arrays is the *ladder processor* of Fig. A.4. This processor performs a *plane rotation* according to

$$
\begin{bmatrix} x' \\ y' \end{bmatrix} = \begin{bmatrix} 1 & K_m^f \\ K_m^b & 1 \end{bmatrix} \begin{bmatrix} x \\ y \end{bmatrix} , \tag{A.42a}
$$

where $[x, y]^T$ is a vector in the two-dimensional space and $[x', y']^T$ is the *rotated* vector. The "ladder rotor" defined by (A.42a) may be compared with the now widely accepted *Givens rotor* [A.7] discussed in Chaps. 3 and 4. The Givens rotor was defined as [recall (3.33a)]

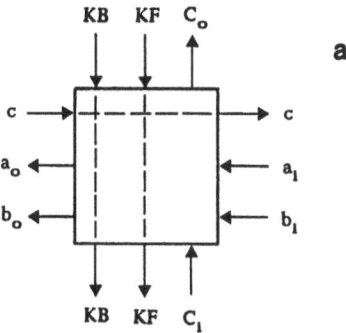

a

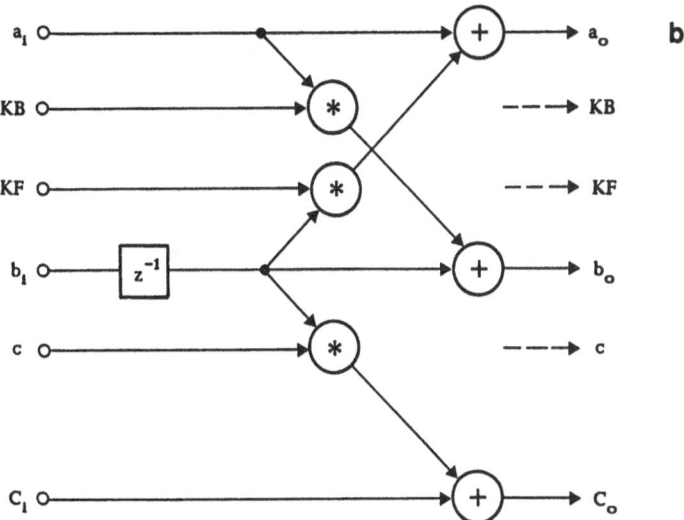

b

Fig. A.3. Internal cell of ARRAYLAD 1. (a) Block symbol. Dotted line means through-connection. (b) Internal structure

$$
\begin{bmatrix} x' \\ y' \end{bmatrix} = \begin{bmatrix} c & s \\ -s & c \end{bmatrix} \begin{bmatrix} x \\ y \end{bmatrix} \quad , \tag{A.42b}
$$

where $[c]$ and $[s]$ denote cosine and sine of a *rotation angle*. The Givens rotation (A.42b) is *circular* whereas the ladder rotor (A.42a) performs a *noncircular* rotation of the two-dimensional vector $[x, y]^T$. Givens rotors are used in the Gentleman and Kung array [A.8, 9] (see also Chap. 4) where one is interested in solving the recursive least-squares problem by direct recursive triangularization of the *observation matrix* $\mathbf{X}_m(t)$. The input in the case of the Gentleman and Kung array based on the Givens rotor (A.42b) is the data sequence $\{x(t)\}$. On the other hand, the input to

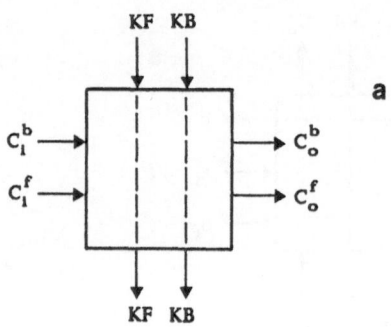

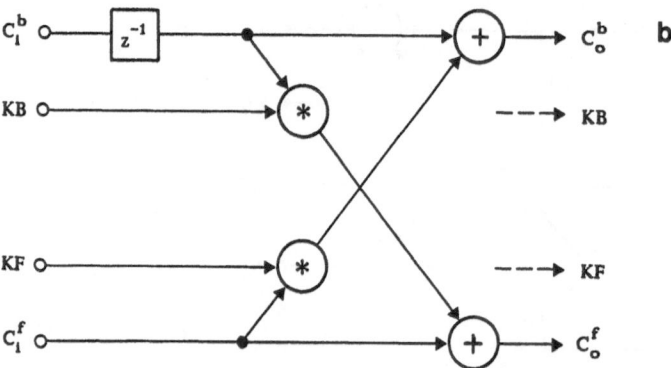

Fig. A.4. Internal cell of ARRAYLAD 2. (a) Block symbol. Dotted line means through-connection. (b) Internal structure

ARRAYLAD 2 based on the ladder rotors is the sequence of autocorrelation coefficients $\{c_i^f(t)\}$. This implies that, in the case of the ladder rotors, one assumes that the Normal Equations are *explicitly* computed. The explicit computation of the Normal Equations *increases* the condition number of the underlying least-squares problem. Hence one can generally state that the Givens processors in the Gentleman and Kung array will require a smaller dynamic range than the ladder processors of ARRAYLAD 2. On the other hand, the computation of the ladder rotor requires only two divisions, whereas the determination of the sines and cosines in the Givens rotor are based on the computation of "strange" operations like $(a^2 + b^2)^{1/2}$. A limitation of the arrays presented in this section is that we assumed that the observation matrix $X_m(t)$ is formed from *subsequent* vectors of observation $x(t-1)$, $x(t-2)$, . . . , $x(t-m)$. The Gentleman and Kung array discussed in Chap. 4 does not make this limiting assumption, and therefore it can be used for the processing of *spatial* data, for example, the prediction of a process $x_0(t)$ from sensor array inputs $x_1(t)$, $x_2(t)$, . . . , $x_m(t)$.

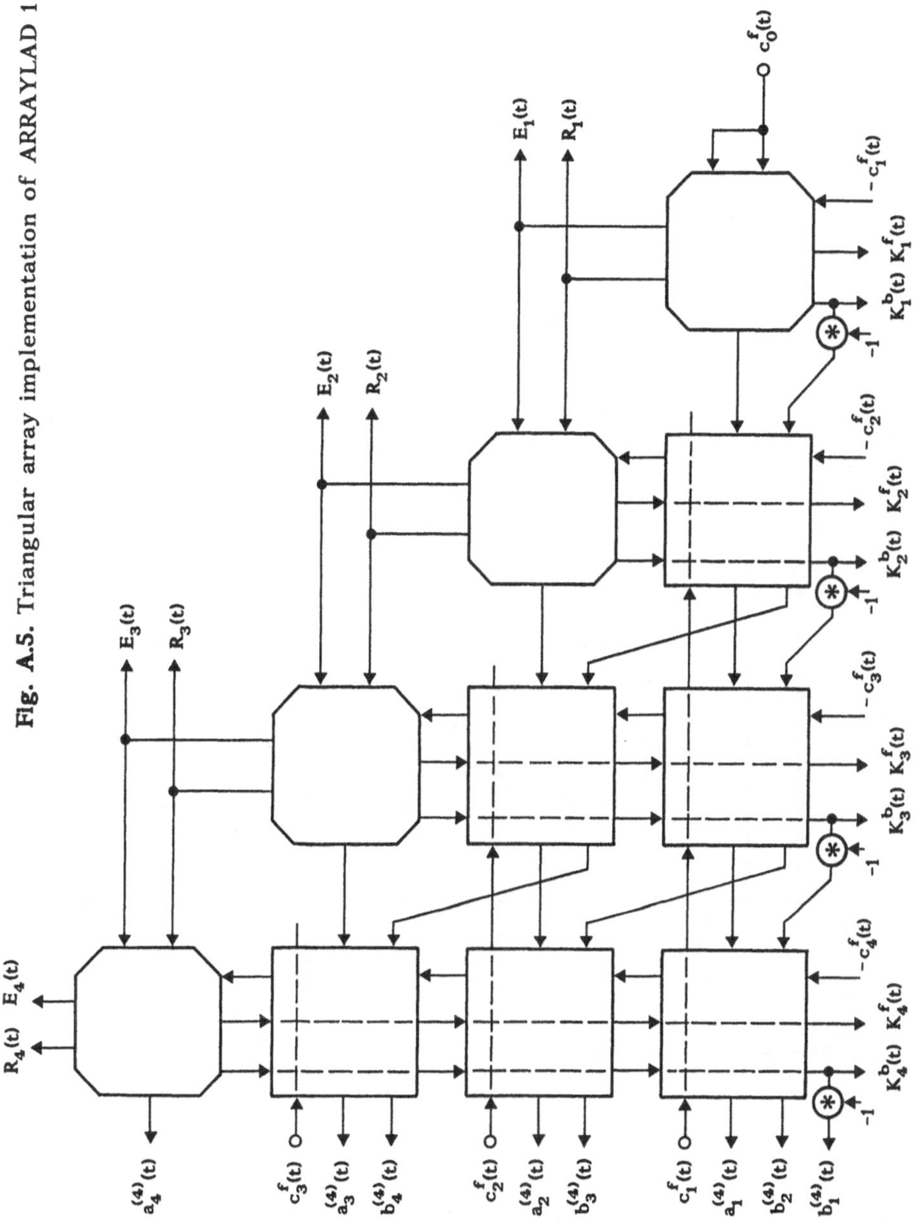

Fig. A.5. Triangular array implementation of ARRAYLAD 1

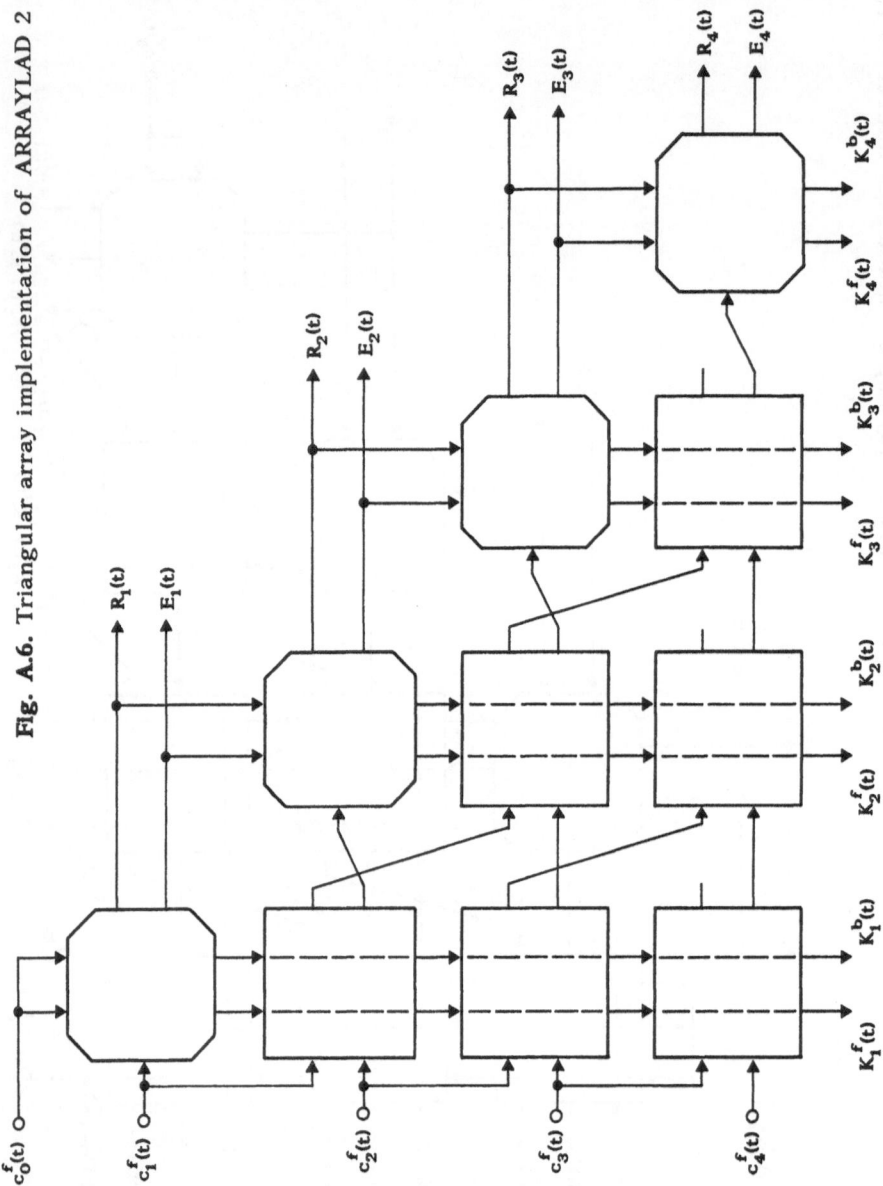

Fig. A.6. Triangular array implementation of ARRAYLAD 2

On the other hand, the Gentleman and Kung array based on the Givens processor requires *twice* as many multiplications as ARRAYLAD 2 at the same order since the Givens processor (A.42b) requires twice as many multiplications as the ladder processor (A.42a). A second hard limitation of the Gentleman and Kung array is that this array allows only a simple *exponential* weighting of past data. The steady-state behavior of simple exponentially weighted least-squares estimation is generally poor due to the strong influence of the actual (incoming) data. The parameters appear to be very noisy due to the *spectral distortion* inherent to the simple exponential window. Another drawback of exponential weighting is that the tracking behavior is also poor. This is a consequence of the slow decay of the exponential window. In fact, Bershad and Macchi [A.10] showed recently that even a simple LMS algorithm can have a better tracking behavior than an exponentially weighted RLS algorithm in some cases.

The ARRAYLAD algorithms of this appendix, however, offer the rather nice and useful feature of higher-order recursive windowing in the computation of the autocorrelation sequence used as input in the arrays. See Chap. 2 for a summary of some useful recursive windowing algorithms. In particular, the structure of the arrays is *not* affected by the particular kind of windowing involved in the autocorrelation sequence computation, since the time-recursive computation of autocorrelation coefficients is completely *decoupled* from the pure order recursive computations in the array.

Another advantage of the ladder rotors is that their parameters, namely, the ladder reflection coefficients, have a *meaningful interpretation* as the coefficients of the ladder form adaptive filter, which provides a useful parametrization of an observed time series and an analogous representation of many physical processes. Numerous techniques have been developed for the analysis of the ladder parameters. The Givens rotors, on the other hand, do not have such a meaningful interpretation. The computation of transversal predictor parameters or ladder reflection coefficients from the Gentleman and Kung array necessitates a lengthy computation of an ill-conditioned back-substitution process, which almost negates the improvements in the accuracy of the rotors.

Both the Gentleman and Kung array and ARRAYLAD 2 are *systolic arrays* [A.5,6] where a computation wavefront propagates through the array in a distinct direction. Moreover, one easily finds that ARRAYLAD 2 can be implemented using Givens processors, where the two direct path multipliers in the processor are simply omitted or a multiplication by 1 is performed. In the case of ARRAYLAD 1, one can decompose the structure into a *systolic* part, which computes the Levinson recursions (A.36d,e), using again the ladder processors of ARRAYLAD 2, and a *parallel* processing part, which computes the covariance $C_m(t)$ according to (A.36c), which is simply the computation of an *inner product* of increasing length when the order of the algorithm increases. Thus, ARRAYLAD 1 and ARRAYLAD 2 present two processing schemes that compute the more familiar transversal and ladder parameters instead of the less meaningful Givens rotors. The structures of the underlying arrays, however, were found to be very similar, and one can even use components of a Gentleman and Kung array to construct, for example, the array of ARRAYLAD 2 in a simple way. Besides its application as a "rotational" cell in the triangular array section, the ladder processor has found numerous applications in the area of sensor array processing and orthogonalization of multichannel processes [A.11].

A.2.5 A Square-Root PORLA Algorithm

In many applications where linear prediction algorithms must be implemented on fixed-point signal processors, one is interested in algorithms with limited or small dynamic range requirements of intermediate variables. A simple way to compress the dynamic range of intermediate algorithm variables is square-root normalization. Square-root normalized algorithms have been discussed in the case of the RLS algorithm (recall Chap. 5) and in the case of Lee's RLS ladder algorithm (recall Chap. 9). This section presents a square-root normalized version of ARRAYLAD 2. For this purpose, consider the following square-root normalization of the generalized covariance (A.37a,b):

$$\rho_{m+1,j}^f(t) = C_{m,j}^f(t)E_m^{-1/2}(t)R_m^{-1/2}(t-1) \quad , \tag{A.43a}$$

$$\rho_{m+1,j}^b(t) = C_{m,j}^b(t)E_m^{-1/2}(t)R_m^{-1/2}(t-1) \quad , \tag{A.43b}$$

where $\rho_{m+1,j}^f(t)$ and $\rho_{m+1,j}^b(t)$ are the normalized generalized forward and normalized generalized backward covariance, respectively. The reflection coefficients $K_m^f(t)$ and $K_m^b(t)$ may then be expressed in terms of normalized variables

$$K_m^f(t) = - C_{m-1,m}^f(t) R_{m-1}^{-1}(t-1) = - \rho_{m,m}^f(t)E_{m-1}^{1/2}(t)R_{m-1}^{-1/2}(t-1), \tag{A.44a}$$

$$K_m^b(t) = - C_{m-1,m}^f(t) E_m^{-1}(t) \quad = - \rho_{m,m}^f(t)E_{m-1}^{-1/2}(t)R_{m-1}^{1/2}(t-1). \tag{A.44b}$$

Furthermore,

$$K_m^f(t) K_m^b(t) = \rho_{m,m}^{f\ 2}(t) \quad . \tag{A.45}$$

A postmultiplication of the inner-loop recursions (A.41c,d) of ARRAYLAD 2 with the square-root product $E_m^{-1/2}(t)R_m^{-1/2}(t-1)$ gives

$$\rho_{m+1,j}^{f}(t) = C_{m-1,j}^{f}(t) E_m^{-1/2}(t) R_m^{-1/2}(t-1)$$

$$+ K_m^{f}(t) C_{m-1,j-1}^{b}(t-1) E_m^{-1/2}(t) R_m^{-1/2}(t-1) \quad , \qquad (A.46a)$$

$$\rho_{m+1,j}^{b}(t) = C_{m-1,j-1}^{b}(t-1) E_m^{-1/2}(t) R_m^{-1/2}(t-1)$$

$$+ K_m^{b}(t) C_{m-1,j}^{f}(t) E_m^{-1/2}(t) R_m^{-1/2}(t-1) \quad . \qquad (A.46b)$$

Introducing the intermediate variables $\alpha_m^{f}(t)$ and $\alpha_m^{b}(t)$

$$\alpha_m^{f}(t) = E_{m-1}^{1/2}(t) E_m^{-1/2}(t) R_{m-1}^{1/2}(t-1) R_m^{-1/2}(t-1) \quad , \qquad (A.47a)$$

$$\alpha_m^{b}(t) = E_{m-1}^{1/2}(t-1) E_m^{-1/2}(t) R_{m-1}^{1/2}(t-2) R_m^{-1/2}(t-1) \quad , \qquad (A.47b)$$

one sees that in the normalized case, the inner-loop recursion of ARRAY-LAD 2 takes on the form

$$\rho_{m+1,j}^{f}(t) = \alpha_m^{f}(t) \rho_{m,j}^{f}(t) + K_m^{f}(t) \alpha_m^{b}(t) \rho_{m,j-1}^{b}(t-1) \quad , \qquad (A.48a)$$

$$\rho_{m+1,j}^{b}(t) = \alpha_m^{b}(t) \rho_{m,j-1}^{b}(t-1) + K_m^{b}(t) \alpha_m^{f}(t) \rho_{m,j}^{f}(t) \quad . \qquad (A.48b)$$

A comparison of (A.45) and (A.21a,b) reveals that the square-root normalized forward and backward residual energies can be updated according to

$$E_m^{1/2}(t) = E_{m-1}^{1/2}(t) \left[1 - \rho_{m,m}^{f\,2}(t) \right]^{1/2} \quad , \qquad (A.49a)$$

$$R_m^{1/2}(t) = R_{m-1}^{1/2}(t-1) \left[1 - \rho_{m,m}^{f\,2}(t) \right]^{1/2} \quad , \qquad (A.49b)$$

which completes the derivation. Table A.3 is a summary of this square-root normalized version of ARRAYLAD 2.

A quick inspection of the algorithm listed in Table A.3 reveals that the inner loop requires twice as many multiplications as the inner-loop recursion of the unnormalized version of ARRAYLAD 2. The structure of the computations is in fact identical to the Givens rotor (A.42b). One can therefore use the rotational cell of a Gentleman and Kung array for the computation of (A.50j,k). The main difference of this algorithm is again that it allows arbitrary windowing of the data sequence in contrast to the recursive Givens reduction (Gentleman and Kung array) which allows only exponential windowing at the same computational cost.

Table A.3. Square-root normalized version of ARRAYLAD 2. The algorithm is initialized from the time-recursively updated sequence of autocorrelation coefficients. This algorithm involves division. Whenever the divisor is small and positive, set $1/x = x$. When the divisor is negative, set $1/x = 0$

FOR $t = 0, 1, 2, \ldots$

Input: forward autocorrelation sequence $c_0^f(t), c_1^f(t), \ldots, c_p^f(t)$

Initialize: $E_0^{1/2}(t) = R_0^{1/2}(t) = \left[c_0^f(t) \right]^{1/2}$

FOR $j = 1, 2, 3, \ldots, p$ *normalize:*

$$\rho_{1,j}^f(t) = \rho_{1,j}^b(t) = c_j^f(t) E_0^{-1/2}(t) R_0^{-1/2}(t-1) \qquad \text{(A.50a)}$$

$$K_1^f(t) = - \rho_{1,1}^f(t) E_0^{1/2}(t) R_0^{-1/2}(t-1) \qquad \text{(A.50b)}$$

$$K_1^b(t) = - \rho_{1,1}^f(t) E_0^{-1/2}(t) R_0^{1/2}(t-1) \qquad \text{(A.50c)}$$

FOR $m = 1, 2, \ldots, p-1$ *compute:*

$$E_m^{1/2}(t) = E_{m-1}^{1/2}(t) \left[1 - \rho_{m,m}^{f\,2}(t) \right]^{1/2} \qquad \text{(A.50d)}$$

$$R_m^{1/2}(t) = R_{m-1}^{1/2}(t-1) \left[1 - \rho_{m,m}^{f\,2}(t) \right]^{1/2} \qquad \text{(A.50e)}$$

$$\alpha_m^f(t) = E_{m-1}^{1/2}(t) E_m^{-1/2}(t) R_{m-1}^{1/2}(t-1) R_m^{-1/2}(t-1) \qquad \text{(A.50f)}$$

$$\alpha_m^b(t) = E_{m-1}^{1/2}(t-1) E_m^{-1/2}(t) R_{m-1}^{1/2}(t-2) R_m^{-1/2}(t-1) \qquad \text{(A.50g)}$$

$$\beta_m^f(t) = K_m^f(t) \alpha_m^b(t) \qquad \text{(A.50h)}$$

$$\beta_m^b(t) = K_m^b(t) \alpha_m^f(t) \qquad \text{(A.50i)}$$

FOR $j = m+1, m+2, \ldots, p$ *compute:*

$$\rho_{m+1,j}^f(t) = \alpha_m^f(t) \rho_{m,j}^f(t) + \beta_m^f(t) \rho_{m,j-1}^b(t-1) \qquad \text{(A.50j)}$$

$$\rho_{m+1,j}^b(t) = \alpha_m^b(t) \rho_{m,j-1}^b(t-1) + \beta_m^b(t) \rho_{m,j}^f(t) \qquad \text{(A.50k)}$$

$$K_{m+1}^f(t) = - \rho_{m+1,m+1}^f(t) E_m^{1/2}(t) R_m^{-1/2}(t-1) \qquad \text{(A.50l)}$$

$$K_{m+1}^b(t) = - \rho_{m+1,m+1}^f(t) E_m^{-1/2}(t) R_m^{1/2}(t-1) \qquad \text{(A.50l)}$$

A.2.6 A Step Towards Toeplitz Systems

We note that ARRAYLAD 2 can be used to derive an algorithm for solving the Yule-Walker equations [A.12]. For this purpose, we simply investigate the algorithm listed in Table A.2 for the case of a time-invariant sequence of autocorrelation coefficients. All quantities will be time invariant. Clearly, in this case, the forward/backward residual energies will attain identical values, hence we may omit the computation of R_m. The forward/backward reflection coefficients will be identical, and hence they may be replaced by a common reflection coefficient K_m. The resulting algorithm is listed in Table A.4 and can be viewed as another "fixed-point implementation" of the Levinson-Durbin algorithm. Interestingly, the algorithm is in some sense more symmetric than the LeRoux-Gueguen algorithm of Table 8.1, since in the new algorithm the right sides of the inner-loop recursions are based on only *two* data variables.

Table A.4. Another "fixed-point computation of partial correlation coefficients" as deduced from ARRAYLAD 2. This algorithm involves division. Whenever the divisor is small and positive, set $1/x = x$. When the divisor is negative, set $1/x = 0$

Input: autocorrelation sequence $c_0^f, c_1^f, \ldots, c_p^f$

Initialize: $E_0 = c_0^f$

$$K_1 = -c_1^f / c_0^f$$

FOR $j = 1, 2, 3, \ldots, p$

$$\left[\; C_{0,j}^f = c_j^f \quad ; \qquad C_{0,j}^b = c_j^f \right.$$

FOR $m = 1, 2, \ldots, p-1$ *compute:*

$$E_m = E_{m-1} + K_m C_{m-1,m}^f \tag{A.51a}$$

FOR $j = m+1, m+2, \ldots, p$ *compute:*

$$C_{m,j}^f = C_{m-1,j}^f + K_m C_{m-1,j-1}^b \tag{A.51b}$$

$$C_{m,j}^b = C_{m-1,j-1}^b + K_m C_{m-1,j}^f \tag{A.51c}$$

$$K_{m+1} = -C_{m,m+1}^f / E_m \tag{A.51d}$$

A.3 Vector Case of New PORLA Algorithms

In this section, we shall derive the recursions of ARRAYLAD 1 and ARRAYLAD 2 for the case of *vector-valued data*. Both *recursive least-squares* and *block processing* versions of the algorithms will be obtained.

We recall the discussion on the estimation of vector autoregressive (VAR) processes given in Sect. 10.3. In the vector-valued case, each observation sample consists of a number q of components. When time progresses, one obtains q interrelated time series $x_1(t)$, $x_2(t)$, . . . , $x_q(t)$. These time series form the *input signal matrix* $S(t)$ [recall (10.52)]

$$S(t) = \left[\ x_1(t), \ x_2(t), \ \ldots \ , \ x_q(t) \right] \ . \tag{A.52}$$

Note again that the input signal matrix $S(t)$ is just the counterpart to the signal vector $x(t)$ in the scalar case. Thus, the *observation block matrix* $\mathbf{X}_m(t)$ is the counterpart to $X_m(t)$ in the scalar case and is therefore defined as

$$\mathbf{X}_m(t) = \left[\ S(t-1), \ S(t-2), \ \ldots \ , \ S(t-m) \right] \ . \tag{A.53}$$

The vector autoregressive process is then completely determined by the *covariance block matrix* $\mathbf{\Phi}^{(m)}(t)$ defined as

$$\mathbf{\Phi}^{(m)}(t) = \left[\Phi_{i,j}(t) \right] \ , \tag{A.54a}$$

$$\Phi_{i,j}(t) = S^T(t-i) \ S(t-j) \ , \tag{A.54b}$$

with the partitioning schemes

$$\mathbf{\Phi}^{(m)}(t) = \begin{bmatrix} S^T(t) \ S(t) & S^T(t) \ \mathbf{X}_m(t) \\ \mathbf{X}_m^T(t) \ S(t) & \mathbf{X}_m^T(t) \ \mathbf{X}_m(t) \end{bmatrix} \ , \tag{A.55a}$$

$$\mathbf{\Phi}^{(m)}(t) = \begin{bmatrix} \mathbf{X}_m^T(t+1) \ \mathbf{X}_m(t+1) & \mathbf{X}_m^T(t+1) \ S(t-m) \\ S^T(t-m) \ \mathbf{X}_m(t+1) & S^T(t-m) \ S(t-m) \end{bmatrix} \ . \tag{A.55b}$$

The *forward and backward prediction error matrices* $e_m(t)$ and $r_m(t)$ are then given by

$$e_m(t) = S(t) - \hat{S}_m^f(t) \ , \tag{A.56a}$$

$$r_m(t) = S(t-m) - \hat{S}_m^b(t) \ . \tag{A.56b}$$

Recall that we assume that the observed data $S(t)$ has been generated by a *vector autoregressive (VAR) process model* of the type

$$S(t) = y(t) + \sum_{j=1}^{m} S(t-j) a_j^{(m)}(t) \quad , \tag{A.57}$$

where $a_j^{(m)}(t)$ is the (matrix-valued) autoregressive parameter of stage j and dimension q×q of the mth order model at time t and $y(t)$ is a vector-valued white Gaussian source.

We shall define the *forward/backward transversal predictor parameter block vectors* $\mathbf{A}^{(m)}(t)$ and $\mathbf{B}^{(m)}(t)$ as follows:

$$\mathbf{A}^{(m)}(t) = \begin{bmatrix} a_1^{(m)}(t) \\ a_2^{(m)}(t) \\ \vdots \\ a_m^{(m)}(t) \end{bmatrix} \quad , \tag{A.58a}$$

$$\mathbf{B}^{(m)}(t) = \begin{bmatrix} b_1^{(m)}(t) \\ b_2^{(m)}(t) \\ \vdots \\ b_m^{(m)}(t) \end{bmatrix} \quad . \tag{A.58b}$$

The predicted process matrices $\hat{S}_m^f(t)$ and $\hat{S}_m^b(t)$ may then be expressed as

$$\hat{S}_m^f(t) = \mathbf{X}_m(t)\,\mathbf{A}^{(m)}(t) \quad , \tag{A.59a}$$

$$\hat{S}_m^b(t) = \mathbf{X}_m(t+1)\,\mathbf{B}^{(m)}(t) \quad . \tag{A.59b}$$

Similarly to the scalar case, one finds the relationships

$$e_m^T(t)\,\hat{S}_m^f(t) = 0 \quad , \tag{A.60a}$$

$$r_m^T(t)\,\hat{S}_m^b(t) = 0 \quad , \tag{A.60b}$$

$$S^T(t)e_m(t) = e_m^T(t)\,e_m(t) = E_m(t) \quad , \tag{A.61a}$$

$$S^T(t-m)r_m(t) = r_m^T(t)\,r_m(t) = R_m(t) \quad , \tag{A.61b}$$

$$\mathbf{e}_m^T(t)\mathbf{S}(t-m-1) = \mathbf{e}_m^T(t)\mathbf{r}_m(t-1) = \mathbf{C}_m(t) \quad , \tag{A.62a}$$

$$\mathbf{S}^T(t)\mathbf{r}_m(t-1) = \mathbf{e}_m^T(t)\mathbf{r}_m(t-1) = \mathbf{C}_m(t) \quad . \tag{A.62b}$$

Consequently, we may describe the *vector-valued transversal forward/backward prediction error filters* as follows:

$$\mathbf{e}_m(t) = \mathbf{S}(t) - \mathbf{X}_m(t)\,\mathbf{A}^{(m)}(t) \quad , \tag{A.63a}$$

$$\mathbf{r}_m(t) = \mathbf{S}(t-m) - \mathbf{X}_m(t+1)\,\mathbf{B}^{(m)}(t) \quad , \tag{A.63b}$$

where $\mathbf{E}_m(t)$ and $\mathbf{R}_m(t)$ are the *forward/backward residual energy matrices* of dimension q×q, and $\mathbf{C}_m(t)$ is the covariance of dimension q×q.

Continuing this line of thought, we shall define the *forward and backward autocorrelation block vectors* consisting of the autocorrelation matrices $\mathbf{C}_i^f(t)$ and $\mathbf{C}_i^b(t)$,

$$\mathbf{C}_i^f(t) = \mathbf{S}^T(t)\mathbf{S}(t-i) \quad , \tag{A.64a}$$

$$\mathbf{C}_i^b(t) = \mathbf{S}^T(t-m)\mathbf{S}(t-m+i) \quad . \tag{A.64b}$$

The forward and backward autocorrelation block vectors can be viewed as the upper row block vector and the reversely ordered lower row block vectors of the covariance block matrix. The *vector-valued ladder form prediction error filter* is then constituted by

$$\mathbf{e}_m(t) = \mathbf{e}_{m-1}(t) + \mathbf{r}_{m-1}(t-1)\mathbf{K}_m^f(t) \quad , \tag{A.65a}$$

$$\mathbf{r}_m(t) = \mathbf{r}_{m-1}(t-1) + \mathbf{e}_{m-1}(t)\mathbf{K}_m^b(t) \quad , \tag{A.65b}$$

where

$$\mathbf{K}_m^f(t) = -\,\mathbf{R}_{m-1}^{-1}(t-1)\mathbf{C}_{m-1}^T(t) \quad , \tag{A.66a}$$

$$\mathbf{K}_m^b(t) = -\,\mathbf{E}_{m-1}^{-1}(t)\mathbf{C}_{m-1}(t) \quad . \tag{A.66b}$$

The quantities $\mathbf{K}_m^f(t)$ and $\mathbf{K}_m^b(t)$ are the *forward/backward ladder reflection matrices* of dimension q×q.

Combining (A.62a,b) and (A.63a,b), one finds

$$\mathbf{C}_m(t) = \left[\, \mathbf{S}^T(t) - \mathbf{A}^{(m)^T}(t)\,\mathbf{X}_m^T(t) \,\right]\mathbf{S}(t-m-1) \quad , \tag{A.67a}$$

$$\mathbf{C}_m(t) = \mathbf{S}^T(t)\left[\, \mathbf{S}(t-m-1) - \mathbf{X}_m(t)\mathbf{B}^{(m)}(t-1) \,\right] \quad . \tag{A.67b}$$

This is an important result since we may now compute the covariance $\mathbf{C}_m(t)$ as the inner product of the transversal predictor parameter block vector and the autocorrelation block vector. Two versions of the computation of $\mathbf{C}_m(t)$, related to forward and backward linear prediction, respectively, can be discovered:

$$
\mathbf{C}_m(t) = - \begin{bmatrix} -\mathbf{I} & \mathbf{A}^{(m)\,T}(t) \end{bmatrix}
\begin{bmatrix}
\mathbf{c}^b_{m+1}(t-1) \\
\mathbf{c}^b_m(t-1) \\
\vdots \\
\mathbf{c}^b_1(t-1)
\end{bmatrix} ,
\tag{A.68a}
$$

$$
\mathbf{C}_m(t) = - \begin{bmatrix} \mathbf{c}^f_1(t), & \mathbf{c}^f_2(t), & \ldots, & \mathbf{c}^f_{m+1}(t) \end{bmatrix}
\begin{bmatrix}
\mathbf{B}^{(m)}(t-1) \\
\\
-\mathbf{I}
\end{bmatrix} .
\tag{A.68b}
$$

In the VAR case, the Levinson type recursions take on the form

$$
\begin{bmatrix} -\mathbf{I} \\ \\ \mathbf{A}^{(m)}(t) \end{bmatrix}
=
\begin{bmatrix} -\mathbf{I} \\ \mathbf{A}^{(m-1)}(t) \\ \mathbf{0} \end{bmatrix}
+
\begin{bmatrix} \mathbf{0} \\ \mathbf{B}^{(m-1)}(t-1) \\ -\mathbf{I} \end{bmatrix} \mathbf{K}^f_m(t) ,
\tag{A.69a}
$$

$$
\begin{bmatrix} \mathbf{B}^{(m)}(t) \\ \\ -\mathbf{I} \end{bmatrix}
=
\begin{bmatrix} \mathbf{0} \\ \mathbf{B}^{(m-1)}(t-1) \\ -\mathbf{I} \end{bmatrix}
+
\begin{bmatrix} -\mathbf{I} \\ \mathbf{A}^{(m-1)}(t) \\ \mathbf{0} \end{bmatrix} \mathbf{K}^b_m(t) .
\tag{A.69b}
$$

A simple combination of (A.65a,b) with (A.61a,b and 62a,b) gives the desired order recursions of forward/backward residual energy matrices as follows:

$$\mathbf{E}_m(t) = \mathbf{E}_{m-1}(t) + \mathbf{C}_{m-1}(t)\mathbf{K}_m^f(t) \quad , \tag{A.70a}$$

$$\mathbf{R}_m(t) = \mathbf{R}_{m-1}(t-1) + \mathbf{C}_{m-1}^T(t)\mathbf{K}_m^b(t) \quad . \tag{A.70b}$$

A.3.1 Vector Case of ARRAYLAD 1

We may now establish an order recursive scheme for computing the forward/backward transversal predictor parameter block vectors and the ladder reflection matrices. This procedure is the vector-valued equivalent to ARRAYLAD 1 presented in Sect. A.2. The vector-valued algorithm is therfore easily obtained from the scalar version (Table A.1). Table A.5 is a complete summary of the vector-valued version of ARRAYLAD 1. In this algorithm, one first computes the order update of the residual energy matrices via (A.70a,b). The order recursive computation of forward/backward transversal predictor parameter block vectors is accomplished by the Levinson type recursion (A.69a,b), while the covariance is computed either using the forward autocorrelation matrices (A.68b), or the backward autocorrelation matrices (A.68a). Without loss of generality, we restrict the discussion to the case using the forward autocorrelation matrices and just mention the fact that very similar algorithms may be constructed when the backward autocorrelation matrices are available.

A.3.2 Computation of Reflection Matrices

Note that the vector-valued algorithms require the explicit computation of expressions of the type $\mathbf{K}^f = -\mathbf{R}^{-1}\mathbf{C}^T$ and $\mathbf{K}^b = -\mathbf{E}^{-1}\mathbf{C}$. In many applications, particularly when the dimension of the involved matrices is large (e.g., for $q = 256$) the forward and backward prediction error energy matrices $\mathbf{E}$ and $\mathbf{R}$ are often *singular*. For this reason, the computation of the reflection coefficients presents a hard subproblem. This author has found, however, that in this case it is particularly useful to employ the concept of the *Penrose pseudoinverse* [A.13,14] in the computation of the reflection matrices.

When using the Penrose pseudoinverse, (A.63f,g) can be rewritten as

$$\mathbf{K}_{m+1}^f(t) = -\mathbf{R}_m^+(t-1)\mathbf{C}_m^T(t) \quad , \tag{A.72a}$$

$$\mathbf{K}_{m+1}^b(t) = -\mathbf{E}_m^+(t)\mathbf{C}_m(t) \quad . \tag{A.72b}$$

See Sects. 3.3.3 and 3.3.4 for an introductory discussion of the pseudo-inverse concept in this context. A FORTRAN coded routine for computing expressions of the type (A.72a,b) may be found in the book by Lawson and Hanson [A.13]. This is the subroutine SVDRS, which calls QRBD, H12, G1, G2, and the function DIFF. In the subroutine QRBD one may replace statement 118

Table A.5. Vector case of ARRAYLAD 1. This algorithm involves matrix inversion. Use Lawson and Hanson's subroutine SVDRS [A.13] for computation of (A.71f,g). See also the comments in Sect. A.3.2

FOR t = 0, 1, 2, . . .

Input: forward autocorrelation sequence $C_0^f(t), C_1^f(t), \ldots, C_p^f(t)$

Initialize: $E_0(t) = R_0(t) = C_0^f(t)$; $C_0(t) = C_1^f(t)$

$\qquad\qquad K_1^f(t) = - C_0^{f^{-1}}(t-1)\, C_1^{f^T}(t)$; $K_1^b(t) = - C_0^{f^{-1}}(t)\, C_1^f(t)$

FOR m = 1, 2, . . . , p − 1 *compute:*

$$E_m(t) = E_{m-1}(t) + C_{m-1}(t)K_m^f(t) \tag{A.71a}$$

$$R_m(t) = R_{m-1}(t-1) + C_{m-1}^T(t)K_m^b(t) \tag{A.71b}$$

$$\begin{bmatrix} -I \\ \\ A^{(m)}(t) \end{bmatrix} = \begin{bmatrix} -I \\ \\ A^{(m-1)}(t) \\ \\ 0 \end{bmatrix} + \begin{bmatrix} 0 \\ \\ B^{(m-1)}(t-1) \\ \\ -I \end{bmatrix} K_m^f(t) \tag{A.71c}$$

$$\begin{bmatrix} B^{(m)}(t) \\ \\ -I \end{bmatrix} = \begin{bmatrix} 0 \\ \\ B^{(m-1)}(t-1) \\ \\ -I \end{bmatrix} + \begin{bmatrix} -I \\ \\ A^{(m-1)}(t) \\ \\ 0 \end{bmatrix} K_m^b(t) \tag{A.71d}$$

$$C_m(t) = - \begin{bmatrix} C_1^f(t), & C_2^f(t), & \ldots, & C_{m+1}^f(t) \end{bmatrix} \begin{bmatrix} B^{(m)}(t-1) \\ \\ -I \end{bmatrix} \tag{A.71e}$$

$$K_{m+1}^f(t) = - R_m^{-1}(t-1)C_m^T(t) \tag{A.71f}$$

$$K_{m+1}^b(t) = - E_m^{-1}(t)C_m(t) \tag{A.71g}$$

118   G = SQRT(ONE + F**2)

by the following chain of statements:

118-1   IF (ABS (F) .GT. 1E10) THEN
118-2   G = ABS(F)
118-3   ELSE
118-4   G = SQRT(ONE + F**2)
118-5   ENDIF

A.3.3 Multichannel Prediction Error Filters

After estimating the transversal predictor parameter matrices, or the ladder reflection matrices, one is sometimes interested in prediction error filtering of the observed process. This section presents the explicit structures of the associated prediction error filters in the vector-valued (multichannel) case.

Considering (A.63a), one obtains the calculation of the top component of the residual matrix $\mathbf{e}_m(t)$ in the case of forward prediction error filtering as follows:

$$
\left[e_1(t), e_2(t), \ldots, e_q(t) \right] = \left[x_1(t), x_2(t), \ldots, x_q(t) \right]
$$

$$
+ \left[x_1(t-1), \ldots, x_q(t-1); x_1(t-2), \ldots, x_q(t-2); \ldots \right.
$$

$$
\left. \ldots; x_1(t-m), \ldots, x_q(t-m) \right] \mathbf{A}^{(m)}(t) . \tag{A.73}
$$

Using (A.58a,b), we find the multichannel forward prediction error filter is given by

$$
\begin{bmatrix} e_1(t) \\ e_2(t) \\ \vdots \\ e_q(t) \end{bmatrix} = \begin{bmatrix} x_1(t) \\ x_2(t) \\ \vdots \\ x_q(t) \end{bmatrix} + \sum_{j=1}^{m} \mathbf{a}_j^{(m)^T}(t) \begin{bmatrix} x_1(t-j) \\ x_2(t-j) \\ \vdots \\ x_q(t-j) \end{bmatrix} . \tag{A.74}
$$

The ladder form prediction error filter in the multichannel case is then constituted by the relations

$$
\begin{bmatrix} e_1^{(m)}(t) \\ e_2^{(m)}(t) \\ \vdots \\ e_q^{(m)}(t) \end{bmatrix} = \begin{bmatrix} e_1^{(m-1)}(t) \\ e_2^{(m-1)}(t) \\ \vdots \\ e_q^{(m-1)}(t) \end{bmatrix} + \mathbf{K}_m^{f^T}(t) \begin{bmatrix} r_1^{(m-1)}(t-1) \\ r_2^{(m-1)}(t-1) \\ \vdots \\ r_q^{(m-1)}(t-1) \end{bmatrix} , \qquad (A.75a)
$$

$$
\begin{bmatrix} r_1^{(m)}(t) \\ r_2^{(m)}(t) \\ \vdots \\ r_q^{(m)}(t) \end{bmatrix} = \begin{bmatrix} r_1^{(m-1)}(t-1) \\ r_2^{(m-1)}(t-1) \\ \vdots \\ r_q^{(m-1)}(t-1) \end{bmatrix} + \mathbf{K}_m^{b^T}(t) \begin{bmatrix} e_1^{(m-1)}(t) \\ e_2^{(m-1)}(t) \\ \vdots \\ e_q^{(m-1)}(t) \end{bmatrix} . \qquad (A.75b)
$$

A.3.4 Vector Case of ARRAYLAD 2

Sometimes, one is only interested in computing the ladder reflection matrices. In those cases, the vector-valued Levinson type recursion in the vector-valued version of ARRAYLAD 1 (Table A.5) can be replaced by a more efficient computation scheme involving an intermediate recursion variable, the so-called *generalized forward/backward covariance*.

By comparing the definition of the covariance block matrix of the observed process (A.54a,b) with the definition of the forward/backward autocorrelation block vectors (A.64a,b), we can write

$$
\mathbf{C}_i^f(t) = \boldsymbol{\Phi}_{0,i}(t) \quad , \qquad (A.76a)
$$

$$
\mathbf{C}_i^b(t) = \boldsymbol{\Phi}_{m,m-i}(t) \quad . \qquad (A.76b)
$$

Using these expressions, and noting that $\boldsymbol{\Phi}_{i,j}(t) = \boldsymbol{\Phi}_{j,i}^T(t)$, the computation of the covariance $\mathbf{C}_m(t)$ may alternatively be expressed as

$$
\mathbf{C}_m(t) = -\left[\boldsymbol{\Phi}_{m+1,0}^T(t), \; \boldsymbol{\Phi}_{m+1,1}^T(t), \; \ldots, \; \boldsymbol{\Phi}_{m+1,m}^T(t) \right] \begin{bmatrix} -\mathbf{I} \\ \\ \\ \mathbf{A}^{(m)}(t) \end{bmatrix} , \qquad (A.77a)
$$

$$\mathbf{C}_m(t) = - \left[\Phi_{0,1}(t), \Phi_{0,2}(t), \ldots, \Phi_{0,m+1}(t) \right] \begin{bmatrix} \mathbf{B}^{(m)}(t-1) \\ \\ -\mathbf{I} \end{bmatrix} . \qquad (A.77b)$$

In the vector-valued case, one defines the generalized forward/backward covariance as

$$\mathbf{C}_{m,j}^{f}(t) = - \left[\Phi_{j,0}^{T}(t), \Phi_{j,1}^{T}(t), \ldots, \Phi_{j,m}^{T}(t) \right] \begin{bmatrix} -\mathbf{I} \\ \\ \mathbf{A}^{(m)}(t) \end{bmatrix} , \qquad (A.78a)$$

$$\mathbf{C}_{m,j}^{b}(t) = - \left[\Phi_{j,0}^{T}(t), \Phi_{j,1}^{T}(t), \ldots, \Phi_{j,m}^{T}(t) \right] \begin{bmatrix} \mathbf{B}^{(m)}(t) \\ \\ -\mathbf{I} \end{bmatrix} . \qquad (A.78b)$$

Substituting the Levinson type recursions (A.69a,b) into the expressions for the generalized covariance (A.78a,b), one obtains a generalized covariance recursion as follows:

$$\mathbf{C}_{m,j}^{f}(t) = \mathbf{C}_{m-1,j}^{f}(t) + \mathbf{C}_{m-1,j-1}^{b}(t-1) \mathbf{K}_{m}^{f}(t) , \qquad (A.79a)$$

$$\mathbf{C}_{m,j}^{b}(t) = \mathbf{C}_{m-1,j-1}^{b}(t-1) + \mathbf{C}_{m-1,j}^{f}(t) \mathbf{K}_{m}^{b}(t) , \qquad (A.79b)$$

which is simply the vector-valued counterpart recursion to (A.41a,b). Finally, we note that the generalized covariance is related to the "conventional" covariance via the expression

$$\mathbf{C}_m(t) = \mathbf{C}_{m,m+1}^{f}(t) . \qquad (A.80)$$

The complete algorithm based on the recursions (A.79a,b) is listed in Table A.6.

Table A.6. Vector case of ARRAYLAD 2. This algorithm involves matrix inversion. Use Lawson and Hanson's subroutine SVDRS [A.13] for computation of (A.81e,f). See also the comments in Sect. A.3.2

FOR t = 0, 1, 2, . . .

Input: forward autocorrelation sequence $C_0^f(t), C_1^f(t), \ldots, C_p^f(t)$

Initialize: $E_0(t) = R_0(t) = C_0^f(t)$

$$K_1^f(t) = - C_0^{f^{-1}}(t-1) \, C_1^{f^T}(t)$$

$$K_1^b(t) = - C_0^{f^{-1}}(t) \, C_1^f(t)$$

FOR j = 1, 2, . . . , p *initialize:*

$$\left[C_{0,j}^f(t) = C_j^f(t) \quad ; \qquad\qquad C_{0,j}^b(t) = C_j^f(t) \right.$$

FOR m = 1, 2, . . . , p − 1 *compute:*

$$E_m(t) = E_{m-1}(t) + C_{m-1,m}^f(t) K_m^f(t) \tag{A.81a}$$

$$R_m(t) = R_{m-1}(t-1) + C_{m-1,m}^{f^T}(t) K_m^b(t) \tag{A.81b}$$

FOR j = m+1, m+2, . . . , p

$$\left[C_{m,j}^f(t) = C_{m-1,j}^f(t) + C_{m-1,j-1}^b(t-1) \, K_m^f(t) \tag{A.81c} \right.$$

$$\left. C_{m,j}^b(t) = C_{m-1,j-1}^b(t-1) + C_{m-1,j}^f(t) \, K_m^b(t) \tag{A.81d} \right.$$

$$K_{m+1}^f(t) = - R_m^{-1}(t-1) C_{m-1,m}^{f^T}(t) \tag{A.81e}$$

$$K_{m+1}^b(t) = - E_m^{-1}(t) C_{m-1,m}^f(t) \tag{A.81f}$$

A.3.5 Stationary Case – Block Processing Algorithms

In the vector-valued case, a true recursive least-squares updating of the matrix-valued parameters seems impractical in many applications due to the $O(p^2 q^3)$ dependence of the computational costs. This fact frequently motivates a *blockwise processing* of the data, in that the autocorrelation block vector is recursively updated for a certain number of measured input data vectors. The parameters are subsequently extracted by a block processing algorithm. Such block processing algorithms for the vector case can be deduced from the recursive algorithms listed in Table A.5 and Table A.6 simply by omitting the time index.

A.3.6 Concluding Remarks

We have presented the triangular array ladder algorithms ARRAYLAD 1 and ARRAYLAD 2, which were discovered by the author shortly after the manuscript of this book was completed. Obviously, the new techniques exhibit the same algorithm structure as Strobach's early PORLA algorithms [A.1], i.e., the new algorithms are initialized from the time-recursively updated covariance (second-order) information of the observed process (recall Fig. 8.3). The true least-squares ladder form is again constructed by a procedure that is entirely based on numerically stable and robust recursions involving residual energies and transversal predictor or ladder predictor parameters. Numerically uncertain residual signals are not used in the recursions. Throughout the tests, ARRAYLAD 2 , in particular, has proven to have outstanding numerical properties. This algorithm can be highly recommended.

It has become apparent that in the stationary case ARRAYLAD 1 is equivalent to the classical Levinson-Durbin algorithm. But, interestingly, in the case of stationary data, ARRAYLAD 2 is not equivalent to the LeRoux-Gueguen algorithm. We were able to derive a similar fixed-point implementation of the Levinson-Durbin algorithm by simply using ARRAYLAD 2 as a basis. Similarly, it is also possible to derive PORLA type algorithms for the general recursive least-squares case from a generalization of the LeRoux-Gueguen algorithm. This was recently demonstrated by Sokat [A.15, 16]. See Sect. 8.3.4.

Besides these interesting theoretical ideas, we have demonstrated the systolic array structures of the new algorithms. Particularly the structure of ARRAYLAD 2 deserves some attention. In this case, it turns out that the internal ("rotational") cell has exactly the same structure as a ladder section. This is an important result, since now the ladder form prediction error filter and the triangular array section can be constructed by using exactly *the same* type of processor cells. The rotational cell in the case of ARRAYLAD 2 requires only two multipliers, in contrast to the Gentleman and Kung array, which requires four multipliers per cell while being much less flexible in terms of data windowing. These considerations show that the previously unpublished scheme ARRAYLAD 2 is likely to be one of the most attractive methods for solving the recursive least-squares problem stably and efficiently. A comparison of the structures of the ARRAYLAD algorithms and the Givens rotor underlying the Gentleman and Kung array revealed that the basic design of a Gentleman and Kung array can easily be modified to obtain the structure of an ARRAYLAD algorithm.

In Sect. A.3, we extended the algorithms ARRAYLAD 1 and ARRAYLAD 2 to the more general vector autoregressive case. Explicit listings of the algorithms were presented and we discussed special problems appearing in the VAR case, such as the computation of $\mathbf{K} = \mathbf{X}^{-1}\mathbf{Y}$, a central problem in the case of VAR estimation algorithms. Concluding this appendix, we showed how the algorithms can be used in the stationary (block processing) case.

Other algorithms with triangular structure were developed by Ling, Manolakis and Proakis [A.17]. This algorithm is an extension of the well-known modified Gram-Schmidt algorithm [A.13] to the recursive case. Most recently, triangular array algorithms were also developed by Kalouptsidis and Theodoridis [A.18], and by Theodoridis, Kalouptsidis and Bakirtzis [A.19]. A pipelined architecture for triangular array algorithms was considered by Theodoridis [A.20]. Ling [A.21] showed how one can obtain a triangular array ladder algorithm based on Givens rotations. He also discusses the relationships between the Gentleman and Kung array (recursive Givens reduction) and the modified recursive Gram-Schmidt algorithm [A.17]. Triangular array algorithms for least-squares FIR filtering, parameter estimation and prediction are becoming increasingly important because of their inherent ability for an easy implementation on highly concurrent systolic structures.

References

Chapter 1

1.1 C.F. Gauss (1809): Theoria Motus Corporum Coelestium. English translation: *Theory of the Motion of Heavenly Bodies* (Dover, New York 1963)

1.2 G.E.P. Box, G.M. Jenkins: *Time Series Analysis: Forecasting and Control* (Holden Day, San Francisco 1970)

1.3 G.M. Jenkins, D.G. Watts: *Spectral Analysis and its Applications* (Holden Day, San Francisco 1968)

1.4 G.C. Goodwin, R.L. Payne: *Dynamic System Identification, Experimental Design and Data Analysis* (Academic, New York 1977)

1.5 A.S. Willsky: *Digital Signal Processing and Control and Estimation Theory* (MIT Press, Cambridge, MA 1979)

1.6 C.L. Lawson, R.J. Hanson: *Solving Least-Squares Problems* (Prentice Hall, Englewood Cliffs, NJ 1974)

1.7 N. Wiener: *Extrapolation, Interpolation, and Smoothing of Stationary Time Series* (MIT Press, Cambridge, MA 1942)

1.8 R.E. Kalman: A new approach to linear filtering and prediction problems. J. Basic Eng. **82**, 35–45 (1960)

1.9 S. Haykin: *Adaptive Filter Theory* (Prentiçe Hall, Englewood Cliffs, NJ 1985)

1.10 D. Godard: Channel equalization using a Kalman filter for fast data transmission. IBM J. Res. Dev. **18**, 267–273 (1974)

1.11 R.L. Plackett: Some theorems in least squares. Biometrika **37**, 149 (1950)

1.12 J. Sherman, W.J. Morrison: Adjustment of an inverse matrix corresponding to the changes in the elements of a given column or a given row of the original matrix. Ann. Math. Stat. **20**, 621 (1949)

1.13 M.D. Srinath, P.K. Rajasekaran: *An Introduction to Statistical Signal Processing with Applications* (Wiley, New York 1979)

1.14 L. Ljung, M. Morf, D.D. Falconer: Fast calculation of gain matrices for recursive estimation schemes. Int. J. Control **27**, 1–19 (1978)

1.15 G. Carayannis, D. Manolakis, N. Kalouptsidis: A fast sequential algorithm for least-squares filtering and prediction. IEEE Trans. ASSP **31**, 1394–1402 (1983)

378   References

1.16   J.M. Cioffi: The fast adaptive rotors RLS algorithm. IEEE Trans. ASSP, to appear (1989)

1.17   W.M. Gentleman: Least-squares computation by Givens transformations without square roots. J. Inst. Math. Appl. **12**, 329-336 (1973)

1.18   H.T. Kung, W.M. Gentleman: Matrix triangularization by systolic arrays. Proc. SPIE, Real-Time Signal Processing IV, **298**, 16 (1981)

1.19   J. McWhirter: Recursive least-squares minimization using a systolic array. Proc. SPIE, Real-Time Signal Processing VI, **431**, 18-26 (1983)

1.20   J.W. Givens: Numerical computation of the characteristic values of a real symmetric matrix. Oak Ridge National Laboratory, ORNR 1574, Internal Report (1954)

1.21   G. Kubin: Stabilization of the RLS algorithm in the absence of persistent excitation. Proc. IEEE Int. Conf. on ASSP, New York (1988) pp. 1369-1372

1.22   B. Widrow, M.E. Hoff, Jr.: Adaptive switching circuits. Weston Conf., IRE, Part 4, 96-104 (1960)

1.23   M. Morf: Ladder forms in estimation and system identification. Proc. 11th Asilomar Conf. Circuits Systems and Computers, Monterey, CA (1977) pp. 424-429

1.24   J. Heinhold: *Einführung in die Höhere Mathematik* (Carl Hanser, München 1976)

1.25   N. Levinson: The Wiener RMS (root-mean-square) error criterion in filter design and prediction. J. Math. Phys. **25**, 261-278 (1947)

1.26   J. Durbin: The fitting of time series models. Rev. Int. Stat. **28**, 233-244 (1960)

1.27   A. Cumani: On a covariance lattice algorithm for linear prediction. Proc. IEEE Int. Conf. on ASSP, Paris (1982) pp. 651-654

1.28   P. Strobach: Pure order recursive least-squares ladder algorithms. IEEE Trans. ASSP **34**, 880-897 (1986)

1.29   D.T.L. Lee: Canonical Ladder Form Realizations and Fast Estimation Algorithms. Ph.D. Dissertation, Stanford University, Stanford, CA (1980)

1.30   T. Kato: *Perturbation Theory for Linear Operators 2nd ed.* (Springer, Berlin, Heidelberg 1976)

1.31   J.M. Cioffi, T. Kailath: Fast, recursive least-squares transversal filters for adaptive filtering. IEEE Trans. ASSP **32**, 304-337 (1984)

1.32   M.L. Honig , D.G. Messerschmitt: *Adaptive Filters* (Cluwer, Boston, MA 1984)

1.33   L. Ljung, T. Söderström: *Theory and Practice of Recursive Identification* (MIT Press, Cambridge, MA 1983)

1.34   H.J. Larson, B.O. Shubert: *Probabilistic Models in Engineering Sciences, Volume II: Random Noise, Signals, and Dynamic Systems* (Wiley, New York 1979)

1.35 G.U. Yule (1907): "On the Theory of Correlation for Any Number of Variables, Treated by a New System of Notation", in *Statistical Papers of George Udny Yule* (Griffin, London 1971)

1.36 U. Grenander, G. Szegö: *Toeplitz Forms and Their Applications* (California Press, Berkeley, CA 1958)

1.37 J.D. Markel, A.H. Gray: *Linear Prediction of Speech,* Communications and Cybernetics, Vol. 12 (Springer, Berlin, Heidelberg 1976)

1.38 B.S. Atal, S.L. Hanauer: Speech analysis and synthesis by linear prediction of the speech wave. J. Acoust. Soc. Am. **50**, 637–655 (1971)

1.39 F. Itakura, S. Saito: Digital filtering techniques for speech analysis and synthesis. Proc. 7th Int. Conf. Acoust. **25–C–1**, Budapest (1971) pp. 261–264

1.40 J. LeRoux, C. Gueguen: A fixed-point computation of partial correlation coefficients. IEEE Trans. ASSP **25**, 257–259 (1977)

1.41 P. Delsarte, Y. Genin: The split Levinson algorithm. IEEE Trans. ASSP **34**, 470–478 (1986)

1.42 H. Krishna: New split Levinson, Schur, and lattice algorithms for digital signal processing. IEEE Int. Conf. on ASSP, New York (1988) pp. 1640–1642

1.43 H. Krishna, S.D. Morgera: The Levinson recurrence and fast algorithms for solving Toeplitz systems of linear equations. IEEE Trans. ASSP **35**, 839–848 (1987)

1.44 P. Delsarte, Y. Genin: On the splitting of classical algorithms in linear prediction theory. IEEE Trans. ASSP **35**, 645–653 (1987)

1.45 I. Schur (1917): Über Potenzreihen die im Innern des Einheitskreises beschränkt sind. J. Reine Angew. Math. **147**, 205–232; ibid. **148**, 122–145 (1917)

1.46 U. Appel, A. von Brandt: Recursive lattice algorithms with finite duration windows. Proc. IEEE Int. Conf. on ASSP, Paris (1982) pp. 647–650

1.47 N.J. Bershad, O. Macchi: Comparison of RLS and LMS algorithms for tracking a chirped signal. Proc. IEEE Int. Conf. on ASSP, Glasgow (1989) pp. 896–899

1.48 J. Makhoul: Stable and efficient lattice methods for linear prediction. IEEE Trans. ASSP **25**, 423–428 (1977)

1.49 J.P. Burg: Maximum Entropy Spectral Analysis. Ph.D. Dissertation, Stanford University, Stanford, CA (1975)

1.50 J. Sokat, P. Strobach: A new class of pure order recursive ladder algorithms. Proc. IEEE Int. Workshop on Digital Signal Processing, Stanford Sierra Lodge, Tahoe, CA (1988)

1.51 S.-Y. Kung: VLSI array processors. IEEE ASSP Magazine **2**, 4–22 (1985)

Chapter 2

2.1   N. Levinson: The Wiener RMS (root-mean-square) error criterion in filter design and prediction. J. Math. Phys. **25**, 261-278 (1947)

2.2   N. Wiener: *Extrapolation, Interpolation, and Smoothing of Stationary Time Series* (MIT Press, Cambridge, MA 1942)

2.3   H.J. Larson, B.O. Shubert: *Probabilistic Models in Engineering Sciences, Volume II: Random Noise, Signals, and Dynamic Systems* (Wiley, New York 1979)

2.4   T.C. Hsia: *System Identification* (Lexington Books, Lexington, MA 1977)

2.5   L. Ljung, T. Söderström: *Theory and Practice of Recursive Identification* (MIT Press, Cambridge, MA 1983)

2.6   G.U. Yule (1907): "On the Theory of Correlation for Any Number of Variables, Treated by a New System of Notation", in *Statistical Papers of George Udny Yule* (Griffin, London 1971)

2.7   J.D. Markel, A.H. Gray: *Linear Prediction of Speech,* Communications and Cybernetics, Vol. 12 (Springer, Berlin, Heidelberg 1976)

2.8   L.R. Rabiner, B. Gold: *Theory and Application of Digital Signal Processing* (Prentice-Hall, Englewood Cliffs, NJ 1975)

2.9   U. Appel, A. von Brandt: Recursive lattice algorithms with finite duration windows. Proc. IEEE Int. Conf. on ASSP, Paris (1982) pp. 647-650

2.10 P. Strobach: Efficient covariance ladder algorithms for finite arithmetic applications. Signal Processing **13**, 29-70 (1987)

2.11 T.P. Barnwell III: Recursive windowing for generating autocorrelation coefficients for LPC analysis. IEEE Trans. ASSP **29**, 1062-1066 (1981)

Chapter 3

3.1   C.L. Lawson, R.J. Hanson: *Solving Least-Squares Problems* (Prentice Hall, Englewood Cliffs, NJ 1974)

3.2   H.R. Schwarz, H. Rutishauser, E. Stiefel: *Numerical Analysis of Symmetric Matrices* (Prentice-Hall, Englewood Cliffs, NJ 1973)

3.3   F.S. Acton: *Numerical Methods that Work* (Harper and Row, New York 1970)

3.4   G.H. Golub: Numerical methods for solving linear least-squares problems. Numer. Math. **7**, 206-216 (1965)

3.5   B.W. Dickinson: Estimation of partial correlation matrices using Cholesky decomposition. IEEE Trans. Autom. Control **24**, 302-305 (1979)

3.6   B.W. Dickinson, J.M. Turner: Reflection coefficient estimation using Cholesky decomposition. IEEE Trans. ASSP **27**, 146-149 (1979)

3.7   D. Gibson: On reflection coefficients and the Cholesky decomposition. IEEE Trans. ASSP **25**, 93–96 (1977)

3.8   J.W. Givens: Numerical computation of the characteristic values of a real symmetric matrix. Oak Ridge National Laboratory, ORNR 1574, Internal Report (1954)

3.9   W.M. Gentleman: Least-squares computation by Givens transformations without square roots. J. Inst. Math. Appl. **12**, 329–336 (1973)

3.10   S. Hammarling: A note on modifications to the Givens plane rotation. J. Inst. Math. Appl. **13**, 215–218 (1974)

3.11   J.G. Nash, S. Hansen: Modified Faddeeva algorithm for concurrent execution of linear algebraic operations. IEEE Trans. Comput. **37**, 129–137 (1988)

3.12   J. Götze, U. Schwiegelshohn: An orthogonal method for solving systems of linear equations without square roots and with few divisions. Proc. Int. Conf. on ASSP, Glasgow (1989) pp. 1298–1300

3.13   A.S. Householder: *The Theory of Matrices in Numerical Analysis* (Blaisdell, New York 1964)

3.14   A.O. Steinhardt: Householder transforms in signal processing. IEEE ASSP Magazine **5**, 4–12 (1988)

3.15   C. Rader, A. Steinhardt: Hyperbolic Householder transformations. IEEE Trans. ASSP **34**, 1589–1602 (1986)

3.16   E.F. Deprettere (ed.): *Singular Value Decomposition and Signal Processing: Algorithms, Applications, and Architectures* (North-Holland, Amsterdam 1989)

3.17   R. Penrose: A general inverse for matrices. Proc. Cambridge Philos. Soc. **51**, 406–413 (1955)

3.18   G.H. Golub, C. Reinsch: "Singular Value Decomposition and Least-Squares Solutions", in *Handbook for Automatic Computation II, Linear Algebra,* ed. by J.H. Wilkinson, C. Reinsch (Springer, New York 1971)

Chapter 4

4.1   C.L. Lawson, R.J. Hanson: *Solving Least-Squares Problems* (Prentice Hall, Englewood Cliffs, NJ 1974)

4.2   W.M. Gentleman: Least-squares computation by Givens transformations without square roots. J. Inst. Math. Appl. **12**, 329–336 (1973)

4.3   J.W. Givens: Numerical computation of the characteristic values of a real symmetric matrix. Oak Ridge National Laboratory, ORNR 1574, Internal Report (1954)

4.4   H.T. Kung, W.M. Gentleman: Matrix triangularization by systolic arrays. Proc. SPIE, Real-Time Signal Processing IV, **298**, 16 (1981)

4.5   J. McWhirter: Recursive least-squares minimization using a systolic array. Proc. SPIE, Real-Time Signal Processing VI, **431**, 18–26 (1983)

4.6   H.J. Larson, B.O. Shubert: *Probabilistic Models in Engineering Sciences, Volume II: Random Noise, Signals, and Dynamic Systems* (Wiley, New York 1979)

4.7   H.T. Kung: Why systolic architectures. IEEE Computer **15**, 37-46 (1982)

4.8   S.-Y. Kung, H.J. Whitehouse, T. Kailath (eds.): *VLSI and Modern Signal Processing* (Prentice-Hall, Englewood Cliffs, NJ 1985)

4.9   J.E. Valder: The CORDIC trigonometric computing technique. IEEE Trans. Electromagn. Compat. **9**, 227-231 (1960)

4.10   K. Hwang: *Computer Arithmetic: Principles, Architectures, and Design* (Wiley, New York 1979)

4.11   P. Strobach: Efficient covariance ladder algorithms for finite arithmetic applications. Signal Processing **13**, 29-70 (1987)

4.12   T.P. Barnwell III: Recursive windowing for generating autocorrelation coefficients for LPC analysis. IEEE Trans. ASSP **29**, 1062-1066 (1981)

4.13   J.M. Cioffi: High-speed systolic implementation of fast QR adaptive filters. Proc. Int. Conf. on ASSP, New York (1988) pp. 1584-1587

4.14   J.M. Cioffi: The fast adaptive rotors RLS algorithm. IEEE Trans. ASSP, to appear (1989)

4.15   G.H. Golub: Numerical methods for solving linear least-squares problems. Numer. Math. **7**, 206-216 (1965)

4.16   M. Bellanger: The potential of QR adaptive filter variables for signal analysis. Proc. Int. Conf. on ASSP, Glasgow (1989) pp. 2166-2169

Chapter 5

5.1   D. Godard: Channel equalization using a Kalman filter for fast data transmission. IBM J. Res. Dev. **18**, 267-273 (1974)

5.2   R.L. Plackett: Some theorems in least squares. Biometrika **37**, 149 (1950)

5.3   R.E. Kalman: A new approach to linear filtering and prediction problems. J. Basic Eng. **82**, 35-45 (1960)

5.4   M.D. Srinath, P.K. Rajasekaran: *An Introduction to Statistical Signal Processing with Applications* (Wiley, New York 1979)

5.5   L. Ljung, M. Morf, D.D. Falconer: Fast calculation of gain matrices for recursive estimation schemes. Int. J. Control **27**, 1-19 (1978)

5.6   T.C. Hsia: *System Identification* (Lexington Books, Lexington, MA 1977)

5.7   L. Ljung, T. Söderström: *Theory and Practice of Recursive Identification* (MIT Press, Cambridge, MA 1983)

5.8   J. Sherman, W.J. Morrison: Adjustment of an inverse matrix corresponding to the changes in the elements of a given column or a given row of the original matrix. Ann. Math. Stat. **20**, 621 (1949)

5.9   J.M. Cioffi, T. Kailath: Fast, recursive least-squares transversal filters for adaptive filtering. IEEE Trans. ASSP **32**, 304–337 (1984)

5.10 G. Carayannis, D. Manolakis, N. Kalouptsidis: A fast sequential algorithm for least-squares filtering and prediction. IEEE Trans. ASSP **31**, 1394–1402 (1983)

5.11 N. Kalouptsidis, G. Carayannis, D. Manolakis: A fast covariance type algorithm for sequential LS filtering and prediction. IEEE Trans. Autom. Control **29**, 752–755 (1984)

5.12 B. Widrow, M.E. Hoff, Jr.: Adaptive switching circuits. Weston Conf. Rec., IRE, Part 4, 96–104 (1960)

5.13 B. Widrow, J.M. McCool, M.G. Larimore, C.R. Johnson: Stationary and nonstationary learning characteristics of the LMS adaptive filter. Proc. IEEE **64**, 1151–1161 (1976)

5.14 B. Widrow, S.D. Stearns: *Adaptive Signal Processing* (Prentice-Hall, Englewood Cliffs, NJ 1985)

5.15 J.E. Potter: *New Statistical Formulas*. Memo **40**, Instrumentation Laboratory, Massachusetts Institute of Technology (1963)

5.16 G. Kubin: Stabilization of the RLS algorithm in the absence of persistent excitation. Proc. Int. Conf. on ASSP, New York (1988) pp. 1369–1372

5.17 G.J. Bierman: *Factorization Methods for Discrete Sequential Estimation* (Academic, New York 1977)

5.18 M. Morf: Fast Algorithms for Multivariable Systems. Ph.D. Dissertation, Stanford University, Stanford, CA (1974)

5.19 M. Morf, T. Kailath, L. Ljung: Fast algorithms for recursive identification. Proc. IEEE Conf. Dec. Control, 916–921 (1976)

5.20 E. Eleftheriou, D.D. Falconer: Tracking properties and steady-state performance of RLS adaptive filter algorithms. IEEE Trans. ASSP **34**, 1097–1110 (1986)

5.21 J.M. Cioffi: Fast Transversal Filters for Communications Applications. Ph.D. Dissertation, Stanford University, Stanford, CA (1984)

5.22 J.M. Cioffi, T. Kailath: Windowed fast transversal filters adaptive algorithms with normalization. IEEE Trans. ASSP **33**, 607–625 (1985)

5.23 C. Samson: A unified treatment of fast Kalman algorithms for identification. Int. J. Control **35**, 909–934 (1982)

5.24 M.L. Honig, D.G. Messerschmitt: *Adaptive Filters* (Cluwer, Boston, MA 1984)

5.25 S.T. Alexander: *Adaptive Signal Processing: Theory and Applications* (Springer, New York 1986)

5.26 D.T.L. Lee: Canonical Ladder Form Realizations and Fast Estimation Algorithms. Ph.D. Dissertation, Stanford University, Stanford, CA (1980)

5.27 N.J. Bershad: On the optimum gain parameter in LMS adaptation. IEEE Trans. ASSP **35**, 1065–1067 (1987)

5.28 N.J. Bershad: Behavior of ε-normalized LMS algorithm with Gaussian input. IEEE Trans. ASSP **35**, 636-644 (1987)

5.29 N.J. Bershad, P.L. Feintuch: A normalized frequency domain LMS adaptive algorithm. IEEE Trans. ASSP **34**, 452-461 (1986)

5.30 C.F. Cowan, P.M. Grant: *Adaptive Filters* (Prentice-Hall, Englewood Cliffs, NJ 1985)

Chapter 6

6.1   J.D. Markel, A.H. Gray: *Linear Prediction of Speech*, Communications and Cybernetics, Vol. 12 (Springer, Berlin, Heidelberg 1976)

6.2   J. Heinhold: *Einführung in die Höhere Mathematik, Teil 1 - 4* (Carl-Hanser-Verlag, München 1976)

6.3   J. Makhoul: Stable and efficient lattice methods for linear prediction. IEEE Trans. ASSP **25**, 423-428 (1977)

6.4   J.P. Burg: Maximum Entropy Spectral Analysis. Ph.D. Dissertation, Stanford University, Stanford, CA (1975)

6.5   B.S. Atal: Speech analysis and synthesis by linear prediction of the speech wave. J. Acoust. Soc. Am. **47**, 65 (1970)

6.6   B.S. Atal and S.L. Hanauer: Speech analysis and synthesis by linear prediction of the speech wave. J. Acoust. Soc. Am. **50**, 637-655 (1971)

6.7   F. Itakura, S. Saito: Digital filtering techniques for speech analysis and synthesis. Proc. 7th Int. Conf. Acoust., Vol. **25-C-1**, 261-264, Budapest (1971)

6.8   B.S. Atal, M.R. Schroeder: Adaptive predictive coding of speech signals. Bell Syst. Tech. J. **49**, 1973-1986 (1970)

6.9   S.J. Mason: Feedback theory: further properties of signal flow graphs. Proc. IRE **44**, 920-926 (1956)

6.10 J.P. Burg: Maximum entropy spectral analysis. in Proc. 37th Ann. Int. Meeting, Soc. Explor. Geophysics, Oklahoma City (1967)

6.11 S.L. Marple: *Digital Spectral Analysis* (Prentice-Hall, Englewood Cliffs, NJ 1987)

6.12 S. Haykin (ed.): *Nonlinear Methods of Spectral Analysis*, Top. Appl. Phys., Vol. 34 (Springer, Berlin, Heidelberg 1979)

6.13 D.E. Boekee, H.H. Buss: "Order Estimation of Autoregressive Models" in 4th Aachener Kolloquium "Theorie und Anwendung der Signalverarbeitung", ed. by H.D. Lüke RWTH Aachen (1981) pp. 126-130

6.14 A. Fettweis: Digital filter structures related to classical filter networks. Archiv Elektr. Übertragung **25**, 78-89 (1971)

6.15 A. Fettweis: On sensitivity and round-off noise in wave digital filters. IEEE Trans. ASSP **22**, 383-384 (1974)

6.16 A. Sedlmeyer, A. Fettweis: Digital filters with true ladder configuration. Int. J. Circuit Theory Appl. **1**, 5-10 (1973)

6.17 L.T. Bruton: Low-sensitivity digital ladder filters. IEEE Trans. Circuits Syst. **22**, 168-176 (1975)

6.18 S. Kay, S.L. Marple: Spectrum analysis - A modern perspective. Proc. IEEE **69**, 1380-1416 (1981)

6.19 P.F. Fougere: A solution to the problem of spontaneous line splitting in maximum entropy power spectrum analysis. J. Geophys. Res. **82**, 1051-1054 (1977)

Chapter 7

7.1 N. Levinson: The Wiener RMS (root-mean-square) error criterion in filter design and prediction. J. Math. Phys. **25**, 261-278 (1947)

7.2 J. Durbin: The fitting of time series models. Rev. Int. Stat. **28**, 233-244 (1960)

7.3 J. LeRoux, C. Gueguen: A fixed-point computation of partial correlation coefficients. IEEE Trans. ASSP **25**, 257-259 (1977)

7.4 P. Delsarte, Y. Genin: The split Levinson algorithm. IEEE Trans. ASSP **34**, 470-478 (1986)

7.5 J. Makhoul: Stable and efficient lattice methods for linear prediction. IEEE Trans. ASSP **25**, 423-428 (1977)

7.6 G.U. Yule (1907): "On the Theory of Correlation for Any Number of Variables, Treated by a New System of Notation", in *Statistical Papers of George Udny Yule* (Griffin, London 1971)

7.7 G.U. Yule: On a method of investigating periodicities in disturbed series, with special reference to Wölfer's sunspot numbers. Philos. Trans. R. Soc. London **A226**, 267-298 (1927)

7.8 J.D. Markel, A.H. Gray: *Linear Prediction of Speech*, Communications and Cybernetics, Vol. 12 (Springer, Berlin, Heidelberg 1976)

7.9 H. Krishna: New split Levinson, Schur, and lattice algorithms for digital signal processing. in IEEE Int. Conf. on ASSP, New York (1988) pp. 1640-1642

7.10 U. Grenander, G. Szegö: *Toeplitz Forms and Their Applications* (California Press, Berkeley, CA 1958)

7.11 H. Krishna, S.D. Morgera: The Levinson recurrence and fast algorithms for solving Toeplitz systems of linear equations. IEEE Trans. ASSP **35**, 839-848 (1987)

7.12 J. Makhoul: Linear prediction: A tutorial review. Proc. IEEE **63**, 561-580 (1975)

Chapter 8

8.1   J. LeRoux, C. Gueguen: A fixed-point computation of partial correlation coefficients. IEEE Trans. ASSP **25**, 257–259 (1977)

8.2   A. Cumani: On a covariance lattice algorithm for linear prediction. Proc. IEEE Int. Conf. on ASSP, Paris (1982) pp. 651–654

8.3   J. Makhoul: Stable and efficient lattice methods for linear prediction. IEEE Trans. ASSP **25**, 423–428 (1977)

8.4   P. Strobach: Pure order recursive least-squares ladder algorithms. IEEE Trans. ASSP **34**, 880–897 (1986)

8.5   P. Strobach: A new approach to the least-squares ladder estimation algorithm. Proc. IASTED Int. Symp. Appl. Signal Proc. Digital Filtering, Paris (1985) pp. 37–40

8.6   J. Sokat, P. Strobach: A new class of pure order recursive ladder algorithms. Proc. IEEE Int. Workshop on Digital Signal Processing, Stanford Sierra Lodge, Tahoe, CA (1988)

8.7   J. Sokat: The split generalized LeRoux-Gueguen algorithm for least-squares problems. Proc. Int. Symp. on Signals, Systems, and Electronics, URSI, Erlangen (Sept. 1989) to appear

8.8   P. Strobach: Efficient covariance ladder algorithms for finite arithmetic applications. Signal Processing **13**, 29–70 (1987)

8.9   N. Levinson: The Wiener RMS (root-mean-square) error criterion in filter design and prediction. J. Math. Phys. **25**, 261–278 (1947)

8.10   J. Durbin: The fitting of time series models. Rev. Int. Stat. **28**, 233–244 (1960)

8.11   J. Heinhold: *Einführung in die Höhere Mathematik* (Carl Hanser, München 1976)

8.12   J.P. Burg: Maximum Entropy Spectral Analysis. Ph.D. Dissertation, Stanford University, Stanford, CA (1975)

8.13   J.D. Markel, A.H. Gray: *Linear Prediction of Speech,* Communications and Cybernetics, Vol. 12 (Springer, Berlin, Heidelberg 1976)

8.14   U. Grenander, G. Szegö: *Toeplitz Forms and Their Applications* (California Press, Berkeley, CA 1958)

8.15   P. Delsarte, Y. Genin: On the splitting of classical algorithms in linear prediction theory. IEEE Trans. ASSP **35**, 645–653 (1987)

8.16   H. Krishna: New split Levinson, Schur, and lattice algorithms for digital signal processing. IEEE Int. Conf. on ASSP, New York (1988) pp. 1640–1642

8.17   I. Schur (1917): Über Potenzreihen die im Innern des Einheitskreises beschränkt sind. J. Reine Angew. Math. **147**, 205–232; ibid. **148**, 122–145 (1917)

8.18   P. Strobach: A fast recursive approach to FIR system identification. Proc. EUSIPCO-86, The Hague (1986) pp. 1055–1058

8.19   P. Strobach: Recursive covariance ladder algorithms for ARMA system identification. IEEE Trans. ASSP **36**, 560–580 (1988)

8.20 S.Y. Kung, Y.H. Hu: A highly concurrent algorithm and pipelined architecture for solving Toeplitz systems. IEEE Trans. ASSP **31**, 66-76 (1983)

8.21 I.C. You, Y.H. Hu, W.S. Feng: A novel implementation of pipelined Toeplitz system solver. Proc. IEEE **74**, 1463-1464 (1986)

Chapter 9

9.1   D.T.L. Lee: Canonical Ladder Form Realizations and Fast Estimation Algorithms. Ph.D. Dissertation, Stanford University, Stanford, CA (1980)

9.2   G.W. Stewart: Error and perturbation bounds for subspaces associated with certain eigenvalue problems. SIAM Rev. **15**, 727-764 (1973)

9.3   A. Björck, G.H. Golub: Numerical methods for computing angles between linear subspaces. Math. Comp. **27**, 579-594 (1973)

9.4   P.A. Wedin: Perturbation theory for pseudoinverses. B.I.T. **13**, 217-232 (1973)

9.5   S. Afriat: Orthogonal and oblique projections and the characteristics of pairs of vector spaces. Proc. Cambridge Philos. Soc. **53**, 800-816 (1957)

9.6   H. Zassenhaus: Angles of inclination in correlation theory. Am. Math. Mon. **71**, 218-219 (1964)

9.7   T.N.E. Greville: Solutions of the matrix equation **XAX** = **X**, and relations between oblique and orthogonal projectors. SIAM J. Appl. Math. **26**, 828-832 (1974)

9.8   H.J. Landau, H.O. Pollak: Prolate spheroidal wave functions, Fourier analysis and uncertainty - II. Bell Syst. Tech. J. **XI**, 65-84 (1961)

9.9   J.E. Mazo: On the angle between two Fourier subspaces. Bell Syst. Tech. J. **56**, 411-426 (1977)

9.10 I.C. Gohberg, M.G. Krein: *Introduction to the Theory of linear Nonselfadjoint Operators* (American Math. Soc., Providence, RI 1969)

9.11 H. Bart, I.C. Gohberg, M.A. Kaashoek: *Minimal Factorization of Matrix and Operator Functions* (Birkhäuser, Basel 1979)

9.12 P. Van Dooren, P. Dewilde: Minimal factorization of rational matrices. Proc. 17th IEEE Conf. Dec. Control, San Diego (1979) pp. 170-171

9.13 J.M. Cioffi, T. Kailath: Fast, recursive least-squares transversal filters for adaptive filtering. IEEE Trans. ASSP **32**, 304-337 (1984)

9.14 M.L. Honig, D.G. Messerschmitt: *Adaptive Filters* (Cluwer, Boston, MA 1984)

9.15 S.T. Alexander: *Adaptive Signal Processing: Theory and Applications* (Springer, New York 1986)

9.16 D.T.L. Lee, M. Morf, B. Friedlander: Recursive least-squares ladder estimation algorithms. IEEE Trans. ASSP **29**, 627-641 (1981)

9.17 F. Ling, D. Manolakis, J.G. Proakis: Numerically robust least-squares lattice-ladder algorithms with direct updating of the reflection coefficients. IEEE Trans. ASSP **34**, 837-845 (1986)

9.18 L.J. Griffiths: A continuously adaptive filter implemented as a lattice structure. Proc. IEEE Int. Conf. ASSP, Hartford (1977) pp. 683-686

9.19 L.J. Griffiths: An adaptive lattice structure for noise cancelling applications. Proc. IEEE Int. Conf. ASSP, Tulsa (1978) pp. 87-90

9.20 J. Makhoul, R. Viswanathan: Adaptive lattice methods for linear prediction. Proc. IEEE Int. Conf. ASSP, Tulsa (1978) pp. 83-86

9.21 B. Friedlander: Lattice filters for adaptive processing. Proc. IEEE **70**, 829-867 (1982)

9.22 C. Samson, V.U. Reddy: Fixed-point error analysis of the normalized ladder algorithm. IEEE Trans. ASSP **31**, 1177-1191 (1983)

9.23 B. Porat, B. Friedlander, M. Morf: Square-root covariance ladder algorithms. IEEE Trans. Autom. Control **27**, 813-829 (1982)

9.24 B. Porat, T. Kailath: Normalized lattice algorithms for least-squares FIR system identification. IEEE Trans. ASSP **31**, 122-128 (1983)

9.25 D.T.L. Lee, B. Friedlander, M. Morf: Recursive ladder algorithms for ARMA modeling. IEEE Trans. Autom. Control **27**, 753-764 (1981)

9.26 N. Ahmed, M. Morf, D.T.L. Lee, P.H. Ang: A VLSI speech analysis chip set based on square-root normalized ladder forms. Proc. IEEE Int. Conf. ASSP, Atlanta (1981) pp. 648-653

9.27 D.T.L. Lee, M. Morf: Generalized CORDIC for digital signal processing. Proc. IEEE Int. Conf. ASSP, Paris (1982) pp. 1748-1751

Chapter 10

10.1 B. Friedlander: Lattice filters for adaptive processing. Proc. IEEE **70**, 829-867 (1982)

10.2 B. Porat, T. Kailath: Normalized lattice algorithms for least-squares FIR system identification. IEEE Trans. ASSP **31**, 122-128 (1983)

10.3 P. Strobach: A fast recursive approach to FIR system identification. Proc. EUSIPCO-86, The Hague (1986) pp. 1055-1058

10.4 S.L. Marple: Efficient least-squares FIR system identification. IEEE Trans. ASSP **29**, 62-73 (1981)

10.5 B. Widrow, J.R. Glover, J.M. McCool, J. Kaunitz, C.S. Williams, R.H. Hearn, J.R. Zeidler, E. Dong, R.C. Goodlin: Adaptive noise cancelling: Principles and applications. Proc. IEEE **63**, 1692-1716 (1975)

10.6 D.T.L. Lee, B. Friedlander, M. Morf: Recursive ladder algorithms for ARMA modeling. IEEE Trans. Autom. Control **27**, 753-764 (1981)

10.7 P. Strobach: Recursive covariance ladder algorithms for ARMA system identification. IEEE Trans. ASSP **36**, 560-580 (1988)

10.8 R.L. Moses, J.A. Cadzow, A.A. (Louis) Beex: A recursive procedure for ARMA modeling. IEEE Trans. ASSP **33**, 1188-1196 (1985)

10.9 S. Li, B.W. Dickinson: Application of the lattice filter to robust estimation of AR and ARMA models. IEEE Trans. ASSP **36**, 502-512 (1988)

10.10 E. Karlsson and M.H. Hayes: ARMA modeling of linear time-varying systems: Lattice filter structures and fast RLS algorithms. IEEE Trans. ASSP **35**, 994-1015 (1987)

10.11 P. Whittle: On the fitting of multivariate auto-regressions and the approximate canonical factorization of a spectral density matrix. Biometrika **50**, 129-134 (1963)

10.12 R.A. Wiggins, E.A. Robinson: Recursive solution to the multichannel filtering problem. J. Geophys. Res. **70**, 1885-1891 (1965)

10.13 J.A. Cadzow: High performance spectral estimation - A new ARMA method. IEEE Trans. ASSP **28**, 524-529 (1980)

10.14 S.M. Kay: A new ARMA spectral estimator. IEEE Trans. ASSP **28**, 585-588 (1980)

10.15 S.L. Marple: *Digital Spectral Analysis* (Prentice-Hall, Englewood Cliffs, NJ 1987)

10.16 A.S. Willsky: *Digital Signal Processing and Control and Estimation Theory* (MIT Press, Cambridge, MA 1979)

10.17 M.D. Srinath, P.K. Rajasekaran: *Statistical Signal Processing with Applications* (Wiley, New York 1979)

Chapter 11

11.1 B. Widrow, S.D. Stearns: *Adaptive Signal Processing* (Prentice-Hall, Englewood Cliffs, NJ 1985)

11.2 M.G. Bellanger: New applications of digital signal processing in communications. IEEE ASSP Magazine **3**, 6-11 (1986)

11.3 T. Claasen, W. Mecklenbräuker: Adaptive techniques for signal processing in communications. IEEE Commun. Mag. **23**, No. 11 (1985)

11.4 U. Appel: Detection of parameter jumps in linear systems by means of a joint process lattice algorithm. Proc. Int. Conf. on Digital Signal Proc., Florence (1984) pp. 334-338

11.5 C.T. Mullis, R.A. Roberts: The use of second-order information in the approximation of discrete time linear systems. IEEE Trans. ASSP **24**, 226-238 (1976)

11.6 D.T.L. Lee, B.Friedlander, M. Morf: Recursive ladder algorithms for ARMA modeling. IEEE Trans. Autom. Control **27**, 753-764 (1981)

11.7 P. Strobach: Recursive covariance ladder algorithms for ARMA system identification. IEEE Trans. ASSP **36**, 560-580 (1988)

11.8 S.J. Campanella, H. Sayderhoud, M. Onufry: Analysis of adaptive impulse response echo canceller. COMSAT Tech. Rev. **2**, (1972)

11.9   J.M. Cioffi, T. Kailath: An efficient exact least-squares fractionally spaced equalizer using intersymbol interpolation. IEEE J. SAC **2**, 743-756 (1984)

11.10 L.J. Griffiths: Adaptive structures for multiple-input noise cancelling applications. Proc. IEEE Int. Conf. on ASSP, Washington, DC (1979) pp. 925-928

11.11 M. Honig: Echo cancellation of voiceband data signals using RLS and gradient algorithms. IEEE Trans. Commun. **33**, 65-73 (1985)

11.12 R.W. Lucky: Automatic equalization for digital communications. Bell Syst. Tech. J. **44**, 547-588 (1965)

11.13 M.S. Mueller: Least-squares algorithms for adaptive equalizers. Bell Syst. Tech. J. **60**, 1905-1926 (1981)

11.14 S. Qureski: Adaptive equalization. IEEE Commun. Mag. **20**, 9-16 (1982)

11.15 B.S. Atal, S.L. Hanauer: Speech analysis and synthesis by linear prediction of the speech wave. J. Acoust. Soc. Am. **50**, 637-655 (1971)

11.16 M.R. Schroeder: Vocoders: Analysis and synthesis of speech. Proc. IEEE **54**, 720-734 (1966)

11.17 M.R. Schroeder: Linear predictive coding of speech: Review and current directions. IEEE Commun. Mag. **23**, 54-61 (1985)

11.18 B.S. Atal: High-quality speech at low bit rates: Multipulse and stochastically excited linear predictive coders. Proc. IEEE Int. Conf. on ASSP, Tokyo (1986) pp. 1681-1684

11.19 B.S. Atal: Predictive coding of speech at low bit rates. IEEE Trans. Commun. **30**, 600-614 (1982)

11.20 B.S. Atal, M.R. Schroeder: Stochastic coding of speech signals at very low bit-rates. Proc. Int. Conf. on Commun. (1984) pp. 1610-1613

11.21 J.D. Markel, A.H. Gray: *Linear Prediction of Speech*, Communications and Cybernetics, Vol. 12 (Springer, Berlin, Heidelberg 1976)

11.22 N. Ahmed, M. Morf, D.T.L. Lee, P.H. Ang: A VLSI speech analysis chip set based on square-root normalized ladder forms. Proc. Int. Conf. on ASSP, Atlanta (1981) pp. 648-653

11.23 S.T. Alexander: Transient weight misadjustment properties for the finite precision LMS algorithm. IEEE Trans. on ASSP **35**, 1250-1258 (1987)

11.24 C. Caraiscos, B. Liu: A roundoff error analysis of the LMS adaptive algorithm. IEEE Trans. ASSP **32**, 34-41 (1984)

11.25 S.H. Ardalan, S.T. Alexander: Fixed-point round-off error analysis of the exponentially windowed RLS algorithm for time-varying systems. IEEE Trans. ASSP **35**, 770-783 (1987)

11.26 S. Ljung, L. Ljung: Error propagation properties of recursive least-squares adaptation algorithms. Automatica **21**, 157-167 (1985)

11.27 D.W. Lin: On digital implementation of the fast Kalman algorithms. IEEE Trans. ASSP **32**, 998-1005 (1984)

11.28 F. Ling, J.G. Proakis: Numerical accuracy and stability: Two problems caused by round-off error. Proc. IEEE Int. Conf. on ASSP, San Diego (1985) pp. 30.3.1 - 30.3.4

11.29 J.D. Markel, A.H. Gray, Jr.: Roundoff noise characteristics of a class of orthogonal polynomial structures. IEEE Trans. ASSP **23**, 486-494 (1975)

11.30 C. Samson, V.U. Reddy: Fixed-point error analysis of the normalized ladder algorithm. IEEE Trans. ASSP **31**, 1177-1191 (1983)

11.31 P. Strobach: New forms of least-squares lattice algorithms and a comparison of their round-off error characteristics. Proc. IEEE Int. Conf. on ASSP, Tokyo (1986) pp. 573-576

11.32 P. Strobach: Efficient covariance ladder algorithms for finite arithmetic applications. Signal Processing **13**, 29-70 (1987)

11.33 E.H. Satorius, S.W. Larisch, S.C. Lee, L.J. Griffiths: Fixed-point implementation of adaptive digital filters. Proc. IEEE Int. Conf. on ASSP, Boston (1983) pp. 33-36

11.34 R.A. Pepe, J.D. Rogers: Simulation of fixed-point operations with high level languages. IEEE Trans. on ASSP **35**, 116-117 (1987)

11.35 A. von Brandt: On-line segmentation of time series using a pilot segment. Proc. Int. Conf. on Digital Signal Processing, Florence (1981) pp. 1111-1118

11.36 A. von Brandt: Detecting and estimating parameter jumps using ladder algorithms and likelihood ratio tests. Proc. IEEE Int. Conf. on ASSP, Boston (1983) pp. 1017-1020

11.37 Y. Azenkot, I. Gertner: The least-squares estimation of time-delay between two signals with unknown relative phase shift. IEEE Trans. ASSP **33**, 308-309 (1985)

11.38 J.A. Cadzow: High performance spectral estimation - A new ARMA method. IEEE Trans. ASSP **28**, 524-529 (1980)

11.39 S.M. Kay: A new ARMA spectral estimator. IEEE Trans. ASSP **28**, 585-588 (1980)

11.40 P. Strobach, U. Appel: "Ein Signalprozessor mit Wirt-Gast-Kopplung über gemeinsame Speicherbereiche" in *Architektur und Betrieb von Rechensystemen*, ed. by H. Wettstein, Informatik-Fachberichte, Vol. 78 (Springer, Berlin, Heidelberg 1984)

11.41 R. Gitlin, S. Weinstein: On the required tap-weight precision for digitally implemented adaptive mean-squared equalizers. Bell Syst. Tech. J. **58**, 301-321 (1979)

11.42 E.A. Lee, D.G. Messerschmitt: Pipeline interleaved programmable DSP's: Architecture. IEEE Trans. ASSP **35**, 1320-1333 (1987)

11.43 Y.T. Chan, J.M. Lavoie, J.B. Plant: A parameter estimation approach to the estimation of frequencies of sinusoids. IEEE Trans. ASSP **29**, 214-219 (1981)

11.44 A.J.E.M. Janssen, R.N.J. Veldhuis, L.B. Vries: Adaptive interpolation of discrete time signals that can be modeled as autoregressive processes. IEEE Trans. ASSP **34**, 317–330 (1986)

11.45 A.M. Tekalp, H. Kaufman, J.W. Woods: Identification of image and blur parameters for the restoration of non-causal blur. IEEE Trans. ASSP **34**, 963–969 (1986)

11.46 J.R. Caldas-Pinto, P.E. Wellstedt: Self-tuning filters and predictors for two-dimensional systems, Part 2: Smoothing applications. Int. J. Control **42**, 479–496 (1985)

11.47 J.R. Caldas-Pinto, P.E. Wellstedt: Self-tuning filters and predictors for two-dimensional systems, Part 3: Prediction applications. Int. J. Control **42**, 497–515 (1985)

11.48 T.L. Marzetta: Two-dimensional linear prediction: Autocorrelation arrays, minimum phase prediction error filters and reflection coefficients. IEEE Trans. ASSP **28**, 725–733 (1980)

11.49 S. Ranagath, A.K. Jain: Two-dimensional linear prediction models – Part I: Spectral factorization and realization. IEEE Trans. ASSP **33**, 280–293 (1985)

11.50 G.R. Wagner, P.E. Wellsteadt: Two-dimensional self-tuning predictors and filters. Proc. Multidimensional Signal Processing Workshop, Noordwijkerhout (1987)

11.51 P.E. Wellsteadt, J.R. Caldas-Pinto: Self-tuning filters and predictors for two-dimensional systems, Part 1: Algorithms. Int. J. Control **42**, 457–478 (1985)

11.52 P.A. Maragos, R.M. Mersereau, R.W. Schafer: Two-dimensional linear predictive analysis of arbitrarily shaped regions. Proc. Int. Conf. on ASSP, Boston (1983) pp. 104–107

11.53 T.F. Quatieri: Object detection by two-dimensional linear prediction. Proc. IEEE Int. Conf. on ASSP, Boston (1983) pp. 108–111

11.54 P.A. Maragos, R.W. Schafer, R.M. Mersereau: Two-dimensional linear prediction and its application to adaptive predictive coding of images. IEEE Trans. ASSP **32**, 1213–1228 (1984)

11.55 P. Strobach: Quadtree-structured linear prediction models for image sequence processing. IEEE Trans. Pattern Anal. Machine Intell. **11**, 742–748 (1989)

11.56 T. Kohonen: *Self Organization and Associative Memory, 2nd ed.* (Springer, Berlin, Heidelberg 1988)

11.57 J. Makhoul, R. Schwartz, A. El-Jaroudi: Classification capabilities of two-layer neural nets. Proc. IEEE Int. Conf. on ASSP, Glasgow (1989) pp. 635–638

11.58 D.E. Rumelhart, J.L. McClelland: *Parallel Distributed Processing, Vols. I and II* (MIT Press, Cambridge, MA 1986)

11.59 D.E. Rumelhart, G.E. Hinton, R.J. Williams: Learning internal representations by error propagation. ICS Report 8506, Institute for Cognitive Science, University of California, San Diego (1985)

11.60 S.S. Singhal, L. Wu: Training feed-forward networks with the extended Kalman algorithm. Proc, Int. Conf. on ASSP, Glasgow (1989) pp. 1187-1190

11.61 M.D. Srinath, P.K. Rajasekaran: *An Introduction to Statistical Signal Processing with Applications* (Wiley, New York 1979)

11.62 A.M. Bruckstein, Y.Y. Zeevi: An adaptive stochastic model for the neural coding process. IEEE Trans. Syst., Man and Cybern. **15**, 343-351 (1985)

11.63 J.J. Hopfield: Neural networks and physical systems with emergent collective computational abilities. Proc. Nat. Acad. Sci. USA **79**, 2554 (1982)

11.64 J.J. Hopfield, D.W. Tank: Neural computation of decisions in optimization problems. Biol. Cybern. **52**, 192 (1985)

11.65 D.B. Parker: Learning logic; Tech. Rep. TR-47, Center for Comput. Res. Econ. and Manag. Sci., MIT, Cambridge, MA (1985)

11.66 D.B. Parker: "A Comparison of Algorithms for Neuron-like Cells", in AIP Conf. Proc., Vol. **151** (AIP, New York 1986) pp. 327-332

11.67 B. Widrow: "Generalization and Information Storage in Networks of Adaline Neurons", in *Self-Organizing Systems 1962*, ed. by M.C. Yovitz, G.T. Jacobi, G.D. Goldstein (Spartan Books, Washington, DC 1962) pp. 435-461

11.68 L.J. Griffiths (ed.): Special Section on Neural Networks. IEEE Trans. ASSP **36**, 1107-1190 (1988)

Appendix

A.1   P. Strobach: Pure order recursive least-squares ladder algorithms. IEEE Trans. ASSP **34**, 880-897 (1986)

A.2   J. LeRoux, C. Gueguen: A fixed-point computation of partial correlation coefficients. IEEE Trans. ASSP **25**, 257-259 (1977)

A.3   N. Levinson: The Wiener RMS (root-mean-square) error criterion in filter design and prediction. J. Math. Phys. **25**, 261-278 (1947)

A.4   J. Durbin: The fitting of time series models. Rev. Int. Stat. **28**, 233-244 (1960)

A.5   S.-Y. Kung: VLSI array processors. IEEE ASSP Magazine **2**, 4-22, (1985)

A.6   S.-Y. Kung, H.J. Whitehouse, T. Kailath (eds.): *VLSI and Modern Signal Processing* (Prentice-Hall, Englewood Cliffs, NJ 1985)

A.7   J.W. Givens: Numerical computation of the characteristic values of a real symmetric matrix. Oak Ridge National Laboratory, ORNR 1574, Internal Report (1954)

A.8   W.M. Gentleman: Least-squares computation by Givens transformations without square roots. J. Inst. Math. Appl. **12**, 329-336 (1973)

A.9 H.T. Kung, W.M. Gentleman: Matrix triangularization by systolic arrays. Proc. SPIE, Real-Time Signal Processing IV, **298**, 16 (1981)

A.10 N.J. Bershad, O. Macchi: Comparison of RLS and LMS algorithms for tracking a chirped signal. Proc. IEEE Int. Conf. on ASSP, Glasgow (1989) pp. 896-899

A.11 K.C. Sharman, T.S. Durrani: A triangular adaptive lattice filter for spatial signal processing. Proc. IEEE Int. Conf. on ASSP, Boston (1983) pp. 348-351

A.12 G.U. Yule (1907): "On the Theory of Correlation for Any Number of Variables, Treated by a New System of Notation", in *Statistical Papers of George Udny Yule* (Griffin, London 1971)

A.13 C.L. Lawson, R.J. Hanson: *Solving Least-Squares Problems* (Prentice Hall, Englewood Cliffs, NJ 1974)

A.14 R. Penrose: A general inverse for matrices. Proc. Cambridge Philos. Soc. **51**, 406-413 (1955)

A.15 J. Sokat, P. Strobach: A new class of pure order recursive ladder algorithms. Proc. IEEE Int. Workshop on Digital Signal Processing, Stanford Sierra Lodge, Tahoe, CA (1988)

A.16 J. Sokat: The split generalized LeRoux-Gueguen algorithm for least-squares problems. Proc. Int. Symp. on Signals, Systems, and Electronics, URSI, Erlangen (Sept. 1989)

A.17 F. Ling, D. Manolakis, J.G. Proakis: A recursive modified Gram-Schmidt algorithm for least-squares estimation. IEEE Trans. on ASSP **34**, 829-835 (1986)

A.18 N. Kalouptsidis, S. Theodoridis: Parallel implementation of efficient LS algorithms for filtering and prediction. IEEE Trans. ASSP **35**, 1565-1569 (1987)

A.19 S. Theodoridis, N. Kalouptsidis, D. Bakirtzis: Pipelined algorithm for LS FIR filters with symmetric impulse response. IEEE Trans. ASSP, to appear (1990)

A.20 S. Theodoridis: Pipeline architecture for block adaptive LS FIR filtering and prediction, IEEE Trans. on ASSP, to appear (1990)

A.21 F. Ling: Efficient least-squares lattice algorithms based on Givens rotation with systolic array implementations, Proc. IEEE Int. Conf. on ASSP, Glasgow (1989) pp. 1290-1293

Bibliography

In addition to the references cited in each chapter, we provide the following list of other books, reports and papers related to the material in the text. The list is arranged alphabetically under each heading.

Books

M. Abramowitz, I.A. Stegun: *Handbook of Mathematical Functions with Formulas, Graphs and Mathematical Tables* (Dover, New York 1965)

B.D.O. Anderson, J.B. Moore: *Linear Optimal Control* (Prentice Hall, Englewood Cliffs, NJ 1979)

R.B. Ash: *Real Analysis and Probability* (Academic, New York 1972)

B. Belevitch: *Classical Network Theory* (Holden Day, San Francisco 1969)

M. Bellanger: *Adaptive Digital Filters and Signal Analysis* (Marcel Deccer, New York 1987)

R. Bellman: *Introduction to Matrix Analysis* (McGraw-Hill, New York 1960)

R.E. Blahut: *Fast Algorithms for Digital Signal Processing* (Addison-Wesley, Reading, MA 1985)

G.E.P. Box, G.M. Jenkins: *Time Series Analysis: Forecasting and Control* (Holden Day, San Francisco 1970)

K. Brammer, G. Siffling: *Kalman Bucy Filter* (Oldenbourg, München 1975)

W.L. Brogan: *Modern Control Theory*, 2nd ed. (Prentice-Hall, Englewood Cliffs, NJ 1985)

J.R. Bunch, J.D. Rose (eds.): *Sparse Matrix Computations* (Academic, New York 1976)

C.H. Chen (ed.): *Signal Processing Handbook* (Marcel Dekker, New York 1988)

D.G. Childers (ed.): *Modern Spectral Analysis* (IEEE, New York 1978)

H. Cramer: *Mathematical Methods of Statistics* (Princeton University Press, Princeton, NJ 1951)

R.E. Crochiere, L.R. Rabiner: *Multivariate Digital Signal Processing* (Prentice-Hall, Englewood Cliffs, NJ 1983)

G. Dahlquist, A. Björck: *Numerical Methods* (Prentice-Hall, Englewood Cliffs, NJ 1974)

G.B. Dantzig: *Linear Programming and Extensions* (Princeton University Press, Princeton, NJ 1963)

R. Deutsch: *Estimation Theory* (Prentice-Hall, Englewood Cliffs, NJ 1965)

L.C.W. Dixon: *Nonlinear Optimization* (English Universities Press, London 1972)

L.J. Doob: *Stochastic Processes* (Wiley, New York 1953)

P. Eykhoff: *System Parameter and State Estimation* (Wiley, London 1974)

D.T. Finkbeiner: *Introduction to Matrices and Linear Transformations* (Freeman, San Francisco 1966)

J.L. Flanagan: *Speech Analysis, 2nd ed.*, Communications and Cybernetics, Vol. 3 (Springer, Berlin, Heidelberg 1972)

G.E. Forsythe, M.A. Malcolm, C.B. Moler: *Computer Methods for Mathematical Computations* (Prentice-Hall, Englewood Cliffs, NJ 1977)

R.G. Gallagher: *Information Theory and Reliable Communications* (Wiley, New York 1968)

A. Gelb: *Applied Optimal Estimation* (MIT Press, Cambridge, MA 1974)

L. Gerominus: *Orthogonal Polynomials*, Int. Series on Appl. Mathematics (Pergamon, New York 1960)

A.A. Giordano, F.M. Hsu: *Least Square Estimation with Applications to Digital Signal Processing* (Wiley, New York 1985)

G.H. Golub, C.F. Van Loan: *Matrix Computations* (The Johns Hopkins University Press, Baltimore, MB 1983)

G.C. Goodwin, K.S. Sin: *Adaptive Filtering, Prediction, and Control* (Prentice-Hall, Englewood Cliffs, NJ 1984)

D. Graupe: *Identification of Systems* (Van Nostrand, New York 1972)

S. Haykin: *Communication Systems*, 2nd ed. (Wiley, New York 1983)

S. Haykin: *Introduction to Adaptive Filters* (Macmillan, New York 1984)

S. Haykin (ed.): *Array Signal Processing* (Prentice-Hall, Englewood Cliffs, NJ 1984)

J. Heinhold, K.W. Gaede: *Ingenieur-Statistik* (Carl-Hanser, München 1976)

J. Heinhold, B. Riedmüller: *Lineare Algebra und Analytische Geometrie* (Carl-Hanser, München 1975)

W. Hess: *Pitch Determination of Speech Signals*, Springer Ser. Info. Sci., Vol. 3 (Springer, Berlin, Heidelberg 1983)

A.S. Householder: *Principles of Numerical Analysis* (McGraw-Hill, New York 1953)

A.S. Householder: *The Theory of Matrices in Numerical Analysis* (Blaisdell, Waltham, MA 1964)

R. Isermann: *Prozessidentifikation* (Springer, Berlin, Heidelberg 1974)

N.S Jayant, P. Noll: *Digital Coding of Waveforms* (Prentice-Hall, Englewood Cliffs, NJ 1984)

E.I. Jury: *Theory and Application of the z-Transform Method* (Wiley, New York 1964)

T. Kailath (ed.): *Linear Least-Squares Estimation,* Benchmark Papers in Electrical Engineering and Computer Science (Dowden, Hutchison and Ross, Stroudsburg, PA 1977)

T. Kailath: *Linear Systems* (Prentice-Hall, Englewood Cliffs, NJ 1980)

T. Kailath: *Lectures on Wiener and Kalman Filtering,* CISM Courses, Vol. 140 (Springer, New York 1981)

R.E. Kalman, P.H. Falb, M.A. Arbib: *Topics in Mathematical System Theory* (McGraw-Hill, New York 1969)

K. Kroschel: *Statistische Nachrichtentheorie, 2. Teil: Signalschätzung,* 2. Aufl. (Springer, Berlin, Heidelberg 1988)

D. Luenberger: *Optimization by Vector Space Methods* (Wiley, New York 1969)

C. Mead, L. Conway: *Introduction to VLSI Systems* (Addison-Wesley, Reading, MA 1980)

D. Middleton: *An Introduction to Statistical Communication Theory* (McGraw-Hill, New York 1960)

K.S. Miller: *Complex Stochastic Processes: An Introduction to Theory and Application* (Addison-Wesley, Reading, MA 1974)

R.A. Monzingo, T.W. Miller: *Introduction to Adaptive Arrays* (Wiley, New York 1980)

W. Murray (ed.): *Numerical Methods for Unconstrained Optimization* (Academic, New York 1972)

J.C. Nash: *Compact Numerical Methods for Computers: Linear Algebra and Function Minimization* (Adam Hilger, Bristol 1979).

N. Nilsson: *Learning Machines* (McGraw-Hill, New York 1965)

A.V. Oppenheim, R.W. Schafer: *Digital Signal Processing* (Prentice-Hall, Englewood Cliffs, NJ 1975)

M.R. Osborne: *Finite Algorithms in Optimization and Data Analysis* (Wiley, New York 1985).

J.G. Proakis: *Digital Communications* (McGraw-Hill, New York 1983)

E.A. Robinson: *Multichannel Time-Series Analysis with Digital Computer Programs* (Holden-Day, San Francisco 1967)

E.A. Robinson, S. Treitel: *Geophysical Signal Analysis* (Prentice-Hall, Englewood Cliffs, NJ 1980)

H.W. Schüssler: *Digitale Systeme zur Signalverarbeitung* (Springer, Berlin, Heidelberg, 1973)

H.R. Schwarz: *Numerik Symmetrischer Matrizen* (Teubner, Stuttgart, 1972)

H.R. Schwarz: *Zeitdiskrete Regelungssysteme, Einführung* (Vieweg, Braunschweig 1979)

H.R. Schwarz, J. Waldvogel: *Numerische Mathematik* (Teubner, Stuttgart 1986)

D. O'Shaughnessy: *Speech Communication: Human and Machine* (Addison-Wesley, Reading, MA 1987)

D.L. Snyder: *Random Point Processes* (Wiley Interscience, New York 1975)

T. Söderström, P. Stoica: *System Identification* (Prentice-Hall Int., Hemel Hempstead 1989)

G.W. Stewart: *Introduction to Matrix Computation* (Academic, New York 1973)

J. Stoer: *Einführung in die numerische Mathematik I,* 4. Aufl., Heidelberger Taschenbücher, Band 105 (Springer, Berlin, Heidelberg 1983)

J. Stoer: *Einführung in die numerische Mathematik II,* 2. Aufl., Heidelberger Taschenbücher, Band 114 (Springer, Berlin, Heidelberg 1978)

G. Strang: *Introduction to Applied Mathematics* (Wellesley, Cambridge, MA 1986)

S.A. Tretter: *Introduction to Discrete-Time Signal Processing* (Wiley, New York 1976)

H.L. Van Trees: *Detection, Estimation, and Modulation Theory, Part 1,* (Wiley, New York 1968)

A. Wald: *Sequential Analysis* (Wiley, New York 1947)

S. Weisberg: *Applied Linear Regression* (Wiley, New York 1980)

A.D. Whalen: *Detection of Signals in Noise* (Academic, New York 1971)

N. Wiener: *Extrapolation, Interpolation, and Smoothing of Stationary Time Series* (MIT Press, Cambridge, MA 1942)

J.H. Wilkinson: *Rounding Errors in Algebraic Processes* (Prentice-Hall, Englewood Cliffs 1963)

J.H. Wilkinson: *The Algebraic Eigenvalue Problem* (Oxford University Press, Oxford 1965)

P.C. Young: *Recursive Estimation and Time Series Analysis: An Introduction* (Springer, New York 1984)

Principles of Linear Prediction and Least-Squares

H.M. Ahmed: Signal Processing Algorithms and Architectures; Ph.D. Dissertation, Stanford University, Stanford, CA (1982)

H.M. Ahmed, J.M. Delosome, M. Morf: Highly concurrent computation structures for matrix arithmetic and signal processing. Comput. Mag. 65-82 (1982)

H. Akaike: A new look at the statistical model identification. IEEE Trans. AC-19, 716-723 (1974)

K.J. Aström, P. Eykhoff: System identification - a survey. Automatica 7, 123-162 (1971)

Automatica: *Special Issue on Identification and System Parameter Estimation,* Vol. 17 (1981)

Y. Bistriz: Z-domain continued fraction expansions for stable discrete systems polynomials. IEEE Trans. CAS-32, 1162-1166 (1985)

J.M. Cioffi: A pipelined fast QR-RLS structure for high-speed VLSI implementation of adaptive equalizers. Proc. IEEE Int. Conf. on Communications, Seattle, WA (1987) pp. 853-857

M.A. Clements, S.H. Isabelle: Reconstruction of a positive definite Toeplitz matrix from its sequence of minimum eigenvalues. IEEE Trans. ASSP-36, 1784-1786 (1988)

P. Comon, Y. Robert: A systolic array for computing BA^{-1}. IEEE Trans. ASSP-35, 717-723 (1987)

C.J. Demeure, C.T. Mullis: The Euclid algorithm and the fast computation of cross-covariance and autocorrelation sequences. IEEE Trans. ASSP-37, 545-552 (1989)

C.J. Demeure, L.L. Scharf: Linear statistical models for stationary sequences and related algorithms for Cholesky factorization of Toeplitz matrices. IEEE Trans. ASSP-35, 29-42 (1987)

A. Einstein: Method for the determination of the statistical values of observations concerning quantities subject to statistical fluctuations. Arch. Sci. Phys. et Natur. 37, 254-256 (1914)

J.W. Givens: Computation of plane unitary rotations transforming a general matrix to triangular form. J. Soc. Ind. Appl. Math. 6, 26-50 (1958)

G.H. Golub, C. Reinsch: Singular value decomposition and least-squares problems. Numer. Math. 14, 403-420 (1970)

M.J.R. Healy: Triangular decompostiton of a symmetric matrix. Appl. Stat. 17, 195-197 (1968)

A.S. Householder, F.L. Bauer: On certain methods for expanding the characteristic polynomial. Numer. Math. 1, 29-37 (1959)

T. Kailath: A view of three decades of linear filtering theory. IEEE Trans. IT-20, 145-181 (1974)

T. Kailath, S.-Y. Kung, M. Morf: Displacement ranks of matrices and linear equations. J. Math. Anal. Appl. 68, 395-407 (1979)

A. Khinchin: Correlation theory of stationary stochastic processes. Math. Ann. 109, 604-615 (1934)

S.Y. Kung: Multivariable and Multidimensional Systems: Analysis and Design; Ph.D. Dissertation, Stanford University, Stanford, CA (1977)

V.B. Lawrence and S.K. Tewksbury: Multiprocessor implementation of adaptive digital filters. IEEE Trans. COM-31, 826-835 (1983)

H. Lev-Ari, T. Kailath, J. Cioffi: Least-squares adaptive lattice and transversal filters: A unified geometrical theory. IEEE Trans. IT-30, 222-236 (1984)

F. Ling, D. Manolakis, J.G. Proakis: A flexible, numerically robust array processing algorithm and its relationship to the Givens transformation. Proc. IEEE Int. Conf. on ASSP, Tokyo (1986) pp. 2127-2130

L. Ljung: Analysis of recursive stochastic algorithms. IEEE Trans. AC-22, 551-575 (1977)

S. Ljung: Fast Algorithms for Integral Equations and Least-Squares Identification Problems; Ph.D. Dissertation, Linköping University, Sweden (1983)

O. Macchi, E. Eweda: Second-order convergence analysis of stochastic adaptive linear filtering. IEEE Trans. AC-**28**, 76-85 (1983)

D. Manolakis, N. Kalouptsidis, G. Carayannis: Fast algorithms for discrete-time Wiener filters with optimum lag . IEEE Trans. ASSP-31, 168-179 (1983)

P.E. Mantey, L.J. Griffiths: Iterative least-squares algorithm for signal extraction. Proc. Second Hawaii Int. Conf. Syst. Sci., Western Periodicals Co. (1969) pp. 767-770

R.S. Medaugh, L.J. Griffiths: A comparison of two fast linear prediction algorithms. Proc. IEEE Int. Conf. on ASSP, Atlanta (1981) pp. 293-296

R.S. Medaugh, L.J. Griffiths: Further results of a least-squares and a gradient adaptive lattice algorithm comparison. Proc. IEEE Int. Conf. on ASSP, Paris (1982) pp. 1412-1415

M. Morf, B.W. Dickinson, T. Kailath, A. Vieira: Efficient solution of covariance equations for linear prediction. IEEE Trans. ASSP-**25**, 429-435 (1977)

S.D. Morgera: On bi-orthogonality of Hermitian and skew-Hermitian Szegö/Levinson polynomials. IEEE Trans. ASSP-**37**, 436-439 (1989)

T. Ozaki, H. Tong: On the fitting of nonstationary AR models in time series analysis. Proc. 8th Hawaii Int. Conf. on System Sci. (1975) pp. 224-226

A. Papoulis: Levinson algorithms, Wold's decomposition, and spectral estimation. SIAM Rev. **27**, 405-411 (1985)

V.F. Pisarenko: The retrieval of harmonics from a covariance function. Geophys. J. R. Astron. Soc. **33**, 347-366 (1973)

C.P. Rialan, L.L. Scharf: Fast algorithms for computing QR and Cholesky factors of Toeplitz operators. IEEE Trans. ASSP-**36**, 1740-1948 (1988)

S.A. Ruzinsky, E.T. Olsen: L_1 and L_∞ minimization via a variant of Karmarkar's algorithm. IEEE Trans. ASSP-**37**, 245-253 (1989)

P. Stoica, A. Nehorai: On stability and root location of linear prediction models. IEEE Trans. ASSP-**35**, 582-583 (1987)

D.R. Sweet: Fast Toeplitz orthogonalization. Numer. Math. **43**, 1-21 (1984)

B. Widrow, E. Walach: On the statistical efficiency of the LMS algorithm with nonstationary inputs. IEEE Trans. IT-**30**, Special Issue on Adaptive Filtering, 211-221 (1984)

N. Wiener: Generalized harmonic analysis. Acta Math. **55**, 117-258 (1930)

N. Wiener, E. Hopf: On a class of singular integral equations. Proc. Russian Acad., Math.-Phys. Ser. 696 (1931)

A.S. Willsky: Relationships between digital signal processing and control and estimation theory. Proc. IEEE **66**, 996-1017 (1978)

Transversal Algorithms

S.H. Ardalan, L.J. Faber: A fast ARMA transversal RLS filter algorithm. IEEE Trans. ASSP-**36**, 349–358 (1988)

N.J. Bershad: On error-saturation nonlinearities in LMS adaptation. IEEE Trans. ASSP-**36**, 440–452 (1988)

R.R. Bitmead, B.D.O. Anderson: Lyapunov techniques for the exponential stability of linear difference equations with random coefficients. IEEE Trans. AC-**25**, 782–787 (1980)

G. Carayannis, N. Kalouptsidis, D. Manolakis: Fast recursive algorithms for a class of linear equations. IEEE Trans. ASSP-**20**, 227–239 (1982)

T.A.C.M. Claasen, W.F.G. Mecklenbräuker: Comparison of the convergence of two algorithms for adaptive FIR digital filters. IEEE Trans. ASSP-**29**, 670–678 (1981)

P.M. Clarkson, P.R. White: Simplified analysis of the LMS adaptive filter using a transfer function approximation. IEEE Trans. ASSP–**35**, 987–993 (1987)

D.D. Falconer, L. Ljung: Application of fast Kalman estimation to adaptive equalization. IEEE Trans. COM-**26**, 1439–1446 (1978)

A. Feuer, E. Weinstein: Convergence analysis of LMS filters with uncorrelated data. IEEE Trans. ASSP-**33**, 222–230 (1985)

C. Halkias, G. Carayannis, J. Dologlou, D. Emmanolopoulos: A new generalized recursion for the fast computation of the Kalman gain to solve the covariance equations. Proc. IEEE Int. Conf. on ASSP, Paris (1982) pp. 1760–1763

L.L. Horowitz, K.D. Senne: Performance advantage of complex LMS for controlling narrow-band adaptive arrays. IEEE Trans. ASSP-**29**, 722–735 (1981)

H.J. Kallmann: Transversal filters. Proc. IRE **28**, 302–310 (1940)

M. Morf: Fast Algorithms for Multivariable Systems; Ph.D. Dissertation, Stanford University, Stanford, CA (1974)

C.T. Mullis, R.A. Roberts: Synthesis of minimum round-off noise fixed point digital filters. IEEE Trans. CAS-**23**, 551–562 (1976)

B. Toplis, S. Pasupathy: Tracking improvements in fast RLS algorithms using a variable forgetting factor. IEEE Trans. ASSP-**36**, 206–227 (1988)

B. Widrow, J.M. McCool: A comparison of adaptive algorithms based on the method of steepest descent and random search. IEEE Trans. AP-**24**, 615–637 (1976)

D. Williamson: Delay replacement in direct form structures. IEEE Trans. ASSP-**36**, 453–460 (1988)

Ladder Algorithms

J.G. Ackenhusen: Regular form of Durbin's recursion for programmable signal processors. IEEE Trans. ASSP-**35**, 1628-1631 (1987)

P.L. Chu, D.G. Messerschmitt: Zero-sensitivity of the digital lattice filter. Proc. IEEE Int. Conf. on ASSP, Denver, CO (1980) pp. 89-93

J.M. Cioffi: An unwindowed RLS adaptive lattice algorithm. IEEE Trans. ASSP-**36**, 365-371 (1988)

R.E. Crochiere: Digital ladder structures and coefficient sensitivity. IEEE Trans. AU-**20**, 240-246 (1972)

J.M. Delosome, M. Morf: Mixed and minimal representations for Toeplitz and related systems. Proc. 14th Ann. Asilomar Conf. (1980) pp. 19-24

E.F. Deprettere: Orthogonal Digital Cascade Filters and Recursive Construction Algorithms for Stationary and Nonstationary Fitting of Orthogonal Models; Ph.D. Dissertation, Delft University of Technology, Delft (1980)

P. Dewilde, A.C. Viera, T. Kailath: On a generalized Szegö-Levinson realization algorithm for optimal linear predictors based on a network synthesis approach. IEEE Trans. CAS-**25**, 663-675 (1978)

B.W. Dickinson: Autoregressive estimation using residual energy ratios. IEEE Trans. IT-**24**, 503-506 (1978)

A. Elfishawy, S. Kesler, A. Aboutaleb: Least squares moving object detection in cluttered images. Proc. 31st Midwest Symposium on Circuits and Systems, University of Missouri, Rolla, Missouri (1988)

A. Fettweis, H. Levin, A. Sedlmeyer: Wave digital lattice filters. Int. J. Circuit Theory Appl. **2**, 203-211 (1974)

B. Friedlander: Recursive lattice forms for spectral estimation and adaptive control. Proc. 19th IEEE Int. Conf. Decision Control, Albuquerque, NM (1980) pp. 466-471

B. Friedlander: A modified lattice algorithm for deconvolving filtered impulsive processes. Proc. IEEE Int. Conf. on ASSP, Atlanta (1981) pp. 865-868

B. Friedlander: Lattice methods for spectral estimation. Proc. IEEE **70**, 990-1017 (1982)

B. Friedlander: A lattice algorithm for factoring the spectrum of a moving average process. Proc. Conf. on Information Sciences and Systems, Princeton (1982)

B. Friedlander: Lattice implementation of some recursive parameter estimation algorithms. Int. J. Control **37**, 661-684 (1983)

B. Friedlander, T. Kailath, M. Morf, L. Ljung: Extended Levinson and Chandrasekhar equations for discrete-time linear estimation problems. IEEE Trans. AC-**23**, 653-659 (1978)

B. Friedlander, L. Ljung, M. Morf: Lattice implementation of the recursive maximum likelihood algorithm. Proc. 20th IEEE Conf. Decision Control, San Diego (1981) pp. 1083-1084

B. Friedlander, M. Morf: Efficient inversion formulas for sums of products for Toeplitz and Hankel matrices. Proc. 18th Ann. Allerton Conf. on Communications, Control and Computing (1980)

C. Gibson, S. Haykin: Learning characteristics of adaptive lattice algorithms. IEEE Trans. ASSP-28, 681-691 (1980)

D.F. Gingras: Asymptotic properties of high-order Yule-Walker estimates of the AR parameters of an ARMA time series. IEEE Trans. ASSP-33, 1095-1101 (1985)

A.H. Gray, Jr. , J.D. Markel: Digital lattice and ladder filter synthesis. IEEE Trans. AU-21, 491-500 (1973)

M.L. Honig, D.G. Messerschmitt: Convergence properties of adaptive digital lattice filters. IEEE Trans. ASSP-29, 642-653 (1981)

T. Kailath: Time-variant and time-invariant lattice filters for nonstationary processes. Proc. Fast Algorithms for Linear Dynamic Systems, Aussois, France (1981) pp. 417-464

G.S. Kang, S.S. Everett: Improvement of the LPC analysis. Proc. IEEE Int. Conf. on ASSP, Boston (1983) pp. 89-92

T. Kawase, H. Saki, H. Tokumara: Recursive least-squares circular lattice and escalator estimation algorithms. IEEE Trans. ASSP-31, 228-231 (1983)

H. Kimura, T. Osada: Canonical pipelining of lattice filters. IEEE Trans. ASSP-35, 878-887 (1987)

D.T.L. Lee, M. Morf: Recursive square-root estimation algorithms. Proc. IEEE Int. Conf. on ASSP, Denver (1980) pp. 1005-1017

D.T.L. Lee, M. Morf, B. Friedlander: Recursive least-squares ladder estimation algorithms. IEEE Trans. CAS-28, 467-481 (1981)

H. Leib, M. Eizenman, S. Pashpathy, J. Krolik: Adaptive lattice filter for multiple sinusoids in white noise. IEEE Trans. ASSP-35, 1015-1023 (1987)

H. Lev-Ari: Parametrization and Modeling of Nonstationary Processes; Ph.D. Dissertation, Stanford University, Stanford, CA (1982)

H. Lev-Ari, T. Kailath: Lattice filter parametrization and modeling of nonstationary processes. IEEE Trans. IT-30, 2-15 (1984)

F. Ling, J.G. Proakis: A generalized multichannel least-squares lattice algorithm with sequential processing stages. IEEE Trans. ASSP-32, 381-389 (1984)

J. Makhoul, L.K. Cossell: Adaptive lattice analysis of speech. IEEE Trans. CAS-28, 494-499 (1981)

D.G. Messerschmitt: A class of generalized lattice filters. IEEE Trans. ASSP-28, 198-204 (1980)

K.S. Mitra, R.J. Sherwood: Canonical realizations of digital filters using the continued fraction expansion. IEEE Trans. AU-20, 185-194 (1972)

K.S. Mitra, R.J. Sherwood: Digital ladder networks. IEEE Trans. AU-21, 30-36 (1973)

M. Morf: Ladder forms in estimation and system identification. Proc. 11th Ann. Asilomar Conf., Monterey (1977) pp. 424-429

M. Morf, D.T.L. Lee: Recursive least-squares ladder forms for fast parameter tracking. Proc. IEEE Int. Conf. Decision Control, San Diego (1979) pp. 1362-1367

M. Morf, D.T.L. Lee: State-space structures of ladder canonical forms. Proc. IEEE Conf. Decision Control, Albuquerque (1980) pp. 1221-1224

M. Morf, C.H. Muravchik, D.T.L. Lee: Hilbert space array methods for finite rank process estimation and ladder realizations for adaptive signal processing. Proc. IEEE Int. Conf. on ASSP, Atlanta, GA (1981) pp. 856-859

M. Morf, A. Viera, D.T.L. Lee: Ladder forms for identification and speech processing. Proc. IEEE Int. Conf. Decision Control, New Orleans (1977) pp. 1074-1078

D. Pavlikh, N. Ahmed, S.D. Stearns: An adaptive lattice algorithm for recursive filters. IEEE Trans. ASSP-28, 110-111 (1980)

F.A. Perry, S.R. Parker: Recursive solutions for zero-pole modeling. Proc. 13th Ann. Asilomar Conf., Monterey (1979) pp. 509-512

B. Porat: Contributions to the Theory and Applications of Lattice Filters; Ph.D. Dissertation, Stanford University, Stanford, CA (1982)

B.M. Sallam, S.B. Kesler: Comparison of two modified RLS lattice equalizers in nonstationary environments. Proc. 31st Midwest Symposium on Circuits and Systems, Rolla (1988)

B.M. Sallam, S.B. Kesler: The performance characteristics of the order recursive adaptive lattice equalizer. Proc. 26th Allerton Conf. on Commun., Control and Computing (1989)

E.H. Satorius, S.T. Alexander: Rapid equalization of highly dispersive channels using lattice algorithms. Naval Oceans Systems Center Tech. Report 249, San Diego, CA (1978)

E.H. Satorius, M.J. Shensa: Recursive lattice filters - a brief overview. Proc. 19th IEEE Conf. Decision Control, Albuquerque (1980) pp. 955-959

M.J. Shensa: Recursive least-squares lattice algorithms: A geometrical approach. IEEE Trans. AC-26, 695-702 (1981)

E. Sichor: Fast recursive estimation using the lattice structure. Bell Syst. Tech. J. 61, 97-105 (1982)

G.S. Sidhu, T. Kailath: A shift-invariance approach to Chandrasekhar, Cholesky and Levinson type algorithms. Proc. Johns Hopkins Conf. Inform. Sci. Syst., Baltimore (1975) pp. 324-327

P. Strobach, "Schnelle adaptive Algorithmen zur ordnungsrekursiven Kleinste-Quadrate-Schätzung autoregressiver Parameter", Ph.D. Dissertation, Bundeswehr Univ. Munich, Munich, Germany, 1985.

P. Strobach: A new class of least-squares covariance ladder estimation algorithms. Proc. 19th Ann. Asilomar Conf., Monterey (1985) pp. 116-120

P. Strobach: Efficient and robust covariance ladder algorithms for linear prediction. Proc. European Signal Processing Conf. (EUSIPCO-86), The Hague, The Netherlands (1986) pp. 1025-1028

P. Strobach: New forms of least-squares lattice algorithms and a comparison of their round-off error characteristics. Proc. IEEE Int. Conf. on ASSP, Tokyo, Japan (1986) pp. 573-576

Extensions and Applications

S.H. Adalan, L.J. Faber: A fast ARMA transversal RLS filter algorithm. IEEE Trans. ASSP-36, 349-358 (1988)

H. Akaike: Maximum likelihood identification of Gaussian autoregressive moving average models. Biometrika 60, 255-265 (1973)

S.T. Alexander, Z.M. Rhee: Analytical finite precision results for Burg's algorithm and the autocorrelation method for linear prediction. IEEE Trans. ASSP-35, 626-635 (1987)

S.P. Applebaum, D.J. Chapman: Adaptive arrays with main beam constraints. IEEE Trans. AP-24, 650-662 (1976)

K.J. Aström, P. Eykhoff: System identification - A survey. Automatica 7, 123-162 (1971)

M. Basseville, A. Benveniste: Sequential detection of abrupt changes in spectral characteristics of digital signals. IEEE Trans. IT-29, 709-724 (1983)

A.A. (Louis) Beex, MD. A. Rahman: On averaging Burg spectral estimators for segments. IEEE Trans. ASSP-34, 1473-1484 (1986)

M. Bendir, B. Picinbono: Extensions of the stability criterion for ARMA filters. IEEE Trans. ASSP-35, 425-432 (1987)

A. Benveniste, C. Chaure: AR and ARMA identification algorithms of Levinson-type: An innovations approach. IEEE Trans. AC-26, 1243-1261 (1981)

N. Benvenuto, L.E. Franks, F.S. Hill, Jr.: Dynamic programming methods for designing FIR filters using coefficients -1, 0, +1. IEEE Trans. ASSP-34, 785-792 (1986)

S. Bruzzone, M. Kaveh: Statistical efficiency of correlation based ARMA parameter estimates. Proc. IEEE Int. Conf. on ASSP, Paris (1982) pp. 240-243

D. Chazan, Y. Medan, U. Shvadron: Noise cancellation for hearing aids. IEEE Trans. ASSP-36, 1697-1705 (1988)

R. Chellappa, R.L. Kashyap: Texture synthesis using 2-D noncausal autoregressive models. IEEE Trans. ASSP-33, 194-203 (1985)

A. Cohn: Über die Anzahl der Wurzeln einer algebraischen Gleichung in einem Kreise. Math. Z., 14, 110-148 (1922)

G. Cybenko: The numerical stability of the Levinson-Durbin algorithm for Toeplitz systems of equations. SIAM J. Sci. Stat. Comput. 1, 303-310 (1980)

J.R. Deller, Jr. , G.P. Picache: Advantages of a Givens rotation approach to temporally recursive linear prediction analysis of speech. IEEE Trans. ASSP-37, 429–431 (1989)

S. Dimolitsas: An adaptive IIR filter based on the Hurwitz stability properties and a Chebychev system function approximation. Signal Processing 17, 39–50 (1989)

D.L. Duttweiler, Y.S. Chen: A single-chip VLSI echo canceller. Bell Syst. Tech. J. 59, 149–160 (1980)

S.J. Elliott, J.M. Stothers, P.A. Nelson: A multiple error LMS algorithm and its application to the active control of sound and vibration. IEEE Trans. ASSP-35, 1423–1434 (1987)

L.J. Eriksson, M.C. Allie, R.A. Greiner: The selection and application of an IIR adaptive filter for use in active sound attenuation. IEEE Trans. ASSP-35, 433–437 (1987)

M. Feder, E. Weinstein: Parameter estimation of superimposed signals using the EM algorithm. IEEE Trans. ASSP-36, 477–489 (1988)

B. Friedlander: Instrumental variable methods for ARMA spectral estimation. IEEE Trans. ASSP-31, 404–415 (1983)

B. Friedlander: On the computation of the Cramer-Rao bound for ARMA parameter estimation. IEEE Trans. ASSP-32, 721–727 (1984)

W.F. Gabriel: Adaptive arrays: An introduction. Proc. IEEE 64, 239–272 (1976)

K.A. Gallivan, C.E. Leiserson: High-performance architectures for adaptive filtering based on the Gram-Schmidt algorithm. Proc. SPIE 495, 30–38 (1984)

W.A. Gardner, W.A. Brown, III: A new algorithm for adaptive arrays. IEEE Trans. ASSP-35, 1314–1319 (1987)

W. Gersch: Estimation of the autoregressive parameters of a mixed autoregressive moving-average time series. IEEE Trans. AC-15, 583–588 (1970)

J.D. Gibson: Adaptive prediction in speech differential encoding systems. Proc. IEEE 68, 488–525 (1980)

D.F. Gingras, E. Masry: Autoregressive spectral estimation in additive noise. IEEE Trans. ASSP-36, 490–501 (1988)

D. Graupe, D.J. Krause, J.B. Moore: Identification of autoregressive moving-average parameters of time series. IEEE Trans. AC-20, 104–107 (1975)

Y. Grenier: Time-dependent ARMA modeling of nonstationary signals. IEEE Trans. ASSP-31, 899–911 (1983)

M. Hadidi, M. Morf, B. Porat: Efficient construction of canonical ladder forms for vector autoregressive processes. IEEE Trans. AC-27, 1222–1233 (1982)

P. Hagander, B. Wittenmark: Self-tuning filter for fixed lag smoothing. IEEE Trans. IT-23, 377–384, (1977)

S. Haykin: Radar signal processing. IEEE ASSP Magazine **2**, 2-18 (1985)

W.S. Hodgkiss, J.A. Presley: Adaptive tracking of multiple sinusoids whose power levels are widely separated. IEEE Trans. ASSP-**29**, 710-727 (1981)

M.L. Honig: Convergence models for lattice joint process estimators and least-squares algorithms. IEEE Trans. ASSP-**31**, 415-425 (1983)

F. Itakura, S. Saito: On the optimum quantization of feature parameters in the PARCOR speech synthesizer. IEEE Conf. Speech Communications and Processing, New York (1972) pp. 434-437

C.R. Johnson, Jr.: Adaptive IIR filtering: Current results and open issues. IEEE Trans. IT-**30**, 237-250 (1984)

T. Kailath: A generalized likelihood ratio formula for random signals in Gaussian noise. IEEE Trans. IT-**15**, 350-361 (1969)

N. Kalouptsidis, G. Carayannis, D. Manolakis: Efficient recursive-in-order least-squares FIR filtering and prediction. IEEE Trans. ASSP-**33**, 1175-1187 (1985)

E. Karlsson, M.H. Hayes: ARMA modeling of time-varying systems with lattice filters. Proc. IEEE Int. Conf. on ASSP, Tokyo, Japan (1986) pp. 2335-2338

T. Katayama, T. Hirai, K. Okamura: A fast Kalman filter approach to restoration of blurred images. Signal Processing **14**, 165-175 (1988)

M. Kaveh, G.A. Lippert: An optimum tapered Burg method for linear prediction and spectral analysis. IEEE Trans. ASSP-**31**, 438-444 (1983)

S.M. Kay: Recursive maximum likelihood estimation of autoregressive processes. IEEE Trans. ASSP-**31**, 56-65 (1983)

S.M. Kay, J. Makhoul: On the statistics of the estimated reflection coefficients of an autoregressive process. IEEE Trans. ASSP-**31**, 1447-1456 (1983)

L. Knockaert: An order-recursive algorithm for estimating pole-zero models. IEEE Trans. ASSP-**35**, 154-157 (1987)

A.K. Krishnamurthy, D.G. Childers: Two-channel speech analysis. IEEE Trans. ASSP-**34**, 730-743 (1986)

S.Y. Kung, K.S. Arun, D.V. Bhaskar Rao: State space and SVD based approximation methods for the harmonic retrieval problem. J. Opt. Soc. Am. **73**, 1799-1811 (1983)

I.D. Landau: A feedback system approach to adaptive filtering. IEEE Trans. IT-**30**, 251-262 (1984)

E.A. Lee: Programmable DSP architectures: Part II. IEEE ASSP Mag. **6**, 4-14 (1989)

H. Leung, S. Haykin: Stability of recursive QRD-LS algorithms using finite-precision systolic array implementation. IEEE Trans. ASSP-**37**, 760-763 (1989)

H. Lev-Ari: Modular architectures for adaptive multichannel lattice algorithms. IEEE Trans. ASSP-**35**, 543-552 (1987)

R.P. Lippman: An introduction to computing with neural nets. IEEE ASSP Mag. **4**, 4-22 (1987)

Y. Miyoshi, K. Yamamoto, R. Mizoguchi, M. Yanagida, O. Kakusho: Analysis of speech signals of short pitch period by a sample-selective linear prediction. IEEE Trans. ASSP-**35**, 1233-1240 (1987)

M. Morf, D.T.L. Lee, J. Nicholls, A. Vieira: A classification of algorithms for ARMA models and ladder realizations. Proc IEEE Int. Conf. on ASSP, Hartford, CT (1977) pp. 13-19

A. Nehorai, G. Su, M. Morf: Estimation of time difference of arrival for multiple ARMA sources by pole decomposition. IEEE Trans. ASSP-**31**, 1478-1491 (1983)

M. Ohsmann: An iterative method for the identification of MA-systems. IEEE Trans. ASSP-**36**, 106-109 (1988)

S.J. Orfandis, L.M. Vail: Zero-tracking adaptive filters. IEEE Trans. ASSP-**34**, 1566-1572 (1986)

G. Panda, B. Mulgrew, C.F.N. Cowan, P.M. Grant: A self-orthogonalizing efficient block adaptive filter. IEEE Trans. ASSP-**34**, 1573-1582 (1986)

B. Picinbono: Some properties of prediction and interpolation errors. IEEE Trans. ASSP-**36**, 525-531 (1988)

L. Rabiner, R. Crochiere, J. Allen: FIR system modeling and identification in the presence of noise and with band-limited inputs. IEEE Trans. ASSP-**26**, 319-333 (1978)

M.D.A. Rahman, K.-B. Yu: Total least-squares approach for frequency estimation using linear prediction. IEEE Trans. ASSP-**35**, 1440-1454 (1987)

B.D. Rao: Perturbation analysis of an SVD-based linear prediction method for estimating the frequencies of multiple sinusoids. IEEE Trans. ASSP-**37**, 1026-1035 (1988)

E.A. Robinson, "A historical perspective of spectrum estimation", Proc. IEEE, Vol. 70, Special Issue on Spectral Estimation, pp. 885-907, 1982.

J. Schroeder, R. Yarlagadda: Linear predictive spectral estimation via the L_1 norm. Signal Processing **17**, 19-29 (1989)

J.J. Shynk: Adaptive IIR filtering using parallel-form realizations. IEEE Trans. ASSP-**37**, 519-533 (1989)

C.R. South, C.E. Hoppit, A.V. Lewis: Adaptive filters to improve loudspeaker telephone. Electron. Lett. **15**, 673-674 (1979)

Special Issue on System Identification and Time Series Analysis, IEEE Trans. AC-**19**, 638-951 (1974)

Special Issue on Adaptive Systems, Proc. IEEE **64**, 1123-1240 (1976)

Special Issue on Adaptive Antennas, IEEE Trans. AP-**24**, 585-779 (1976)

Special Issue on Adaptive Signal Processing, IEEE Trans. CAS-**28**, 465-602 (1981)

Special Issue on Spectral Estimation, Proc. IEEE **70**, 883-1125 (1982)

Special Issue on Adaptive Arrays, IEE Proc. on Commun., Radar Signal Proc. **130**, 1-151 (1983)

Special Issue on Linear Adaptive Filtering, IEEE Trans. IT-**30**, 131-295 (1984)

Special Issue on Adaptive Processing Antenna Systems, IEEE Trans. AP-**34**, 276-464 (1986)

P. Stoica, A. Nehorai: On linear prediction models constrained to have unit-modulus poles and their use for sinusoidal frequency estimation. IEEE Trans. ASSP-**36**, 940-942 (1988)

P. Stoica, T. Söderström, F.-N. Ti: Overdetermined Yule-Walker estimation of the frequencies of multiple sinusoids: Accuracy aspects. Signal Processing **16**, 155-174 (1989)

J.A. Stuller: Maximum-likelihood estimation of time-varying delay - Part 1. IEEE Trans. ASSP-**35**, 300-313 (1987)

D. Swanson, F.W. Symons: A fixed-point implementation of the un-normalized least-squares lattice using scaled-integer arithmetic. IEEE Trans. ASSP-**35**, 1781-1782 (1987)

D.N. Swingler, R.S. Walker: Line-array beamforming using linear prediction for aperture interpolation and extrapolation. IEEE Trans. ASSP-**37**, 16-30 (1989)

D.W. Tufts, R. Kumaresan: Estimation of frequencies of multiple sinusoids: Making linear prediction perform like maximum likelihood. Proc. IEEE **70**, 975-989 (1982)

G. Ungerboeck: Fractional tap-spacing equalizer and consequences for clock recovery in data modems. IEEE Trans. COM-**24**, 856-864 (1976)

B.D. Van Veen, K.M. Buckley: Beamforming: A versatile approach to spatial filtering. IEEE ASSP Mag. **5**, 4-24 (1988)

A.M. Walker: Large sample estimation of parameters for autoregressive processes with moving-average residuals. Biometrika **49**, 117-131 (1962)

B. Widrow, J. McCool, M. Ball: The complex LMS algorithm. Proc. IEEE **63**, 719-720 (1975)

Z. Wu: Multidimensional state-space model Kalman filtering with application to image restoration. IEEE Trans. ASSP-**33**, 1576-1592 (1985)

List of Algorithm Summaries

Subject Index